Plant-Derived Feed Additives and Animal-Based Foods

Plant-Derived Feed Additives and Animal-Based Foods

Linking Animal Health, Product Quality, and Consumer Perceptions

Edited by

Mohamed E. Abd El-Hack
Department of Industrial Pharmacy,
College of Pharmaceutical Sciences and Drug Manufacturing,
Misr University for Science and Technology (MUST),
Giza, Egypt

ELSEVIER

ACADEMIC PRESS
An imprint of Elsevier
elsevier.com/books-and-journals

Academic Press is an imprint of Elsevier

125 London Wall, London EC2Y 5AS, United Kingdom
50 Hampshire Street, 5th Floor, Cambridge, MA 02139, United States

Publisher's note: Elsevier takes a neutral position with respect to territorial disputes or jurisdictional claims in its published content, including in maps and institutional affiliations.

Books and Journals published by Elsevier comply with applicable product safety requirements. For any product safety concerns or queries, please contact our authorised representative, Elsevier B.V., at productsafety@elsevier.com.

For accessibility purposes, images in electronic versions of this book are accompanied by alt text descriptions provided by Elsevier. For more information, see https://www.elsevier.com/about/accessibility.

Notices
Knowledge and best practice in this field are constantly changing. As new research and experience broaden our understanding, changes in research methods, professional practices, or medical treatment may become necessary.

Practitioners and researchers must always rely on their own experience and knowledge in evaluating and using any information, methods, compounds, or experiments described herein. In using such information or methods they should be mindful of their own safety and the safety of others, including parties for whom they have a professional responsibility.

To the fullest extent of the law, neither the Publisher nor the authors, contributors, or editors, assume any liability for any injury and/or damage to persons or property as a matter of products liability, negligence or otherwise, or from any use or operation of any methods, products, instructions, or ideas contained in the material herein.

ISBN: 978-0-443-45446-2

For Information on all Academic Press publications visit our website at https://www.elsevier.com/books-and-journals

Publisher: Mara Conner
Acquisitions Editor: Kelsey Connors
Editorial Project Manager: Avni Gupta
Production Project Manager: Rachel E. G. Michael
Cover Designer: Bakyalakshmi S

Typeset by Aptara, New Delhi, India

Contents

Contributors

Mohamed E. Abd El-Hack, Department of Industrial Pharmacy, College of Pharmaceutical Sciences and Drug Manufacturing, Misr University for Science and Technology (MUST), Giza, Egypt

Ibnath Haque Abony, Faculty of Fisheries, Bangladesh Agricultural University, Mymensingh, Bangladesh

Sheikh Adil, Faculty of Veterinary Sciences and Animal Husbandry, Division of Livestock Production and Management, Sher-e-Kashmir University of Agricultural Sciences and Technology of Kashmir (SKUAST-Kashmir), Srinagar, Jammu and Kashmir, India

Sohail Ahmad, Department of Poultry Production, Faculty of Animal Production and Technology, University of Veterinary and Animal Sciences, Lahore, Pakistan

Chaudhry Ahsan Akram, Department of Livestock and Poultry Production, Faculty of Veterinary Sciences, Bahauddin Zakariya University, Multan, Pakistan

Muhammad Arif, Department of Animal Sciences, College of Agriculture, University of Sargodha, Sargodha, Punjab, Pakistan

Elwy A. Ashour, Poultry Department, Faculty of Agriculture, Zagazig University, Zagazig, Egypt

Zeeshan Ahmad Bhutta, Laboratory of Veterinary Immunology and Biochemistry, College of Veterinary Medicine, Chungbuk National University, Cheongju, Republic of Korea

Abu Darda, Department of Biochemistry and Chemistry, Sylhet Agricultural University, Sylhet, Bangladesh

Ahmed I. Elsherbeni, Animal Production Systems Research Department, Animal Production Research Institute, Agricultural Research Center, Giza, Egypt

Arooj Fatima, Department of Veterinary Medicine, Faculty of Veterinary Sciences, University of Veterinary and Animal Sciences, Lahore, Pakistan

Mahmoud Kamal, Animal Production Systems Research Department, Animal Production Research Institute, Agricultural Research Center, Giza, Egypt

Asmaa Fahmy Khafaga, Department of Pathology, College of Veterinary Medicine, King Faisal University, Al-Ahsa, Saudi Arabia

Ehsaan Ullah Khan, Department of Animal Nutrition, Faculty of Animal Production and Technology, University of Veterinary and Animal Sciences, Lahore, Pakistan

Mohammad Mehedi Hasan Khan, Department of Biochemistry and Chemistry, Sylhet Agricultural University, Sylhet, Bangladesh

Mahmoud Moustafa, Department of Biology, College of Science, King Khalid University, Abha, Kingdom of Saudi Arabia

Muhammad Muneeb, Department of Animal Nutrition, Faculty of Animal Production and Technology, University of Veterinary and Animal Sciences, Lahore, Pakistan

Abrar H. Omar, Department of English, College of Languages and Translation, Misr University for Science and Technology (MUST), Giza, Egypt

Nazifa Rahman, Department of Biochemistry and Chemistry, Sylhet Agricultural University, Sylhet, Bangladesh

Shamima Sultana, Department of Biochemistry and Chemistry, Sylhet Agricultural University, Sylhet, Bangladesh

Ayman A. Swelum, Department of Theriogenology, Faculty of Veterinary Medicine, Zagazig University, Zagazig, Egypt

Ayman El-sayed Taha, Department of Public Health, College of Veterinary Medicine, King Faisal University, Al-Ahsa, Saudi Arabia

Islam Mohamed Youssef, Animal Production Systems Research Department, Animal Production Research Institute, Agricultural Research Center, Giza, Egypt

About the Editor

Mohamed E. Abd El-Hack, PhD, is a Full Professor of Industrial Pharmacy at the College of Pharmaceutical Sciences and Drug Manufacturing, Misr University for Science and Technology (MUST), Egypt, and a Full Professor of Poultry Nutrition at Zagazig University, Egypt. His research spans animal nutrition, functional feed additives, phytogenic bioactive compounds, and sustainable livestock production, with particular emphasis on the intersection between animal-derived foods, nutraceutical science, and human health.

Dr. Abd El-Hack received his PhD in Poultry Production from Zagazig University in 2015. Over the past decade, he has built a globally recognized research portfolio focusing on natural growth promoters, plant-derived bioactives, probiotics, and innovative nutritional interventions aimed at improving animal health, productivity, and the biological value of animal-derived foods.

He has authored and coauthored an extensive body of scholarly work, including numerous peer-reviewed articles and academic book chapters with major international publishers. His scientific influence has led to his inclusion among the world's top 2% most influential scientists in the global ranking compiled by Stanford University. His publications have received tens of thousands of citations and maintain high impact across international scientific databases.

Beyond his research contributions, Dr. Abd El-Hack plays an active role in the international scientific community through editorial services for several peer-reviewed journals and participation in global research and evaluation initiatives in animal science, veterinary sciences, and functional food systems. Through his academic and editorial leadership, he continues to advance integrative approaches that connect animal production, nutrition, and sustainable food systems, with the broader aim of enhancing both animal health and human nutrition.

Preface

In recent years, it has become increasingly clear that the way we feed animals has far-reaching consequences beyond simple growth and productivity. Modern feeding strategies influence not only animal health and welfare but also the quality and functional properties of the foods we consume, as well as how these products are perceived by consumers. This book is rooted in the understanding that plant-derived feed additives are more than just supplements—they are powerful tools that can support health, enhance performance, and contribute to sustainable production systems.

Among these natural compounds, certain herbal bioactives stand out for their neuroactive potential, influencing both physiological and neurological functions in animals. This approach highlights the close connection between nutrition, behavior, and overall well-being, showing that what animals consume can have subtle yet significant effects on their biology and, ultimately, on the quality of animal-derived foods.

Throughout the book, the discussion moves from foundational concepts of conventional feeding to the growing role of functional phyto-feed strategies. Topics such as oxidative stress, inflammation, gut health, and the mechanisms of bioactive compounds are explored in an integrated manner, emphasizing how these elements work together to support animal resilience and productivity. The practical applications of these compounds, including their effects on gut microbiota, overall health, and product quality, are highlighted to provide a clear link between scientific knowledge and real-world feeding practices.

The book also considers the broader perspective, examining the indirect benefits of functional feeding for human health, consumer perceptions, and market behavior. Ethical considerations, environmental sustainability, and regulatory frameworks are discussed to provide a complete picture of the opportunities and challenges associated with implementing these strategies effectively. Looking forward, the book encourages readers to think about the future of animal nutrition at the intersection of feed, food, and health, emphasizing innovation that benefits animals, producers, and consumers alike.

I am deeply grateful to all contributors for their expertise and dedication, as well as to the editorial and production teams whose support has been invaluable. My hope is that this book will serve as a practical and inspiring resource for researchers, students, and professionals, promoting a more connected understanding of animal nutrition and its broader impacts on food quality and consumer well-being.

Prof. Dr. Mohamed E. Abd El-Hack

Chapter 1

Introduction—From conventional feeding to functional phytofeed approaches

Islam Mohamed Youssef[a], **Sheikh Adil**[b] **and Mohamed E. Abd El-Hack**[c]

[a]*Animal Production Systems Research Department, Animal Production Research Institute, Agricultural Research Center, Giza, Egypt,* [b]*Faculty of Veterinary Sciences and Animal Husbandry, Division of Livestock Production and Management, Sher-e-Kashmir University of Agricultural Sciences and Technology of Kashmir (SKUAST-Kashmir), Srinagar, Jammu and Kashmir, India,* [c]*Department of Industrial Pharmacy, College of Pharmaceutical Sciences and Drug Manufacturing, Misr University for Science and Technology (MUST), Giza, Egypt*

1.1 Introduction

The emergence of agriculture and human civilization is intricately connected to the development of animal husbandry methods. The initial systematic efforts in livestock nutrition relied on natural grazing and agricultural by-products such as straw and husks, which constituted the primary nutritional resource for early domesticated animals (Guiné et al., 2021). Over millennia, these rudimentary practices were supplanted by more systematic feeding methods, especially during the Industrial Revolution, when complex feed formulations designed to meet nutritional needs—were adopted. This development established the foundation for modern animal husbandry, enabling enhanced control over animal growth and productivity (Coffey et al., 2016).

Traditional feeding methods exhibit significant limits, albeit in their efficacy in enhancing production. The predominant practice involves the widespread application of synthetic chemicals and antibiotics to promote growth and prevent diseases. Historically, these drugs have enhanced feed efficiency and reduced mortality; however, their excessive usage has adversely impacted the environment, consumer health, and contributed to the development of antibiotic resistance (Youssef et al., 2024a; Youssef et al., 2024b). Human health is jeopardized by antibiotic residues in meat and feces, while sustainability problems have emerged owing to the environmental degradation associated with extensive animal agriculture, including water contamination and greenhouse

Plant-Derived Feed Additives and Animal-Based Foods: Linking Animal Health, Product Quality, and Consumer Perceptions. DOI: https://doi.org/10.1016/B978-0-443-45446-2.00008-9

gas emissions (GHGs). Moreover, synthetic chemicals may disrupt the gut microbiome, thereby jeopardizing animal health and welfare, and often lack comprehensive long-term safety evidence (Ikusika and Haruzivi, 2022). The idea of "functional feeding" has come up as a viable option to solve these problems. It is the deliberate inclusion of feeding materials and additives that not only provide the basic nutritional needs of the animals but also provide additional physiological benefits. The need for sustainable, health-promoting options in the industry, along with advances in technology, is driving this paradigm shift (Placha et al., 2022). Probiotics, prebiotics, phytogenics, enzymes, organic acids, etc., are examples of functional feeding additives that are now being used organically to enhance animal performance. These additives improve the efficiency of feed conversion and the incidence of disease by promoting the microbiota, the mucosa, and the immune system (Bedasso, 2021; Youssef et al., 2023a; Youssef et al., 2023b; Youssef et al., 2024).

Targeted feeding, sustainability, and consumer-oriented quality standards are some of the most important developments in animal nutrition, which support the expansion of functional feeding. With an increasing world population, there is an increasing demand for animal protein, which in turn requires an increase in animal production with fewer resources. Functional nutrition offers a means to fulfill these demands while considering moral, environmental, and health-related concerns. The complex interactions between nutrition, physiology, and microbiome, recognized as essential determinants of animal health and productivity, are increasingly comprehended (Varzakas and Antoniadou, 2024).

This chapter will provide the reader with a foundational overview of the concept of functional feeding in animal nutrition within this context. The chapter will attempt to address the scientific basis of functional feeding, the limitations of conventional feeding systems, and the evolution of feeding systems throughout the ages. The various forms of functional feed additives, the way they function, and their role in sustainable animal production systems will also be discussed. The role of the chapter in the broader context of enhancing the understanding of modern animal nutrition will also be discussed. The chapter will be part of a comprehensive volume of chapters that aim to improve our understanding of animal nutrition and feeding systems. The book will attempt to address the gap between conventional systems and emerging technologies by providing the reader with some insights into the subject. The book will attempt to address the way we can improve the move toward more healthful and sustainable animal nutrition systems. The overall aim of the book will be to improve animal welfare, food safety, and the environment.

1.2 Conventional animal feeding systems

1.2.1 Historical overview

The first-time people domesticated animals was in the Fertile Crescent about 12,000 years ago. This is where traditional animal feeding systems come from.

That was when people began to use animals for work, food, and other things. At first, animals ate garbage and leftovers from crops (Hamouda et al., 2025). They did this by scavenging and grazing naturally. As farming changed, so did how people fed their animals. The Industrial Revolution marked a significant transformation with the introduction of compound feed, which consists of carefully crafted mixtures of cereals, protein meals, and additives designed to meet specific nutritional needs. These feeds made livestock production more consistent and effective, especially in systems where animals could not rely on pasture alone (Coffey et al., 2016).

Protein meals, especially soybean meals, gave the body the amino acids it needed to grow and reproduce. Cereals such as corn, wheat, and barley have become important sources of energy. Synthetic additives such as vitamins, minerals, and antibiotic growth promoters (AGPs) were added to make food more effective, stop disease, and improve the quality of the product. These changes turned livestock farming into a very specialized and industrialized field and laid the groundwork for modern animal nutrition (Kumar et al., 2025).

1.2.2 Advantages and achievements

Conventional means of feeding have proved key in securing the world from food crises. Thanks to them, feed formulations have improved, animal physiology has become more science-based, and productivity and efficiency have soared. Producers are now able to make more meat, milk, and eggs with less input because of a noticeable increase in the feed conversion ratio (FCR), which shows the proportion of food intake by animals that is expended for body tissue accumulation (Akintan et al., 2024).

Economic affordability is another major advantage. Common ingredients such as cereal grains and oilseed meals benefit from economies of scale in production and distribution, but many feedstuffs vary regionally. Because of the low prices of animal protein, consumers have cheaper access to it, and livestock worldwide has been increasingly prosperous (Van Middelaar et al., 2014). Agroindustrial by-products such as distillers' grains and wheat bran have also increased the sustainability of the conventional systems, which have been able to reduce waste and feeding cost (Kumar et al., 2025).

Even though conventional feeding systems have worked well in the past, they have a lot of problems. It is a big problem when diets are not made with the needs of different species in mind, because this can lead to nutrient imbalances. Too many or too few nutrients can cause slow growth, poor reproduction, and a higher risk of getting sick (Kumar et al., 2025). Using growth promoters and antibiotics in feed has also raised serious health and environmental issues. In the past, these additives made feed more efficient and lowered death rates, but using them too much has led to antibiotic residues building up in manure and animal products (Cufadar et al., 2024). These residues may enter the environment through waste disposal and runoff, contributing to the dissemination of antimicrobial resistance

(AMR), which poses a significant threat to global public health. Resistant bacteria and genes that stay in soil and water for a long time could be bad for the health of people and animals (Apreja et al., 2022).

Sustainability of the environment is an urgent concern, and as for traditional feed on livestock, it requires a significant amount of energy, water, and land to produce feed (Broom, 2023). The production of crops used for feed, especially soy and corn, affects the environment by causing deforestation, loss of biodiversity, and GHG. Also, manure from animals that have been fed ample amounts of food can contaminate waterways, while simultaneously releasing powerful greenhouse gases such as nitrous oxide and methane gas into our atmosphere (Esteves et al., 2018). This presents challenges; therefore it is necessary to develop healthy and sustainable feeding practices moving forward. Conventional types of systems have been successful in increasing productivity at unprecedented rates, but their future continues to be dependent upon addressing the above-mentioned issues through innovation, regulation, and greater emphasis on animal nutrition.

1.3 Drivers for change in animal nutrition

1.3.1 Consumer demands and perceptions

In recent years, customers have increasingly dictated the direction of animal nutrition, with consumers wanting more natural, less processed foods with greater transparency about what is in the food they eat. This is demonstrated by the rise of the "clean-label" movement (Inguglia et al., 2023). Many customers now prefer foods that are labeled as organic, natural, or sustainably grown when selecting animal products that do not contain artificial ingredients or chemical preservatives, contain no antibiotics, and are not from genetically modified organisms (GMOs). This trend is rooted in concerns about food safety and quality/nutritional value, but there may also be aesthetic considerations. Still, according to several surveys, more than 70% of consumers consider the ingredient list when selecting products, and many will pay a premium for products that reflect their values (Akintan et al., 2024).

In response to the changing consumer preferences regarding food production, producers and feed suppliers are now required to reformulate diets with new ingredients, eliminate the use of artificial growth promoters, and increase the use of natural feed additives. Because of the demand for improved traceability and transparency, several advancements have occurred, including. digital labeling and blockchain supply chain tracking; both technologies allow consumers to verify which animals have been used in the production of animal products and what type of ingredients were provided to those animals (Ellahi et al., 2023).

1.3.2 Animal health and welfare concerns

In the context of sustainable livestock production, animal health and welfare have rapidly gained popularity as one of the critical concerns. Conventionally, animal feeding methods have been causing stress, suppression of the immune system, and increased susceptibility to diseases, particularly in the case of intensive farms. These physical problems can lead to production losses, wastage of the feed, and degradation of the quality of the product (Sicuso et al., 2025).

Increasing awareness of humane treatment, healthier surroundings, and reduced use of unnecessary interventions has led to a greater societal focus on ethical farming practices (Rossi and Garner, 2014). An example of a standard used to assess welfare is called "Five Freedoms," which outlines the freedom to be free from hunger, discomfort, pain, fear, and the ability to express natural behaviors. Poor welfare not only compromises the health of animals; it also limits productivity and contributes to increased environmental impacts of livestock systems because of inefficient use of resources and the increased emissions per unit produced (Llonch et al., 2017). As such, feed strategies are being redesigned to enhance immune function, reduce stress, and increase resilience. More functional feed additives such as probiotics, prebiotics, and phytogenics are being used to support gut health, which reduces the occurrence of disease and improves performance and welfare (Dosoky et al., 2024).

1.3.3 Regulatory and policy pressure

Regulatory systems globally are changing rapidly due to escalating risks related to the use of antibiotics in agriculture (Abou-Kassem et al., 2025). Because of the global rise of AMR, regulatory agencies and various governmental bodies have placed stringent limits on AGP usage throughout the world. Many countries, including Denmark, Sweden, and the Netherlands, have all but banned AGPs. Most notably, the European Union has completely outlawed their use (Zheng et al., 2025).

This action is part of a larger trend toward developing sustainable livestock systems and represents a nonreactive aspect of implementing regulatory policies. Regulatory agencies currently focus on replacement methods that will sustain production while preserving the environment and protecting human health, as well as improving biosecurity and immunization programs. They are recommending that nonantibiotic feed additives (i.e., essential oils, enzymes, and antimicrobial peptides) be used as alternatives to AGPs (Brown et al., 2014).

In addition, regulatory agencies are now incorporating new environmental metrics into their feed approval processes to evaluate how feed ingredients can affect GHGs, as well as land and water degradation. This comprehensive approach reflects a growing recognition that animal feed decisions influence more than animal performance (Garnett, 2009).

1.3.4 One health perspective

The One Health framework strongly supports the idea that animal nutrition should be reevaluated because of the connections between human health, animal health, and the environment, and to integrate these approaches to address the potential risks to all three areas. Evaluating the safety of feed additives not only for human consumers and those who work with animals on farms, but also for the environment, such as ecological systems in agriculture, requires that their nutritional effectiveness for farm animals also be evaluated (Mantovani et al., 2022). An example of the importance of evaluating the safety of additives in relation to human health is that some feed additives can influence how the compounds produced when they are metabolized by animals will be disposed of in the environment. In addition, some feed additives can affect the health of farm workers who are exposed to them during feed handling, and the potential for animal product residues in animal products. When evaluating risk–benefit ratios under the One Health paradigm, it is essential for evaluating feed interventions to ensure that they do not have negative unintended consequences for the ecological balance of the environment and the health of humans (Varzakas and Antoniadou, 2024).

Moreover, the development of feed methods that mitigate zoonotic risks is a point of the One Health paradigm. Functional feeds can lower the chance of disease transmission between humans and animals through animal welfare enhancement and decreasing the number of pathogens in the environment. This factor is very crucial considering the outbreak of new infectious diseases, a great number of which originate in the human, animal, environment nexus (Shurson et al., 2022). Building a more resilient and responsible animal industry is, in fact, one of the results of integrating One Health principles in animal feed formulation and policy design. The approach calls for animal nutrition to be aligned with greater public health and sustainability goals, and thus it stimulates interdisciplinary collaboration, informed decision-making, and a risk-averse attitude (Roy et al., 2025).

1.4 Emergence of functional phytofeed

1.4.1 Conceptual definition

Functional phytofeeds offer significant development in the field of animal nutrition by providing an alternative to both standard feed and functional feeding methods. Traditional food aims to give the animals protein, energy, vitamins, and minerals, but does not specifically target improving health or other physiological effects (Boukid and Rosell, 2025). The main purpose of functional feed is to supplement essential nutrients with additional benefits to support immunity, gut health, and resilience to stress. Phytofeed represents a specific subset of functional feed and is developed using bioactive substances from plants. Various plant extracts, botanicals, and herbs contain secondary metabolites that can modify biological functions of animals (Bedasso et al., 2021; Negm et al., 2025).

Functional feeds produce physiological effects that go beyond simple nutrition; these include improving immune response, decreasing oxidative stress, modifying the gut microbiome, and increasing animal welfare and production.

Functional phytofeeds are particularly attractive as alternatives to synthetic feed additives and AGPs because of their natural origin, low toxicity, and minimal risk of developing AMR (Zuberi et al., 2024).

1.4.2 Role of plant-derived additives

Bioactive agents from plants will have diverse and distinctive chemical structures as well as biological properties, and many of these are contained within plant-based additives used to produce a variety of phytofeed formulations. Polyphenols are one of the most significant types of bioactive substances occurring in plant-based phytofeed additives. These compounds, found in fruits, vegetables, and herbs, exhibit powerful antioxidant activity and help to scavenge free radicals, which can decrease animals' levels of oxidative stress (Bešlo et al., 2023; Abd El-Hack et al., 2025). Flavonoids are a subclass of polyphenols that possess many properties, including antiinflammatory, antimicrobial, and immunomodulatory effects, and contribute to improving gut health and minimizing the risk of disease (Al-Khayri et al., 2022). Some terpenes are constituents of essential oils; these compounds may enhance feed palatability and exhibit antimicrobial and antiparasitic activity (Matté et al., 2023). Alkaloids contain a nitrogen atom and have demonstrated analgesic and antiinflammatory activity as well as effects on metabolic pathways (Gutiérrez-Grijalva et al., 2020). Essential oils can be extracted from aromatic plants such as thyme, oregano, and rosemary, and are associated with many properties, including antimicrobial, antioxidant, and digestive enhancement (Oni and Oke, 2025).

These phytochemicals deliver their actions through multiple routes. For instance, in the case of high-performance animals under stress, antioxidant activity significantly supports the defense of cellular structures from oxidative damage. Antimicrobial characteristics, on the other hand, facilitate the animal's nutrition by enhancing the absorption of nutrients and thus decreasing the chances of infections through the reduction of the pathogenic load in the intestinal tract. Furthermore, the immunomodulatory effects that are beneficial for the innate and adaptive immune responses of animals make them more resistant to diseases. Moreover, different phytochemicals have an impact on the gut microbiome; thus they support the growth of beneficial bacteria and help to keep the harmful ones at bay. As a result, digestion, nutrient absorption, and overall health are improved (Aminullah et al., 2025).

1.4.3 Evidence of effectiveness

According to recent research on functional phytofeed, they have preferred application throughout many different species of livestock, including ruminants, rabbits, and poultry. Phytogenic additives to poultry diets, including basil, lemongrass, peppermint, rosemary, and thyme, were shown to enhance feed conversion rates and improve overall growth performance in poultry as well as improve their immune response to vaccines (Abd El-Hack et al., 2024; Deeb et al., 2024; Deeb et al., 2025). In one trial, chickens receiving rosemary and thyme had enhanced antioxidant enzyme activity in their blood, improved gut morphology, and a greater immune response to infectious bursal disease (Tamir et al., 2023).

Using natural extracts from plants such as cashew nutshell liquid (CNSL) has been proven to reduce methane emissions produced in the rumen of ruminant animals, thereby maintaining their productive capacity while assisting in promoting sustainability. CNSL is rich in natural phenolic compounds, such as anacardic acid, and therefore has a positive influence on rumen fermentation and the rumen microflora, subsequently improving feed efficiency and decreasing GHGs (Branco et al., 2015).

The benefits of using phytofeed with herbal extracts are not solely limited to poultry, as demonstrated in trials conducted in rabbits. Trials utilizing herbal extracts in rabbit diets showed improved meat quality, increased feed intake, and increased growth rate. These benefits can be attributed to increased nutrient absorption, decreased inflammation, and better gut health. Furthermore, phytogenic additives fed to rabbits increased their resistance to gastrointestinal infections and reduced their overall mortality rates (Placha et al., 2022).

1.5 From conventional feeding to functional approaches: A paradigm shift

1.5.1 Comparative overview

The development of functional phytofeed strategies from traditional feeding systems is a revolutionary change in how livestock is managed. Conventional feeding is primarily nutrient-based, utilizing cereals, protein meals, and artificial additives to meet the essential dietary requirements for growth, reproduction, and maintenance (Bedasso, 2021). These feeds focus more on getting enough protein and calories and less on the bigger health and physiological effects. On the other hand, functional phytofeed that is bioactivity-driven has plant-based compounds, such as polyphenols, flavonoids, terpenes, and essential oils that do more than just provide nutrition (Ubbink and Levine, 2025).

Functional phytofeeds are intended to address the following: Improve immunity, alter the gut microbiome, reduce oxidative stress, and promote better animal welfare (Obianwuna et al., 2024). As a result, phytofeeds are also aligned with consumer demand for clean-label products, sustainability, and increasing concerns regarding antibiotic resistance. In contrast with conventional feeds that may leave behind residues of chemicals, phytogenic alternatives offer natural options that promote health and performance while ensuring safe consumption of food products (Williams et al., 2020).

Therefore the transition from a production-focused model of feeding livestock to a whole-animal approach toward health is both a philosophical and technical shift from production to health. With continued evidence supporting the use of phytogenic additives and functional phytofeed in modern commercial feeding operations, the era of ethically based sustainable animal nutrition is underway.

1.5.2 Sustainability and environmental impacts

Sustainability is one of the main reasons for this shift in mindset toward feed. Traditional methods of feeding typically depend upon the use of resource-intensive crops, artificial substances such as pesticides and herbicides, and antibiotics that contribute to harming our environment by emitting greenhouse gases, depleting our soil, and polluting our waterways. Conversely, functional phytofeeding encourages decreased usage of chemical inputs and greater use of naturally occurring bioactive compounds, which are both biodegradable and have lesser impacts on ecosystems (Zhang et al., 2018). Functional feeding practices (also known as functional feed) provide added value to agricultural wastes by converting them into useful animal feeds. Some examples of potential functional feeds include by-products from banana blossoms, citrus peels, grape skins (pomace), and residual herbs that are rich in flavor compounds, including polyphenols, flavonoids, and dietary fiber (Malenica et al., 2023). In addition to the cost savings of the feed ingredient itself, the reduction of waste produced, and the more efficient use of resources also support the development of a biobased economy. Research has confirmed that the addition of commercial fruit and vegetable by-products to the diets of ruminant animals has the potential to increase the quality of the milk they produce, reduce the amount of methane they emit to the atmosphere, and enhance the digestibility of the diet they are given, all without impacting the performance of the animals (Panda et al., 2025). This approach to sustainable development supports the sustainable development goals set by the UN and aligns with other international initiatives to reduce the environmental impact of livestock production (Priya et al., 2023).

1.5.3 Integration into modern feed systems

Phytofeed in functional form must be integrated into existing feed systems, which require advanced formulation and delivery technology to be effective. The addition of phytoadditives to premixes is a widely used method that simplifies compound feed usage and allows for equal distribution. Nevertheless, due to concerns such as volatility, inactivation during processing, and bioavailability, new methods of encapsulation and nanoencapsulation have been created (Manickam et al., 2020).

Encapsulation protects sensitive phytocompounds from inactivation in the gastrointestinal tract, protecting them from oxidation, temperature, and moisture. In addition, nanoformulations improve bioavailability, which allows for more targeted action and reduced dosage. Essential oils and flavonoids are less effective in compound feed because of inactivation during pelleting and storage, and these methods can be applied in such cases (Teng et al., 2023).

Functional phytofeed also has promise as a synergistic agent when used with other bioactive feed additives, such as probiotics, prebiotics, and postbiotics. Prebiotics serve as a growth promoter for the good bacteria that the probiotics provide. The metabolic by-products of the probiotics, referred to as postbiotics, also have health advantages. These feed additives can increase feed efficiency, immunomodulate, and improve gut health when they are used with phytogenics. It has been shown that when strains of probiotics and oregano oil are used together, they reduce pathogenic bacteria and increase growth rates in poultry (Onomu and Okuthe, 2024; Zuberi et al., 2024). The ideas of precision nutrition, in which feed is optimized for physiological responses as well as nutritional requirements, can be seen in this integrative approach. It is a holistic approach that is economically and environmentally sustainable and promotes animal health.

1.6 Opportunities and future perspectives

There are revolutionary possibilities for sustainable livestock production with the introduction of functional feed additives, particularly phyto-based compounds. The bioactive properties of these additives, such as herbs, spices, and agricultural waste, extend beyond providing nutrition to the animal, helping to enhance animal health, prevent diseases, and improve feed efficiency (Kazemi, 2025). Functional feeds, for example, help to reduce chemicals in animal products, thereby addressing the problem of AMR by replacing synthetic growth promoters and antibiotics. Moreover, the reduction of GHGs and the utilization of waste help to promote environmental sustainability through the incorporation of functional feeds. For example, the incorporation of vegetable residues, brewer's yeast, and fruit peels into animal feed formulations reduces waste and depletion of resources, thereby reducing the cost of animal feed and promoting the circular economy (Akintan et al., 2024).

The production of functional foods for animals requires functional feeds to bring consumers multiple benefits, both nutritionally and environmentally. Functional foods, for example, include meat, milk, and eggs that have added immune-reception, omega-3 fatty acids, and antioxidant components because of livestock feed changes (Stephen et al., 2023). Furthermore, the modified feeds have enhanced the nutrient profile and reduced the levels of harmful residues in animals, which benefits public health and satisfies consumers' desire for clean-label, health-promoting products (Zuberi et al., 2024).

Though there are many positive aspects of the potential for utilizing functional phytofeed, several issues inhibit widespread use. The phytochemical profile of any phytobiotic additive is a variable based on many factors, including but not limited to species of plant, the environmental parameters surrounding the growth of that plant, and the method of extraction used for obtaining the plant material (Coffey et al., 2016). These discrepancies create difficulties in the areas of feed formulation/quality control while influencing the ability to replicate and gain efficacy. In addition, producers who work on a relatively small scale may struggle with the costs of acquiring, processing, and/or maintaining bioactive compounds, particularly when employing advanced technologies such as encapsulation or nanoformulation (Hartog et al., 2013). Regulatory frameworks also present major obstacles to the implementation of functional additives into animal feeds. For many regions, the process by which new additives are approved is lengthy and complicated, often requiring a significant amount of data surrounding both their safety and efficacy. The lack of standardized global standards adds further complications to innovation and gaining access to various markets. In addition, a limited history of use in animal feed combined with food safety concerns has resulted in restrictions being placed on certain functional ingredients, such as insect proteins or specific herbal extracts (Pexas et al., 2023).

Future work should investigate standardizing (and utilizing) the phytochemical profile of various plants, developing reliable pathways to develop/enforce regulations which protect both the consumer (safety) and qualified professionals (innovation), and improving methods of extracting and delivering functional feeds at a cost which is acceptable to end users. Development of enabling environments in which to implement functional feed technologies requires collaboration among researchers, industry stakeholders, and legislators. Functional feed solutions are an exceedingly well-thought-out and effective combination of health benefits for livestock, ecological stewardship by livestock operations, and nutrition for humans in meeting the changing needs of the global population because of climate change, consumer preferences, and the effects of pollution on food security.

1.7 Conclusion

This shift has taken place due to poor human/animal relations stemming from agriculture's historically poor performance when utilizing artificial feed additives/antibiotics, subsequently causing an increase in environmental issues (e.g., antibiotic resistance) and reducing human consumer confidence. Functional phytofeeds developed using functional bioactive plant-based products provide animals with nutritional benefits that enhance their overall performance naturally better than traditional methods of feeding animals through the use of artificial additives. The phytogenic additives will improve upon feed efficiency/welfare of livestock while also reducing their negative impacts on the environment, using phytogenic additives to improve oxidative balance, immune modulation, plus gut integrity for livestock.

In addition, using them in feeding programs would be in line with the global move toward environmentally friendly and non-antibiotic livestock production practices. The minimization of waste and resource utilization would also be in line with the push toward the circular bioeconomy. However, standardization, cost-effectiveness, and harmonization of regulations would be some of the challenges to be worked out. For the phytofeeds to be effective and consistent in their use and delivery, research and innovation in the field would be critical. In conclusion, effective phytofeeds would be a forward-thinking and innovative way forward in food safety, sustainability, and productivity, and would be an integral part of future animal nutrition systems.

References

Abd El-Hack, M.E., AboElMaati, M.F., Abusudah, W.F., Awlya, O.F., Almohmadi, N.H., Fouad, W., Mohamed, H.S., Youssef, I.M., Al-Gabri, N.A., Othman, S.I., Allam, A.A., Taha, A.E., Tellez-Isaias, G., Mansour, A.M., 2024. Consequences of dietary cinnamon and ginger oils supplementation on blood biochemical parameters, oxidative status, and tissue histomorphology of growing Japanese quails. Poult. Sci. 103 (2). doi:10.1016/j.psj.2023.103314.

Abd El-Hack, M.E., Rudayni, H.A., Aldhalmi, A.K., Youssef, I.M., Arif, M., Alawam, A.S., Allam, A.A., Khafaga, A.F., Ashour, E.A., Khan, M.M.H., 2025. Beyond traditional uses: unveiling the epigenetic, microbiome-modulating, metabolic, and nutraceutical benefits of Salvia officinalis in human and livestock nutrition. J. Funct. Foods 128. doi:10.1016/j.jff.2025.106843.

Abou-Kassem, D.E., Ashour, E.A., Mahrose, K.M., Youssef, I.M., Tellez-Isaias, G., Swelum, A.A., Abd El-Hack, M.E., 2025. Growth performance, carcass traits, meat composition, digestibility coefficients and caecal microbiota of growing rabbits fed diets supplemented with a herbal mixture. Trop. Anim. Health Prod. 57 (1). doi:10.1007/s11250-024-04249-8.

Akintan, O., Gebremedhin, K.G., Uyeh, D.D., 2024. Animal feed formulation—connecting technologies to build a resilient and sustainable system. Animals 14 (10). doi:10.3390/ani14101497.

Al-Khayri, J.M., Sahana, G.R., Nagella, P., Joseph, B.V., Alessa, F.M., Al-Mssallem, M.Q., 2022. Flavonoids as potential anti-inflammatory molecules: a review. Molecules 27 (9). doi:10.3390/molecules27092901.

Aminullah, N., Mostamand, A., Zahir, A., Mahaq, O., Azizi, M.N., 2025. Phytogenic feed additives as alternatives to antibiotics in poultry production: a review. Vet. World 18 (1), 141–154. doi:10.14202/vetworld.2025.141-154.

Apreja, M., Sharma, A., Balda, S., Kataria, K., Capalash, N., Sharma, P., 2022. Antibiotic residues in environment: antimicrobial resistance development, ecological risks, and bioremediation. Environ. Sci. Pollut. Res. 29 (3), 3355–3371. doi:10.1007/s11356-021-17374-w.

Bedasso, G.T., 2021. The functional feed additives in animal nutrition: the substitute to antibiotics. J. Res. Agric. Anim. Sci. 8, 18–23.

Bešlo, D., Golubić, N., Rastija, V., Agić, D., Karnaš, M., Šubarić, D., Lučić, B., 2023. Antioxidant activity, metabolism, and bioavailability of polyphenols in the diet of animals. Antioxidants 12 (6), 1141. doi:10.3390/antiox12061141.

Boukid, F., Rosell, C.M., 2025. Advances and challenges in breakfast cereals: nutrition, innovation, and sustainability. Crit. Rev. Food Sci. Nutr. 66 (11), 2159–2173. doi:10.1080/10408398.2025.2564889.

Branco, A.F., Giallongo, F., Frederick, T., Weeks, H., Oh, J., Hristov, A.N., 2015. Effect of technical cashew nut shell liquid on rumen methane emission and lactation performance of dairy cows. J. Dairy Sci. 98 (6), 4030–4040. doi:10.3168/jds.2014-9015.

Broom, D.M., 2023. The sustainability of cattle production systems. In: Cattle Welfare in Dairy and Beef Systems: A New Approach to Global Issues. Springer International Publishing, Cham, pp. 351–377.

Brown, V.A., Grootjans, J., Ritchie, J., Townsend, M., Verrinder, G., 2014. Sustainability and Health: Supporting Global Ecological Integrity in Public Health. Taylor and Francis, Australia, pp. 1–327. doi:10.4324/9781315065779.

Coffey, D., Dawson, K., Ferket, P., Connolly, A., 2016. Review of the feed industry from a historical perspective and implications for its future. J. Appl. Anim. Nutr. 4. doi:10.1017/jan.2015.11.

Cufadar, Y., Beyari, E.A., Hassoubah, S.A., Curabay, B., Sevim, B., Aksoy, A., Alfassam, H.E., Rudayni, H.A., Allam, A.A., Youssef, I.M., Moustafa, M., Al-Shahrani, M.O., Algopishi, U., Hassan, H., Abd El-Hack, M.E., 2024. Impacts of Bacillus probiotics on productive performance and egg quality criteria in laying Japanese quails. J. Appl. Poult. Res. 33 (3). doi:10.1016/j.japr.2024.100445.

Deeb, S.E., Ashour, E.A., Abd El-Hack, M.E., El-Maaty, M.A., Youssef, I.M., Adil, S., Elolimy, A.A., Swelum, A.A., 2024. Impacts of dietary different levels of thyme leave powder as a natural growth promoter on growth performance, carcass characteristics, and blood indices of broilers. Poult. Sci. 103 (12). doi:10.1016/j.psj.2024.104396.

Deeb, S.E., Ashour, E.A., Youssef, I.M., Alshehry, G., Abuljadayel, D.A., Aljahdali, N., Albaqami, N.M., Bahnas, M.M., Abd El-Hack, M.E., 2025. Enhancing broiler growth, carcass quality, blood parameters and intestinal microbial load: the role of dietary garlic powder as a natural growth promoter. Ann. Anim. Sci. 25 (4), 1467–1478. doi:10.2478/aoas-2025-0024.

Dosoky, W.M., Farag, S.A., Almuraee, A.A., Youssef, I.M., Awlya, O.F.A., Abusudah, W.F., Qadhi, A., Arbaeen, A.F., Moustafa, M., Hassan, H., Tellez-Isaias, G., 2024. Vitamin C and/or garlic can antagonize the toxic effects of cadmium on growth performance, hematological, and immunological parameters of growing Japanese quail. Poult. Sci. 103 (3). doi:10.1016/j.psj.2024.103457.

Ellahi, R.M., Wood, L.C., Bekhit, A.E.D.A., 2023. Blockchain-based frameworks for food traceability: a systematic review. Foods 12 (16). doi:10.3390/foods12163026.

Esteves, E.M.M., Esteves, V.P.P., Bungenstab, D.J., Araújo, O.Q.F., Morgado, C.R.V., 2018. Greenhouse gas emissions related to biodiesel from traditional soybean farming compared to integrated crop-livestock systems. J. Cleaner Prod. 179, 81–92. doi:10.1016/j.jclepro.2017.12.262.

Garnett, T., 2009. Livestock-related greenhouse gas emissions: impacts and options for policy makers. Environ. Sci. Policy 12 (4), 491–503. doi:10.1016/j.envsci.2009.01.006.

Guiné, R.P.F., Florença, S.G., Barroca, M.J., Anjos, O., 2021. The duality of innovation and food development versus purely traditional foods. Trends Food Sci. Technol. 109, 16–24. doi:10.1016/j.tifs.2021.01.010.

Gutiérrez-Grijalva, E.P., López-Martínez, L.X., Contreras-Angulo, L.A., Elizalde-Romero, C.A., Heredia, J.B., 2020. Plant alkaloids: structures and bioactive properties. In: Plant-derived Bioactives: Chemistry and Mode of Action. Springer Singapore, Mexico, pp. 85–117. doi:10.1007/978-981-15-2361-8_5.

Hamouda, R.E., Youssef, I.M., Gharib, H.B., El-Menawey, M.A., Youssif, M.A.M., Osman, M.A., Abdel-Aziz, Y.A., Rudayni, H.A., Allam, A.A., Alawam, A.S., Khafaga, A.F., Abd El-Hack, M.E., Abdel-Halim, A.A., 2025. Effects of an enhanced feeding model on productivity and sustainability of broilers and hybrid chickens under Egyptian small-scale family systems. Poult. Sci. 104 (11). doi:10.1016/j.psj.2025.105845.

Hartog, D.A.L., Sijtsma, R., 2013. Challenges and opportunities in animal feed and nutrition. In: 11th World Conference on Animal Production, pp. 1–15.

Ikusika, O.O., Haruzivi, C., 2022. Alternatives to the use of antibiotics in animal production. In: Alagawany, M., Abd El-Hack, M.E., Farag, M.R. (Eds.), Antibiotics and Probiotics in Animal Food: Impact and Regulation. Academic Press, Cambridge, MA, USA, pp. 305–321.

Inguglia, E.S., Song, Z., Kerry, J.P., O'Sullivan, M.G., Hamill, R.M., 2023. Addressing clean label trends in commercial meat processing: strategies, challenges and insights from consumer perspectives. Foods 12 (10). doi:10.3390/foods12102062.

Kazemi, M., 2025. Revolutionizing veterinary medicine: the role of nanoparticles in advancing animal health, nutrition and disease management. Vet. Med. Sci. 11 (5). doi:10.1002/vms3.70528.

Kumar, S., Singh, U.K., Ansari, M.H., Kashyap, R., Singh, A.K., 2025. Conventional and non-conventional feed resources enhancing animal production system. In: Kumar, S., Kumar, A., Singh, R. (Eds.), Grazing Strategies and Animal Production Systems. IntechOpen, London, UK https://doi.org/10.5772/intechopen.1006869.

Llonch, P., Haskell, M.J., Dewhurst, R.J., Turner, S.P., 2017. Current available strategies to mitigate greenhouse gas emissions in livestock systems: an animal welfare perspective. Animal 11 (2), 274–284. doi:10.1017/s1751731116001440.

Malenica, D., Kass, M., Bhat, R., 2023. Sustainable management and valorization of agri-food industrial wastes and by-products as animal feed: for ruminants, non-ruminants and as poultry feed. Sustainability 15 (1), 117. doi:10.3390/su15010117.

Manickam, V., Sundar, V., Panchangam, R.L., Amiti, S., K., A., Tamizhselvi, R., 2020. Nanotechnology in delivery and targeting of phytochemicals. In: Nanopharmaceuticals: Principles and Applications, 2. Springer International Publishing, Cham, pp. 211–264.

Mantovani, A., Aquilina, G., Cubadda, F., Marcon, F., 2022. Risk-benefit assessment of feed additives in the One Health perspective. Front. Nutr. 9. doi:10.3389/fnut.2022.843124.

Matté, E.H.C., Luciano, F.B., Evangelista, A.G., 2023. Essential oils and essential oil compounds in animal production as antimicrobials and anthelmintics: An updated review. Anim. Health Res. Rev. 24 (1), 1–11. doi:10.1017/S1466252322000093.

Negm, M.H., Aldhalmi, A.K., Ashour, E.A., Mohamed, L.A., Youssef, I.M., Moustafa, M., Abdelglil, M.I., Abdel-Shafi, S., Mahgoub, S.A., Abd El-Hack, M.E., Swelum, A.A., 2025. The effect of celery (Apium graveolens) seed powder in the diet on growth, immune function, antioxidant status and gut microbial balance in broiler chickens. Trop. Anim. Health Prod. 57 (8). doi:10.1007/s11250-025-04625-y.

Obianwuna, U.E., Chang, X., Oleforuh-Okoleh, V.U., Onu, P.N., Zhang, H., Qiu, K., Wu, S., 2024. Phytobiotics in poultry: revolutionizing broiler chicken nutrition with plant-derived gut health enhancers. J. Anim. Sci. Biotechnol. 15 (1). doi:10.1186/s40104-024-01101-9.

Oni, A.I., Oke, O.E., 2025. Gut health modulation in poultry through phytogenics: mechanisms, benefits, and applications. Front. Vet. Sci. 12, 1616734.

Onomu, A.J., Okuthe, G.E., 2024. The role of functional feed additives in enhancing aquaculture sustainability. Fishes 9 (5). doi:10.3390/fishes9050167.

Panda, J., Amrit, R., Mishra, A.K., Chakraborty, A., Rustagi, S., Nath, P.C., Sarabandi, K., Sarma, H., Wagh, M.S., Mohanta, Y.K., 2025. Sustainable valorization of fruit and vegetable waste for bioactive compounds: advancing functional food and wellness. Waste Biomass Valor. 16 (7), 3357–3386. doi:10.1007/s12649-025-02937-6.

Pexas, G., Doherty, B., Kyriazakis, I., 2023. The future of protein sources in livestock feeds: implications for sustainability and food safety. Front. Sustain. Food Syst. 7. doi:10.3389/fsufs.2023.1188467.

Placha, I., Gai, F., Pogány Simonová, M., 2022. Natural feed additives in animal nutrition—their potential as functional feed. Front. Vet. Sci. 9, 1062724.

Priya, A.K., Alagumalai, A., Balaji, D., Song, H., 2023. Bio-based agricultural products: a sustainable alternative to agrochemicals for promoting a circular economy. RSC Sustain. 1 (4), 746–762. doi:10.1039/d3su00075c.

Rossi, J., Garner, S.A., 2014. Industrial farm animal production: a comprehensive moral critique. J. Agric. Environ. Ethics 27 (3), 479–522. doi:10.1007/s10806-014-9497-8.

Roy, S., Malla, S., Dekari, D., Choudhury, T.G., 2025. One Health Approach: Human, Environment, and Animal Health Management of Fish Diseases. Springer Nature India, India, pp. 281–297. doi:10.1007/978-981-96-0270-4_12.

Shurson, G.C., Urriola, P.E., van de Ligt, J.L.G., 2022. Can we effectively manage parasites, prions, and pathogens in the global feed industry to achieve One Health? Transbound. Emerg. Dis. 69 (1), 4–30. doi:10.1111/tbed.14205.

Sicuso, D.A, Previti, A., Pugliese, M., Passantino, A., 2025. Climate change impacts on livestock and resulting effects on animal health: current challenges in food safety, consumer protection, and animal welfare. J. Consum. Prot. Food Saf. 20 (1), 1–3. doi:10.1007/s00003-025-01544-w.

Stephen, J., Manoharan, D., Radhakrishnan, M., 2023. Immune boosting functional components of natural foods and its health benefits. Food Prod. Process. Nutr. 5 (1). doi:10.1186/s43014-023-00178-5.

Tamir, B., Ayele, M., Waktole, H., Wakjira, B., Regassa, F., Regassa, F., Tufa, T.B., 2023. Effects of dietary supplementation of phytogenic feed additives on broiler feed conversion efficiency and immune response against infectious bursal disease vaccine. J. World's Poult. Res. 13 (2), 180–190. doi:10.36380/jwpr.2023.20.

Teng, H., Zheng, Y., Cao, H., Huang, Q., Xiao, J., Chen, L., 2023. Enhancement of bioavailability and bioactivity of diet-derived flavonoids by application of nanotechnology: a review. Crit. Rev. Food Sci. Nutr. 63 (3), 378–393. doi:10.1080/10408398.2021.1947772.

Ubbink, J., Levine, A.S., 2025. From processed foods to ultraprocessed foods: evolution of an industry model and impact on dietary quality, health, and society. Annu. Rev. Food Sci. Technol. 16 (1), 1–24. doi:10.1146/annurev-food-111523-122028.

Van Middelaar, C.E., Dijkstra, J., Berentsen, P.B.M., De Boer, I.J.M., 2014. Cost-effectiveness of feeding strategies to reduce greenhouse gas emissions from dairy farming. J. Dairy Sci. 97 (4), 2427–2439. doi:10.3168/jds.2013-7648.

Varzakas, T., Antoniadou, M., 2024. A holistic approach for ethics and sustainability in the food chain: the gateway to oral and systemic health. Foods 13 (8), 1224. doi:10.3390/foods13081224.

Williams, A.R., Andersen-Civil, A.I.S., Zhu, L., Blanchard, A., 2020. Dietary phytonutrients and animal health: Regulation of immune function during gastrointestinal infections. J. Anim. Sci. 98 (4). doi:10.1093/jas/skaa030.

Youssef, I.M., Aldhalmi, A.K., Felemban, S.G., Elsherbeni, A.I., Khalil, H.A., Hassan, M.S., El Halim, H.S.A., El-Hack, M.E.A., Youssef, K.M., Swelum, A.A., Tufarelli, V., Abo-Samra, M.A., 2024a. Mannan oligosaccharides as a prebiotic for laying hens: effects on fertility, hatchability, productive performance, and immunity. Transl. Anim. Sci. 8. doi:10.1093/tas/txae123.

Youssef, I.M., Elsherbeni, A.I., Almuraee, A.A., Nass, N.M., Beyari, E.A., Alshammarii, N.M., Abdel-Ghany, A.M., Ahmed, E.S.G., Nasr, S., Youssef, K.M., Salem, H.M., Abd El-Hack, M.E., Saber, H.S., 2024b. Influence of using synbiotics by various routes on Mandarah male chicks: intestinal bacterial counts, gut morphology and histological status. Poult. Sci. 103 (5). doi:10.1016/j.psj.2024.103601.

Youssef, I.M., Khalil, H.A., Jaber, F.A., Alhazzaa, R.A., Alkholy, S.O., Almehmadi, A.M., Alhassani, W.E., Al-Shehri, M., Hassan, H., Hassan, M.S., Abd El Halim, H.S., El-Hack, M.E.A., Youssef, K.M., Abo-Samra, M.A., 2023a. Influence of dietary mannan-oligosaccharides supplementation on hematological characteristics, blood biochemical parameters, immune response and histological state of laying hens. Poult. Sci. 102 (11). doi:10.1016/j.psj.2023.103071.

Youssef, I.M., Khalil, H.A., Shakoori, A.M., Bagadood, R.M., Alyahyawi, A.Y., Alhazzaa, R.A., Fakiha, K.G., Nasr, S., Abo-Samra, M.A., Hassan, M.S., Halim, H.S.A.E., El-Hack, M.E.A., Jaremko, M., Al-Nemi, R., Youssef, K.M., 2023b. Immune response, hematological traits, biochemical blood parameters, and histological status of laying hens influenced by dietary chitosan-oligosaccharides. Poult. Sci. 102 (9). doi:10.1016/j.psj.2023.102834.

Youssef, I.M., Khalil, H.A., Swelum, A.A., Al Garadi, M.A., Balasubramanian, B., Hassan, M.S., Abd El Halim, H.S., Abd El-Hack, M.E., Youssef, K.M., Abo-Samra, M.A., 2024. Influence of dietary chitosan-oligosaccharides supplementation on productive and reproductive performance of laying hens. Ann. Anim. Sci. 24 (2), 491–502. doi:10.2478/aoas-2023-0082.

Zhang, L., Yan, C., Guo, Q., Zhang, J., Ruiz-Menjivar, J., 2018. The impact of agricultural chemical inputs on environment: global evidence from informetrics analysis and visualization. Int. J. Low Carbon Technol. 13 (4), 338–352. doi:10.1093/ijlct/cty039.

Zheng, S., Li, Y., Chen, C., Wang, N., Yang, F., 2025. Solutions to the dilemma of antibiotics use in livestock and poultry farming: regulation policy and alternatives. Toxics 13 (5), 348. doi:10.3390/toxics13050348.

Zuberi, A., Kamran, M., Younus, N., Abdel-Tawwab, M., 2024. Functional feed additives: current trends. Front. Aquac. 3, 1385508.

Chapter 2

Oxidative stress, inflammation, and gut-related health

Islam Mohamed Youssef[a], Muhammad Arif[b], Asmaa Fahmy Khafaga[c] and Mohamed E. Abd El-Hack[d]

[a]*Animal Production Systems Research Department, Animal Production Research Institute, Agricultural Research Center, Giza, Egypt,* [b]*Department of Animal Sciences, College of Agriculture, University of Sargodha, Sargodha, Punjab, Pakistan,* [c]*Department of Pathology, College of Veterinary Medicine, King Faisal University, Al-Ahsa, Saudi Arabia,* [d]*Department of Industrial Pharmacy, College of Pharmaceutical Sciences and Drug Manufacturing, Misr University for Science and Technology (MUST), Giza, Egypt*

2.1 Introduction

Mood disorders such as depression and anxiety are becoming increasingly prevalent in the world, largely due to modern dietary and lifestyle factors. Diets high in sugar, saturated fats, and processed foods are major culprits of neurobiological derangements that undermine emotional regulation (Mottis et al., 2025). Of all the mechanisms by which diet is involved in mental well-being, oxidative stress and inflammation are top contenders. The high oxygen utilization rate, fat content, and low antioxidant activity of the brain make it highly susceptible to oxidative damage, which disrupts neuronal transmission and leads to dysregulation of mood (Agnihotri and Aruoma, 2020).

Oxidative stress results from a mismatch between the formation of reactive oxygen species (ROS) and antioxidant defense in the body. Mitochondria are the main producers of ROS in energy metabolism, but malnutrition, pollution, and chronic stress can also increase ROS production (Corkey, 2023). Oxidative stress is increased when lipids, proteins, and DNA are injured by excessive oxidative stress, leading to dysfunction of neurons. Increased oxidative biomarkers of malondialdehyde (MDA) and decreased antioxidants such as glutathione (GSH) have been consistently found in the instance of depression, indicating a strong association between redox imbalance and mood disorder (Vaváková et al., 2015). Redox homeostasis is maintained through antioxidant defense mechanisms that include enzymes such as superoxide dismutase (SOD), catalase (CAT), and glutathione peroxidase (GPx), and nonenzymatic antioxidants such as vitamins C and E. Vulnerability in these systems enhances sensitivity to oxidative damage

Plant-Derived Feed Additives and Animal-Based Foods: Linking Animal Health, Product Quality, and Consumer Perceptions. DOI: https://doi.org/10.1016/B978-0-443-45446-2.00015-6

and emotional liability. Clinical and animal studies suggest that antioxidant therapies (e.g., *N*-acetylcysteine and coenzyme Q10) can decrease depressive-like behavior, though findings are inconsistent (Majmasanaye et al., 2024).

Another important factor that relates diet, oxidative stress, and depression is inflammation. The blood–brain barrier (BBB) is crossed by proinflammatory cytokines TNF-α, IL-1β, and IL-6, which activate microglia and cause neuroinflammation. The connection between obesity and mood disorders may also be affected by HPA activation that results in high levels of cortisol; both may prevent the formation of new cells in areas of the brain that may contribute to mood disorders (Cheiran Pereira et al., 2022). Chronic low-grade inflammation is common in individuals with obesity and metabolic syndrome and also contributes to increasing levels of oxidative stress and impaired mood (Mancini et al., 2021). The types of food consumed will affect both of these systems. Diets heavy in fat and sugar cause oxidative and inflammatory stress, which inhibits hippocampal plasticity and cognitive function (Li et al., 2025).

Antioxidant and antiinflammatory nutrients such as omega-3 fatty acids, polyphenols, and vitamins, on the other hand, promote neural resilience and mood stability. Similarly, plant extracts such as curcumin, ashwagandha, and rosemary regulate Nrf2 antioxidant response and suppress NF-κB-mediated inflammation and exhibit promising neuroprotection with some limitations in bioavailability and standardization (Sivamaruthi et al., 2023).

Oxidative stress and inflammation are linked biological processes through which diet impacts mental well-being. Interplay between the two has provided a foundation for nutritional and therapeutic interventions to reduce the psychological price of the modern diet. The rest of this chapter elaborates in detail on the molecular roles of oxidative stress and inflammation in mood disorders, diet-induced neuroinflammatory impact, and the potential of animal-derived and plant-derived interventions for the preservation of neural redox homeostasis.

2.2 Role of oxidative stress in mood disorders

2.2.1 Brain susceptibility to oxidative damage

The human brain is particularly vulnerable to oxidative injury from a convergence of anatomical, biochemical, and metabolic forces that amplify the generation and impact of ROS (Houldsworth, 2024). Although oxygen is critical for neuronal function, its own metabolism produces ROS in the shape of superoxide anions (O_2^-), hydrogen peroxide (H_2O_2), and hydroxyl radicals ($OH\bullet$), which can harm lipids, proteins, and nucleic acids (Singh et al., 2025a). The brain consumes approximately 20% of the body's oxygen but only comprises 2% of its mass and is therefore an oxidative metabolic hotspot. Such high oxygen demand enables ATP-depleting activities such as synaptic transmission and ion gradient maintenance but threatens the potential for ROS accumulation, especially in ischemia or mitochondrial dysfunction (Cobley et al., 2018). Neural membranes contain high levels of polyunsaturated fatty acids (PUFAs); this characteristic makes these lipids readily available for oxidative damage. The oxidation of PUFAs produces reactive aldehyde products such as MDA and 4-hydroxynonenal that compromise the integrity of the cell membrane and cause disruptions to the cellular signaling processes (Allowitz et al., 2025). Furthermore, because of their extended axonal shape and high surface area-to-volume ratio, neurons are extremely susceptible to oxidative damage, which can interrupt membrane function. The excitatory neurotransmitter glutamate has the possible to trigger excitotoxicity when it is released in excess, leading to calcium overload and the activation of ROS-producing enzymes such as nitric oxide synthase (NOS) and NADPH oxidase (NOX2) (Ferreiro et al., 2012). Intracellular calcium dysregulation is also a key contributor to oxidative stress in the brain.

Excess intracellular calcium activates phospholipase A_2 and mitochondrial pathways that enhance ROS production. Mitochondria, which are also major producers of ATP, are great producers of ROS (Angelova and Abramov, 2024). Abnormal mitochondria can lose electrons in oxidative phosphorylation to form superoxide radicals. This is augmented by the presence of redox-active transition metals such as iron and copper that catalyze the Fenton reaction for reducing H_2O_2 to extremely reactive hydroxyl radicals. Iron is particularly abundant in brain regions such as the substantia nigra, and its release after injury or neurodegeneration enhances oxidative damage (Jomova et al., 2025). Protective antioxidant levels in the brain are relatively low. Enzymes such as SOD, CAT, and GPx are present but typically insufficient to repress excess ROS (Ighodaro and Akinloye, 2018). GSH, a ubiquitous nonenzymic antioxidant, is usually low in neurodegenerative disorders, which further disturbs redox balance (Cacciatore et al., 2012). Even monoamine neurotransmitters such as dopamine and serotonin can auto-oxidize, generating semiquinone radicals and quinones that redox cycle and create ROS. Redox cycling is catalyzed by metal ions and is the reason behind the induction of Parkinson's and Alzheimer's diseases (Hatziagapiou et al., 2019).

2.2.2 Reactive oxygen species production and sources (mitochondria, diet)

ROS are chemically reactive oxygen-derived radicals, including superoxide anions (O_2^-), hydrogen peroxide (H_2O_2), and hydroxyl radicals (OH •). They are generated as a natural by-product of cellular metabolism and play dual physiological and pathophysiological functions. Mitochondria are the principal endogenous producers of ROS in most cells, especially in metabolically active tissues, including the brain, heart, and liver. Under oxidative phosphorylation, electrons are transferred through complexes I to IV of the mitochondrial electron transport chain (ETC). Under normal conditions, a small number of electrons leak from complexes I and III, converting oxygen early on to superoxide radicals. The radicals are then converted to hydrogen peroxide by the activity of SOD and thereafter detoxified by CAT or GPx to water and oxygen

(Choi and Kim, 2019). Mitochondrial ROS (mROS) are not only poisonous byproducts but also signaling molecules that affect cellular activities such as apoptosis, autophagy, and immune responses.

Yet, oxidative stress, a condition in which ROS outweigh antioxidant defense, is brought on by mROS hyperproduction, which is naturally brought on by mitochondrial dysfunction (Shadel and Horvath, 2015). Numerous illnesses, such as cancer, cardiovascular disorders, and neurodegenerative diseases, have this imbalance as their root reason. Age, ischemia-reperfusion injury, and exposure to toxins and/or inflammatory cytokines are circumstances that promote mitochondria to produce ROS. The regulation of mitochondrial redox state is a dynamic process that could lead to oxidation shifts, which result in a loss in membrane potential, a decline in ATP synthesis, and damage to mitochondrial DNA, proteins, and lipids (Xu et al., 2025). Dietary sources are significant in managing ROS levels, in addition to those produced endogenously.

Increased oxidative stress is common in diets that consist of processed foods, as well as those that have a high amount of refined carbohydrates and fats. This is due to the fact that the diets are typically high in sugar, fat, and calories, which contribute to inflammation and mitochondrial dysfunction (Jiang et al., 2021). A diet made up of fat can have damaging effects on how well a mitochondrion can produce energy, thus resulting in the production of more ROS through increased electron leakage (Ruiz-Ramírez et al., 2011). However, a diet with an abundance of antioxidants such as polyphenols, carotenoids, vitamin C and vitamin E will allow the body to neutralize free radicals and provide additional support for the body's defense mechanism against oxidative stress. The preservation of redox balance depends on certain nutrients, such as zinc and selenium, which are cofactors to antioxidant enzymes (Papas, 2019). Through augmented metabolic flux and inflammation of adipose tissue, which again triggers mROS production, calorie excess and obesity are also related to improved ROS production (Corkey, 2023). Mitochondrial function is directly affected by certain food ingredients. For example, the polyphenol resveratrol, which is found in berries and grapes, stimulates sirtuin pathways that lower ROS levels and promote mitochondrial biogenesis.

Also, omega-3 fatty acids decrease oxidative damage and advance the fluidity of the mitochondrial membrane (Morkovin et al., 2024). Heavy metals, similar to lead and cadmium, which are frequently present in tainted food and water, and alcohol can both disrupt mitochondrial enzymes and raise the production of ROS. So, nutrition can be changed to control mitochondrial function and oxidative stress (Jayaraj et al., 2025).

2.2.3 Biomarkers indicative of oxidative stress in depression

The pathophysiology of depressive disorders (DDs) is well known to be influenced by oxidative stress (OS), and some studies have found specific biomarkers that suggest the level of antioxidant defense and oxidative injury in patients with

DD (Barbosa et al., 2020). The production of ROS and the ability of the body's antioxidant system to counteract them have been associated with depression. Biomolecules such as proteins, lipids, DNA, and RNA are oxidatively modified as a result of this mismatch (Bhatt et al., 2020). MDA, a lipid peroxidation product, and 8-hydroxy-2′-deoxyguanosine (8-OHdG), a marker of oxidative DNA damage, are two of the most studied biomarkers of depression. Patients with major DD have also been found to have higher levels of 8-OHdG and MDA, indicating increased oxidative injury to DNA and cellular membranes (Barbosa et al., 2020). Increased levels of 3-nitrotyrosine (3-NT) and protein carbonyl content (PCC), which indicate oxidative modifications of amino acid residues, are examples of protein oxidation that have also been associated with depression (Bandookwala and Sengupta, 2020).

These altered states also have the potential to create neuroinflammatory changes, as well as damaging neurons and altering the functions of proteins. Furthermore, elevated concentrations of lipid peroxidation markers such as 4-hydroxynonenal (4-HNE) and F2-isoprostanes (F2-iso) have been found in people with depression, indicating a potential causal role of OS in this disorder (Barbosa et al., 2020). For example, depressed individuals have been reported to have lower levels of the activity of the important enzymatic antioxidant enzymes such as GPx, CAT, and SOD, which are critical for maintaining the cellular regulation of the redox balance and detoxifying the reactive oxidative species (Ozcan et al., 2004). The decrease in these critical antioxidant enzymes may lead to an increase in oxidative damage and persistence of depressive symptoms. In addition, depressed individuals have also been found to be particularly susceptible to OS because of deficiencies of nonenzymatic antioxidants, such as vitamins A and C (He et al., 2017).

2.2.4 Defense mechanisms for antioxidants

One of the most important cellular mechanisms for avoiding the damaging effects of ROS, which are continuously produced during normal metabolic processes, especially in the mitochondrial energy-generating pathway, is the antioxidant defense system. First-line enzymatic antioxidants, being part of the most crucial elements of this system, consist of SOD, CAT, and GPx. They cooperate to detoxify ROS and prevent oxidative damage to macromolecules in cells, such as lipids, proteins, and DNA (Ighodaro and Akinloye, 2018). SOD catalyzes the superoxide anion (O_2^-) dismutation to molecular oxygen (O_2) and hydrogen peroxide (H_2O_2).

H_2O_2 is less reactive than superoxide but can also be hazardous if not rapidly eliminated (Batool et al., 2022). This is where CAT and GPx step in. Mostly found in peroxisomes, CAT quickly converts H_2O_2 into oxygen and water, preventing its buildup and subsequent Fenton reaction conversion to the extremely reactive hydroxyl radical (OH•) (Anwar et al., 2024). In contrast, GPx utilizes GSH as a substrate to detoxify H_2O_2 and lipid hydroperoxides, oxidizing glutathione (GSSG) in the process. Tripeptide GSH is a powerful nonenzymatic antioxidant that recycles other antioxidants and directly scavenges free radicals (Csiszár et al., 2016). When combined, SOD, CAT, and GPx provide a potent first line of defense against OS. Together, they preserve redox homeostasis and shield cells from oxidative damage. Several pathological states, such as depression, cardiovascular disease, and neurodegenerative disorders, have been etiologically linked to deficiencies and malfunctions of these enzymes. One potentially helpful treatment to counteract oxidative damage and improve cellular resilience is pharmacological enhancement of these antioxidants' expression or activity (Ighodaro and Akinloye, 2018).

2.2.5 Antioxidant therapy

As more and more research links OS to the pathophysiology of depression, antioxidant therapy has gained attention as a possible adjunctive treatment strategy for DD. OS damages cellular components such as DNA, proteins, and lipids because of an imbalance between the production of ROS and the capacity of antioxidant defenses to detoxify them. Increased oxidative damage is indicated by elevated levels of oxidative biomarkers such as MDA, PCC, and 8-hydroxy-2′-deoxyguanosine (8-OHdG) in depressed patients. At the same time, decreased levels of non-enzymatic antioxidants like vitamins A and C, as well as antioxidant enzymes such as GPx, CAT, and SOD, propose a weakened antioxidant defense system (Barbosa et al., 2020).

Antioxidant supplements have been investigated in clinical trials as a potential treatment for depression symptoms. For the reason that they can scavenge ROS and regenerate other antioxidants, vitamins C and E, for example, are effective in lowering depressive scores (Manosso et al., 2022). It has also been established that vitamin A (retinol palmitate) has mood-stabilizing properties, perhaps through its effects on oxidative pathways and neuroinflammation (Kurowska et al., 2023). In randomized controlled trials, *N*-acetylcysteine (NAC), a precursor to GSH, was found to enhance mood and cognitive function by lowering oxidative damage and restoring GSH levels. These results support the theory that depression's biochemical imbalances can be reversed by restoring antioxidant capacity (Bradlow et al., 2022).

The clinical indication is also supported by experimental models. Antioxidant compounds have been shown to reverse behavioral and biochemical alterations induced by chronic stress or neurotoxic challenges in animal studies. For example, in animal studies, the use of polyphenols, flavonoids, and coenzyme Q10 has led to the reversal of antioxidant enzyme activity, oxidative marker levels, and depressive-like behaviors. These compounds not only reduce ROS levels but also affect signaling pathways such as NF-κB and Nrf2, which are involved in inflammation and neuroplasticity (Winiarska-Mieczan et al., 2023). In summary, antioxidant therapy is an effective adjunct therapy in the management of DD, as evidenced by both clinical and experimental studies.

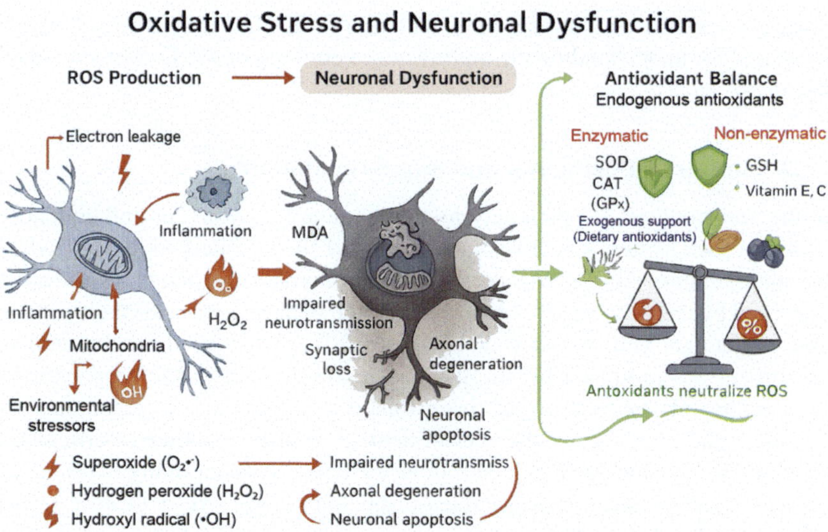

FIGURE 2.1 Schematic illustration of oxidative stress-induced neuronal dysfunction.

Antioxidants offer a biologically plausible and promising approach to improve mental health outcomes by reversing redox imbalance and preventing oxidative damage (Barbosa et al., 2020). Neuronal damage induced by OS is illustrated in Fig. 2.1.

2.3 Inflammatory mechanisms and depression

2.3.1 Peripheral versus central inflammation

The association between peripheral and central immune dysregulation and major depressive disorder (MDD) is increasingly recognized. Systemic sources, including the gut, liver, spleen, and adipose tissue, can induce peripheral inflammation, characterized by the activation of the immune system outside the central nervous system (CNS) (Sun et al., 2022). These peripheral immune signals can affect brain function through humoral pathways, neural pathways (e.g., the vagus nerve), and by breaking down the BBB (Dantzer et al., 2000). Conversely, central inflammation pertains to immune responses occurring within the CNS, predominantly facilitated by astrocytes and microglia. Elevated circulating cytokines, indicative of peripheral immune activation in MDD, can modify BBB permeability, facilitating the ingress of immune cells and inflammatory mediators into the brain (Wu et al., 2022). Neuroinflammation is exacerbated when this infiltration induces the release of proinflammatory cytokines and neurotoxic substances by microglia, the brain's intrinsic immune cells. Bidirectional signaling between the central and peripheral immune systems creates a feedback loop that makes depressive symptoms worse. This crosstalk necessitates therapeutic strategies that address both systemic and neuroinflammatory pathways, which are crucial for understanding the multifactorial etiology of MDD (Müller and Di Benedetto, 2025).

2.3.2 Important cytokines and mood regulation

At the forefront of the pathophysiology of depression are proinflammatory cytokines such as tumor necrosis factor-alpha (TNF-α), interleukin-1β (IL-1β), and interleukin-6 (IL-6). The level of symptoms and treatment resistance is forecasted by these cytokines, which are elevated in the plasma and cerebrospinal fluid of patients with MDD (Farooq et al., 2017). Anhedonia and cognitive impairment are produced by IL-1β-induced dysfunction of the hippocampus and synaptic plasticity. IL-6 has been linked to fatigue and psychomotor retardation, and it regulates the metabolism of neurotransmitters, particularly serotonin and dopamine (Korte and Straub, 2019). Neuronal injury is produced by excitotoxicity and glutamate release modulated by TNF-α (Tolosa et al., 2011). Moreover, cytokines impair neurotrophic pathways and neurogenesis, particularly brain-derived neurotrophic factor (BDNF), which is involved in resilience and mood regulation. The inflammatory hypothesis of depression is validated by the presence of hypercytokinemia in MDD, which also opens the possibility that immune system dysfunction is the underlying cause of the neurobiological profile observed in depression disorders. New therapeutic strategies for mood disorders can emerge from the modulation of these cytokines by pharmacological or nonpharmacological interventions (Jiao et al., 2025).

2.3.3 Activation of the pituitary–adrenal axis caused by inflammation

The hypothalamic–pituitary–adrenal (HPA) axis is a brain system that helps the body deal with stress. It is often not working properly in people with depression. Inflammatory cytokines such as IL-1β, IL-6, and TNF-α can activate the HPA axis by stimulating the hypothalamus to secrete corticotropin-releasing hormone (CRH), which in turn stimulates the pituitary gland to release adrenocorticotropic hormone (ACTH) (Lei et al., 2025). ACTH induces the release of cortisol from the adrenal glands. While cortisol has antiinflammatory actions, chronic elevation from constant inflammation leads to glucocorticoid receptor resistance, causing interference with negative feedback regulation of the HPA axis (Straub and Cutolo, 2016). Prolonged release of cortisol (related to mood disorders), impaired neurogenesis, and reduced size of the hippocampus contribute to the perpetuation of chronic stress and the development of depression (Iob et al., 2020). Chronic stress/interactions with chronic inflammation create a feedback mechanism to maintain depression (Iob et al., 2020). In addition, chronic low-grade inflammation due to elevated levels of peripheral cortisols may reinforce the loop (Iob et al., 2020). Treatment development targeting both the immune and neuroendocrine aspects of depression needs to consider how they interact (Jiao et al., 2025).

2.3.4 Chronic inflammation and hippocampal neurogenesis

Chronic inflammation has profound effects on hippocampal neurogenesis, a process that is very important for learning, memory, and emotional regulation. The hippocampus is the only brain area where adult neurogenesis occurs, and it is highly susceptible to inflammatory injury (Wu and Zhang, 2023). Proinflammatory cytokines IL-1β and TNF-α inhibit the proliferation and differentiation of neural progenitor cells, decrease BDNF expression, and increase the apoptosis of young neurons. These effects lead to structural and functional impairments of the hippocampus, which are normally present in MDD patients (Long et al., 2025). Neuroimaging studies indicated reduced hippocampal volume in depressed patients, associated with disease chronicity and severity. Chronic stress and inflammatory animal models also have compromised neurogenesis and depressive behavior (Kubera et al., 2011). Restoration of hippocampal neurogenesis using antiinflammatory treatment, exercise, or antidepressant therapy has been shown to improve mood and cognitive function. Therapeutic intervention for depression thus has a bright future where neurogenic impairment through inflammation is treated (Azarfarin et al., 2025).

2.3.5 Antiinflammatory drug effects on mood

The therapeutic potential of antiinflammatory drugs in their applications in the treatment of depression has been studied both preclinically and clinically. Antiinflammatory compounds such as cytokine inhibitors, immunomodulatory medications, and NSAIDs are beneficial for reducing depressive symptoms, particularly among patients with elevated inflammatory markers (Adzic et al., 2018). For example, TNF-α inhibitors, such as infliximab, have been noted to have mood-improving effects among treatment-resistant depression with high baseline inflammation (Brymer et al., 2019. Similarly, IL-6 antagonists and minocycline, a tetracycline antibiotic, also showed promise in clinical trials (Garrido-Mesa et al., 2013). Omega-3 fatty acids, with their anti-inflammatory nature, have also been associated with reduced depression symptoms (Giacobbe et al., 2020). Moreover, antidepressants themselves have also been noted to possess antiinflammatory effects through cytokine levels and microglial modulation. However, the heterogeneity of depression and the variability of inflammatory profiles necessitate individualized anti-inflammatory therapy (Drevets et al., 2022). Utilizing biomarkers to screen for inflammation-induced depression may help increase the effectiveness of treatments and lower the number of people who have to undergo an unsuccessful treatment. Antiinflammatory methods of treating people with mood disorders can also be defended as legitimate and biologically sound treatment options (Cătălina et al., 2025).

2.4 Diet-induced neuroinflammation

2.4.1 A diet rich in fats and activation of microglia

A diet that contains high levels of fat (HFDs) is associated with the development of neuroinflammation via microglial activation (the brain's immune cells). Microglial activation has been the signature feature of CNS inflammation and has been observed extensively with the consumption of HFD (Baufeld et al., 2016). Both HFD animal and human models lead to increased circulating levels of saturated fatty acids, which pass through the BBB and stimulate microglia by direct activation through toll-like receptor 4 (TLR4) signaling mechanisms. Activated, they secrete proinflammatory cytokines such as TNF-α, IL-1β, and IL-6, precipitating a neuroinflammatory response (Stathori et al., 2025). The hypothalamus, a major brain area implicated in energy homeostasis, is especially vulnerable to HFD-induced inflammation.

It has been shown by studies that transient exposure to HFD is enough to activate microglia in the mediobasal hypothalamus, fostering gliosis and disruption of appetite- and metabolism-controlling neuronal circuits (Yan and Li, 2022). The inflammatory response is mediated by the NF-κB/IKK-β pathway, which is activated upon dietary fat exposure in microglia. Furthermore, angiotensin II and mitochondrial damage-associated molecular patterns (DAMPs) released during metabolic stress also increase microglial activation by cGAS-STING and TLR signaling (Quan et al., 2025). Experimental evidence from rodent models agrees with these findings. HFD-fed mice exhibit upregulation of ionized calcium-binding adapter molecule 1 (Iba1), a marker of activation of microglia, in the hypothalamus, hippocampus, and cortex. Increases in OS markers and inflammatory cytokines also follow this pattern. It is interesting to note that through excitotoxicity and decreased synaptic plasticity, microglial activation is actively involved in neuronal damage, as opposed to being a bystander effect (Mendes et al., 2018).

2.4.2 The influence of high blood sugar on hippocampal flexibility

Research has demonstrated that the hippocampus, which is involved in the processes of memory (influenced by the formation and utilization of memories), learning (including acquiring new information), and regulating emotion, is adversely affected by a diet high in added sugars, such as sucrose and high-fructose corn syrup (Li et al., 2025). A diet rich in simple sugars affects how the hippocampus's structure and function develop, as well as how it is affected by metabolic or inflammatory insults, as it is particularly susceptible to both types of distress. One of the primary methods by which a high-sugar diet impacts the brain is through the generation of OS and activation of neuroinflammatory processes, both of which interfere with the ability of the brain to communicate through synaptic connections, affecting its ability to regenerate neurons, i.e., neurogenesis. BDNF, which is critical for synaptic plasticity and neuronal survival, is downregulated in animal studies involving excessive sugar consumption (Kaplan et al., 2015). The decrease in BDNF is thought to be associated with abnormal long-term potentiation (LTP), which is the cellular equivalent of memory formation and learning.

It has been shown that rats fed high-sugar diets have lower dendritic spine density in hippocampus neurons and perform poorly in spatial memory tasks such as the Morris water maze test (Olmo et al., 2023). These histopathological changes are accompanied by increased levels of oxidative damage markers such as MDA and proinflammatory cytokines IL-1β and TNF-α. Inhibition of neurotrophic activation and further impairment of synaptic dysfunction are also partial mediators of inflammation (Li et al., 2014). Similarly, excessive sugar intake can affect the signaling of insulin in the brain, resulting in insulin resistance, which further affects the function of the hippocampus.

This cognitive impairment is caused by the dysregulation of insulin, which is also a key element in the release of neurotransmitters and synaptic plasticity (Spinelli et al., 2019). Furthermore, because it is metabolized by astrocytes and is known to generate advanced glycation end products (AGEs), which activate proinflammatory pathways, fructose is specifically found to exacerbate neuroinflammation (Gugliucci, 2017). Of most concern is the fact that this is not dependent on long-term exposure. In fact, a single exposure to a high-sugar diet is found to be sufficient in lowering BDNF levels and modulating memory. This implies that dietary sugars have a quick effect on the brain, especially when exposure occurs at critical developmental periods. As such, adolescents and young adults might be most susceptible because of their brain developmental stage and their sensitivity to external factors (Foulkes and Blakemore, 2018). The effect of fat, sugar, and additives on inflammation markers is depicted in Fig. 2.2.

2.4.3 Neurotoxicity and preservatives

However, while these preservatives are essential for extending the lifespan of food products, preventing microbial contamination, and extending the period for which food remains edible, they have also come under intense scrutiny for their neurotoxicological profiles, especially with regard to excitotoxicity, a type of neurodegenerative pathology involving overstimulation of glutamate receptors, which ultimately causes cell death (Teshome et al., 2022). Monosodium glutamate, sodium nitrite, and artificial food dyes are all types of food preservatives known to exacerbate neuroinflammation/excitotoxicity with high levels of consumption, both acute and chronic. The flavor enhancer MSG can function as an excitatory neurotransmitter agonist and shares structural similarities with glutamate. It has been demonstrated that an acute dietary excess of MSG overstimulates NMDA and AMPA receptors, causing calcium influx, mitochondrial dysfunction, and OS in neurons (Lugrin et al., 2014).

MSG administration causes degeneration of neurons in the hippocampus and hypothalamus, which are important areas for memory and metabolic regulation, according to rodent experimental studies. Increased production

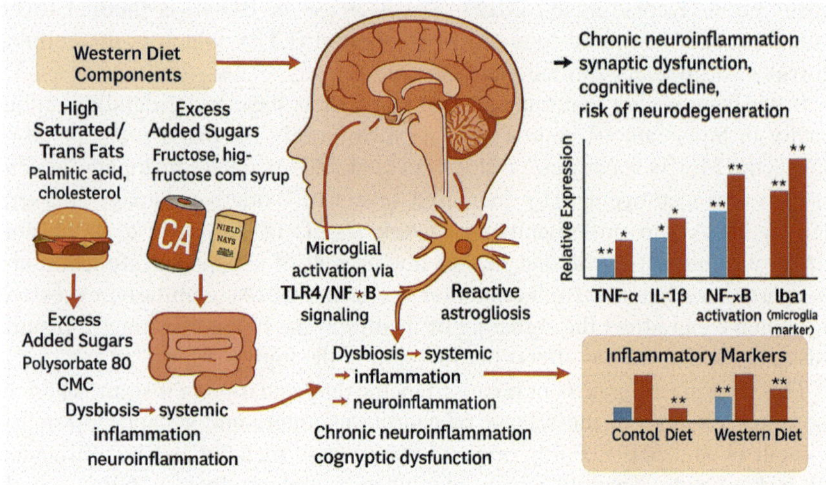

FIGURE 2.2 Western diet promotes neuroinflammation via dietary components.

of lipid peroxidation products, ROS, and proinflammatory cytokines such as TNF-α and IL-1β also accompanies this neurotoxicity (Ankul et al., 2023). Furthermore, exposure to MSG has been linked to astrocyte reactivity and microglial activation, both of which support the inflammatory response. A further common preservative in processed meat is sodium nitrite, which can produce strong neurotoxins called nitrosamines and reactive nitrogen species (RNS). They induce excitotoxic cascades by damaging DNA and interfering with mitochondrial respiration (Albrakati, 2023).

In animal models, artificial food coloring dyes such as sunset yellow and tartrazine have also been linked to behavioral and neurochemical changes. These additives have the potential to promote excitotoxicity by upsetting the balance of neurotransmitters and raising glutamate levels in the brain. Reduced synaptic plasticity, learning impairment, and anxiety-like behavior have all been linked to their long-term consumption. Crucially, early exposure to dietary additives raises concerns because the developing brain is more susceptible to excitotoxic damage (Damotharan et al., 2024).

Usually, the reduced activity of glutamate transporters such as EAAT2 and the inhibition of glutamate uptake by astrocytes are the mechanisms underlying preservative-induced excitotoxicity. Long-term receptor stimulation and extracellular glutamate buildup are the results. By lowering GSH and CAT levels, preservatives can also weaken antioxidant defenses and promote oxidative damage. Together, excitotoxicity and neuroinflammation produce a feedback loop that exacerbates behavioral abnormalities and neuronal dysfunction (Bensid et al., 2022).

2.4.4 Results of behavior in animal models

Determining the behavioral effects of diet-induced neuroinflammation, especially in relation to high-fat and high-sugar diets, has been made possible in large part by animal models. Rodent models consistently show that such diets cause significant behavioral changes, especially raised anxiety-like and depressive-like behavior, impaired cognition, and decreased social interaction, in addition to activating neuroinflammatory reactions. Inflammatory alterations in brain areas essential to emotion regulation and executive function, including the hippocampus, prefrontal cortex, and amygdala, are closely associated with such behavioral changes (Alessandrello et al., 2024). In this case, mice given a high-fat diet for a few weeks show increased immobility in the tail suspension and forced swim tests, which are traditional indicators of depressive-like behavior.

Furthermore, these animals exhibit increased plus maze and decreased exploration in the open field test, both of which point to increased anxiety (Yu et al., 2021). These behavioral alterations are accompanied by decreased expression of neurotrophic factors such as BDNF, increased microglial activation, and higher concentrations of proinflammatory cytokines (Patterson, 2015). According to Tran and Westbrook (2017), high-sugar diets also worsen performance on tests of spatial memory, such as the Morris water maze and novel object recognition, which is indicative of hippocampus-dependent learning impairment. Neuroinflammation brought on by diet also has an impact on motivation and reward perception.

Rats fed obesogenic diets exhibit anhedonia, a primary symptom of depression, as evidenced by reduced reactions to natural rewards and changed dopamine signaling in the nucleus accumbens (Wallace and Fordahl, 2022). In addition, long-term consumption of processed foods high in sugar and saturated fats deteriorates sleep architecture and circadian rhythms, which exacerbates mood swings. It is possible to reverse some of these behavioral shifts. Neuroinflammation can be decreased, and behavior returned to normal by giving animals antiinflammatory drugs (such as minocycline or omega-3 fatty acids) or a regular diet. It shows the potential for therapeutic intervention and the malleability of diet-induced brain adaptations (Ballaz et al., 2023).

2.4.5 Variances by sex in the effects of diet on inflammation

The neuroinflammatory response to diet is significantly influenced by sex differences, which also affect a person's vulnerability to cognitive and affective disorders brought on by consuming too much fat and sugar. Males and females have different profiles of immune activation, hormone regulation, and behavioral responsiveness to obesogenic diets, according to growing evidence from both human and animal studies (De Paula et al., 2024). Hormones, specifically estrogen and testosterone, regulate these proteins differently in the sexes. These hormones influence protein regulation by adjusting immune cells, BBB function, and neuroplasticity. Estrogen, for example, is antiinflammatory and enhances antioxidant defense, potentially providing females with diet-induced inflammation neuroprotection (Vollbracht and Werner, 2024).

Rodent studies have established that male mice fed a high-fat diet show elevated microglial activation and greater proinflammatory cytokine concentrations (e.g., IL-6 and TNF-α) within the hippocampus and hypothalamus compared to females. Greater inflammatory response is correlated with greater spatial memory impairments and greater depressive-like behavior in males (Santos et al., 2022). In contrast, female rodents are resistant to some impairments of cognition but are possibly more susceptible to anxiety-like behavior with identical dietary conditions. This differential behavior is correlated with sex-specificity in neurotrophic factors such as BDNF and synaptic proteins, suggesting differential mechanisms of diet-induced brain dysfunction (Shobeiri et al., 2022). Sex differences also extend to metabolic responses that influence neuroinflammation. Men have greater visceral fat accumulation, which is actively metabolized and releases a higher level of inflammatory adipokines. Subcutaneous fat deposition is higher in women and less inflammatory. Differentiated fat distribution between males and females has been shown to have different levels of impact on the immune system as well as influence the level of systemic inflammation (Cartier et al., 2009). In addition, the gut microbiome also plays an important role in regulating inflammation based on dietary intake and is differentially organized by sex, leading to differential responses to changes in dietary patterns (Shobeiri et al., 2022).

Vulnerability to diet-induced neuroinflammation is also influenced by changes in hormones throughout life. For example, postmenopausal women have lower estrogen levels, which may increase their susceptibility to metabolic dysregulation and neuroinflammatory disorders (Villa et al., 2016). Similarly, elevated inflammation and cognitive decline have been linked to testosterone deficiency in older men. According to Ketchem et al. (2023), all of these results lend credence to the inclusion of sex and hormone characteristics in research on the relationship between diet and brain function.

2.5 Foods based on animals as oxidative balance modulators

2.5.1 The impacts of saturated and unsaturated fatty acids

Whether dietary fat is unsaturated or saturated, its structure plays a crucial role in changing the redox balance and influencing the risk of chronic diseases. Because they are not double-bonded, saturated fatty acids (SFAs), which are widely present in animal products, such as red meat, butter, and whole-fat dairy products, are resistant to chemical reactions. However, they have a less favorable metabolic action (Bajželj et al., 2021). It has been demonstrated that consuming large amounts of SFAs increases oxidative stress by inducing inflammation and lipid peroxidation. Excessive SFA consumption results in mitochondrial impairment and ROS generation, which is especially noticeable in obesity and cardiovascular disease (Tan and Norhaizan, 2024). Monounsaturated fatty acids (MUFAs) and polyunsaturated fatty acids (PUFAs), on the other hand, are antioxidants that protect against OS.

A few double bonds in these fats, which are present in fish, nuts, and other animal products such as grass-fed meat and eggs, help to make cell membranes fluid and increase the activity of antioxidant enzymes (Hegazy et al., 2025). Particularly, omega-3 PUFAs have been demonstrated to enhance endothelial function, control NF-κB-driven inflammation, and suppress ROS production (Ghosh et al., 2017). Substitution of unsaturated fats in place of SFAs in the diet has been associated with reduced levels of oxidative biomarkers such as MDA and improved GSH status (Asghar et al., 2024). Dietary fat's oxidative potential is also influenced by its metabolic state and interactions with other nutrients. In particular, OS and insulin resistance are increased when high SFA intake is combined with hypercarbohydrate intake (Tamer et al., 2020). In addition, how animal fats are stored and fried may affect how stable they are against oxidation. When fried, SFAs' oxidation will decrease, but their metabolic activity will still be proinflammatory. Unsaturated fats, on account of being better, will be susceptible to oxidation under high heat or prolonged storage conditions and therefore run the risk of the formation of harmful lipid peroxides (Machado et al., 2023).

2.5.2 Omega-3s and neural resilience

Eicosapentaenoic acid (EPA) and docosahexaenoic acid (DHA), two of the omega-3 PUFAs that are crucial for meat products such as fatty fish (such as salmon, mackerel, and sardines), have been thoroughly researched for their neuroprotective and antiinflammatory qualities. They influence fluidity, receptor function, and signal transduction and are an essential part of the structure and operation of neuronal membranes (Zárate et al., 2017). DHA is highly concentrated in the brain and is essential for neurotransmission, neurogenesis, and synaptogenesis via the regulation of OS and inflammation. The omega-3 content of animal diet has been associated with improved neural resilience, or the brain's capacity to adapt to stress and trauma (Khelfi, 2024). Omega-3s exert their effect by suppressing the synthesis of proinflammatory cytokines IL-1β, IL-6, and TNF-α and stimulating the production of antiinflammatory mediators, resolvins, and protectins.

Such lipid-derived molecules facilitate resolution and prevent chronic neuroinflammatory diseases that lead to mood disorders and neurodegenerative diseases (Videla et al., 2023). Moreover, omega-3s enhance antioxidant defense mechanisms by increasing the expression of enzymes such as GPx and SOD, which decreases ROS and lipid peroxidation. The combination of both antiinflammatory and antioxidant properties makes omega-3s advantageous for cognitive function and the survival of neurons during periods of metabolic or psychological stress (Arushi et al., 2025). Omega-3 supplementation has demonstrated positive effects on depressive and anxiety-related disorders and cognitive impairment in both clinical and experimental studies. In animal studies, diets containing DHA and EPA have been shown to improve synaptic plasticity, increase BDNF expression, and protect against excitotoxicity. These positive effects have primarily been observed in regions of the brain associated with memory and emotion, such as the hippocampus and prefrontal cortex (Peng et al., 2020).

2.5.3 Role of cholesterol and oxidation products

A lipophilic compound that is present in significant quantities in food of animal origin, such as eggs, meat, and milk, is essential in brain functions, steroid hormone metabolism, and cell membrane structure. There is, however, an increase in findings that its metabolites, which are produced during oxidation, are responsible for neuroinflammation and OS (Schade et al., 2020). Although it is not toxic, its innate tendency to be oxidized under physiological conditions and in a diet that is high in atherogenic foodstuffs is responsible for the formation of an oxidizable compound that can lead to cell dysfunction. Atherosclerosis, neurodegenerative diseases, and mood fluctuations are associated with its metabolites, such as 7-ketocholesterol and 27-hydroxycholesterol, which are produced enzymatically and nonenzymatically (Samadi et al., 2020).

Since it is essential in brain functions, particularly in synaptic functions and myelin structure, its metabolism is tightly controlled in the brain. An increase in dietary cholesterol and its metabolism leads to an increase in its metabolites, which are responsible for triggering inflammation in the brain after crossing the BBB. These molecules trigger inflammation in the brain, which activates microglia and leads to the release of proinflammatory cytokines (Orth and Bellosta, 2012). Furthermore, oxysterols have the ability to cause mitochondrial dysfunction and raise the production of ROS, which increases oxidative damage. Their mechanism on nuclear receptors such as liver X receptors (LXRs) also modulates lipid metabolism and immune responses in the CNS (Ma and Nelson, 2019).

Animal models of a high-cholesterol diet have been shown to increase oxysterol levels in the brain and plasma, which are associated with learning impairment and altered behavior. The effect is particularly pronounced when the cholesterol is consumed in the oxidized form, such as in processed or cooked meat and animal products. Oxidized cholesterol is more cytotoxic and proinflammatory than its native counterpart, hence the significance that underlying food preparation and storage contribute significantly to modulating oxidative risk (Poli et al., 2022).

2.5.4 Cooking methods and formation of advanced glycation end products

Cooking habits significantly affect the oxidative potential of meat foods, primarily through the formation of AGEs. The Maillard reaction, a nonenzymatic protein–sugar reaction that occurs when sugars and proteins are heated to high temperatures, produces AGEs, a class of reactive products (Stadler and Lineback, 2008). Traditional cooking methods such as grilling, frying, roasting, and broiling promote the formation of AGEs, particularly in foods rich in protein, including fish, poultry, and meat. By binding to cellular receptors and generating ROS, they not only trigger inflammation but also induce oxidative stress (Khelfi, 2024). The receptor for AGEs (RAGE), which is expressed on various cell types, including endothelial cells, macrophages, and neurons, can be triggered by dietary AGEs (dAGEs) that accumulate in tissues after ingestion (Garay-Sevilla et al., 2021).

Activation of RAGE by NF-κB triggers intracellular signaling pathways, leading to the secretion of the proinflammatory cytokines TNF-α and IL-6. Since it compromises antioxidant defenses and allows mitochondrial damage, this inflammatory response is inextricably bound to OS (Guarneri et al., 2021). Since AGEs can impair synaptic activity and trigger neuroinflammation, they have been associated with mood disorders, cognitive impairment, and neurodegenerative processes in the brain (Pizza et al., 2011). Factors that influence AGE formation include temperature, water content, and cooking time.

According to Ramesh et al. (2020), moist-heat preparations such as boiling, stewing, and steaming will produce much less of a toxic compound referred to as AGEs than dry-heat cooking methods such as baking, frying, and barbecuing. For the same quantity of meat, the AGE content in grilled meat was greater than that found in boiled meat. Repeated heating (and restorage) of heat-treated animal products may also lead to higher amounts of AGEs and oxidation (Sobral et al., 2018). Changing the cooking method, using an acid-based marinade (e.g., vinegar or lemon juice), and using high-antioxidant foods (e.g., herbs and spices) will all help reduce the intake of AGEs. These actions will reduce the production of toxic metabolites and slow down the Maillard reaction.

2.5.5 Functional animal-based food concepts

Functional animal food is an emerging field within nutrition science. It includes various products made from conventional livestock, such as meat, fish, eggs, and dairy, but improved with respect to several features: Nutritional and bioactive composition, livestock nutrition management, nutrient enrichment or fortification processes, and processing methods (Vlaicu et al., 2023). Functional animal foods include nutrients such as coenzyme Q10, bioactive peptides, conjugated linoleic acid (CLA), omega-3 fatty acids, and antioxidant vitamins A, D, and E. Dama et al. (2024) state that these compounds may have positive influences on human neurological and cardiovascular health, help improve immune function, and lower oxidative stress (OS).

For example, research on both humans and animals have revealed that eggs fortified with selenium or omega-3 fatty acids have antiinflammatory and antioxidant effects (Usturoi et al., 2025). In addition, milk derived from grazing cows contains vitamin E and CLA, which have been revealed to have antiinflammatory and antioxidant effects. Fermented milk products such as kefir and yogurt contain bioactive peptides that have antioxidant effects through mechanisms such as enhanced GSH synthesis and prevention of lipid peroxidation. These peptides can also influence the gut–brain axis, which can improve moods and memory retention (Pereira et al., 2021).

Adding components to various kinds of animal feeds (e.g., flaxseed or algae), as well as adding antioxidant-rich herbs, has been one recent approach to create functional meat products that contain increased amounts of polyphenols and omega-3 fats (Vlaicu et al., 2023). These developments are intended to not only enhance the health-promoting properties of red meat but also to decrease the prooxidant impact commonly associated with red meat consumption. While fish, particularly oily fish such as salmon and mackerel, may be naturally rich in omega-3 fats, there continues to be interest in methods to enrich aquaculture feed with antioxidants to further enhance their benefits (Pateiro et al., 2021).

Functional animal foods relate to the concepts of personalized dietary plans along with preventive nutrition methods. These engineered animal products, which should have been designed to supply specific bioactive compounds, will assist in reducing the risk of diseases such as metabolic syndrome and neurodegenerative disease, both of which will eventually lead to developing cardiovascular disease because of the effects of OS on them. Functional animal foods are overall a great approach to the maintenance of oxidative balance and good health through strategic enrichment of traditional dietary staples (Khelfi, 2024). Fig. 2.3 shows the comparison between saturated versus omega-3 effects on ROS/inflammation.

2.6 Herbal interventions in reducing oxidative burden

2.6.1 Curcumin, ashwagandha, rosemary—Active compounds and mechanisms

Herbal compounds such as curcumin (*Curcuma longa*), withanolides (*Withania somnifera* or ashwagandha), and rosmarinic acid (*Rosmarinus officinalis*, rosemary) have demonstrated high antioxidant and antiinflammatory activity by regulating key cellular signaling pathways (Bhargavi and Madhan Shankar, 2021). These phytochemicals act on molecular targets involved in OS and inflammation, including nuclear factor erythroid 2–related factor 2 (Nrf2), nuclear factor kappa B (NF-κB), and mitogen-activated protein kinases (MAPKs), which are central to redox homeostasis and immune responses (Surh and Na, 2008). Curcumin is one of the best-studied herbal antioxidants.

By triggering the Nrf2 pathway, it increases the transcription of antioxidant response elements (AREs) and upregulates enzymes such as heme oxygenase-1 (HO-1), SOD, and GPx. This strengthens cellular defense against ROS (Shahcheraghi et al., 2022). Meanwhile, curcumin inhibits NF-κB activation by preventing phosphorylation and IκBα degradation, which reduces the expression of proinflammatory cytokines. It also suppresses MAPK signaling, which

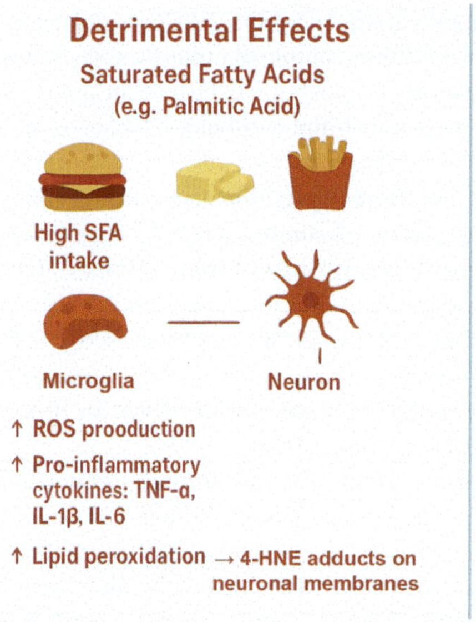

FIGURE 2.3 Comparison between saturated vs omega-3 effects on ROS/inflammation. *ROS*, Reactive oxgen species.

encompasses p38, JNK, and ERK. This further suppresses the expression of inflammatory genes and oxidative damage (Patel et al., 2020). Ashwagandha's bioactive withanolides have similar effects. By inhibiting NF-κB signaling and modifying Nrf2, they lessen OS and inflammation. Ashwagandha also influences the HPA, enhancing neuroprotection and reducing stress-induced ROS production (Basudkar et al., 2024). Rosemary contains rosmarinic acid, which scavenges free radicals and prevents lipid peroxidation. Its neuroprotective and antiinflammatory qualities are further enhanced by its induction of Nrf2 and inhibition of the NF-κB and MAPK pathways (Kosmopoulou et al., 2024). Together, these herbs fight OS and inflammation. Their ability to modulate signaling pathways on multiple targets makes them excellent therapeutic compounds for neurodegenerative disorders, mood disorders, and age-related diseases. Further investigation into the molecular mechanisms and clinical uses of these substances is required before they can be incorporated into evidence-based complementary medicine (Liu et al., 2019).

2.6.2 Behavioral benefits in preclinical models

Behavioral activity was demonstrated by herbal antioxidants such as curcumin, ashwagandha, and rosemary in preclinical models of neuropsychiatric disorders, particularly those related to OS and neuroinflammation (Aswar et al., 2024). Rodent research has demonstrated that these substances may improve cognitive function, increase resilience to stress, and decrease anxiety- and depression-like behavior in rodent models. The mechanisms underlying these effects primarily involve the ability of these compounds to modulate oxidative processes, decrease neuroinflammation, and restore neurochemical homeostasis in certain areas of the brain (i.e., amygdala, prefrontal cortex, and hippocampus) (Cobley et al., 2018).

The effects of curcumin on neurodegeneration, depression, and chronic stress have been well documented. The administration of curcumin to CUS mice also significantly decreased forced swimming immobility and improved open field and elevated plus maze performance (Liu et al., 2014). Increased activity from antioxidant enzymes such as SOD, reduced MDA levels, and increased levels of GSH accompanied the behavioral improvements. Furthermore, curcumin increased levels of the signaling molecule BDNF, which is involved in mood regulation and synaptic plasticity (Radbakhsh et al., 2024).

Ashwagandha withanolides have shown anxiolytic and antidepressant-like effects in various models of behavior. In rodent social defeat stress and corticosterone-induced depression, ashwagandha increased social interaction, reduced anxiety-like behavior, and normalized corticosterone levels. These were linked to reduced oxidative damage and inhibition of proinflammatory cytokines IL-6 and TNF-α. Ashwagandha also modulated the HPA axis, which contributed to superior stress adaptation (Kim et al., 2025).

Rosemary rosmarinic acid has also shown cognition-enhancing and mood-regulating activities in aging and neurotoxicity models. In scopolamine- or lipopolysaccharide (LPS)-treated rats, rosemary extract improved memory performance in the Morris water maze and Y-maze tasks. Such behavioral enhancement was correlated with reduced OS biomarkers and increased acetylcholine levels, suggesting a neuroprotective effect (Hussain et al., 2022). In conclusion, herbal natural products that are prominent for their antioxidant effects have been shown in experimental animals and cultured cells to have potential for preventing symptoms related to depression, anxiety, and cognitive dysfunction. Cobley et al. (2018) described these findings as a strong basis to support moving forward with translational research medicine as well as developing phytotherapeutic interventions aimed at addressing oxidative and inflammatory processes associated with mental illnesses.

2.6.3 Use in functional foods and animal diets

To improve their oxidative resistance and maintain health, herbal antioxidants are increasingly being incorporated into the diet and functional food of animals. In addition to providing food, functional foods are designed to be bioactive, which helps them interact with physiological processes, such as inflammation and OS (Dosoky et al., 2024; Abd El-Hack et al., 2025). Incorporating these spices, which are rich in polyphenols, terpenoids, and alkaloids, into food products and animal feed formulations has been proven to improve their antioxidant, immune, and metabolic responses (Deeb et al., 2025; Negm et al., 2025). The most common functional food applications of curcumin, which is a polyphenol compound found in turmeric, are snack bars, dairy, and beverages.

Curcumin is a powerful antiinflammatory and antioxidant compound with inhibitory effects on NF-κB signaling and activating capabilities of Nrf2 signaling pathways (Rapti et al., 2024). It has been shown to improve immunity, decrease oxidants, and enhance production parameters (growth performance) among some aquaculture and poultry species through increased antioxidant enzyme activity (e.g., SOD and GPx) and decreased lipid peroxidation (Abd El-Hakim et al., 2020). The potential applications of ashwagandha in food are primarily related to its adaptogenic properties, which reduce stress levels and enhance cognition. It is often added to milk, energy bars, and herbal teas as an adaptogen. Research has indicated that adding ashwagandha to feed for animals and poultry increases feed efficiency, stress resilience, and antioxidant status. The means by which ashwagandha helps to promote overall animal and poultry health and productivity is through regulation of the HPA axis and promoting resilience to environmental stressors (Kuruvilla et al., 2025).

Carnosic acid and rosmarinic acid, which function as antioxidants, render rosemary a natural preservative for meat and dairy products. The antioxidant properties of rosemary contribute to the health benefits of humans, the extension of the shelf life of consumables, and the reduction of lipid oxidation. In addition, rosemary contributes to the improvement of gastrointestinal health, the reduction of inflammation markers, and the enhancement of antioxidant status in animal diets. Liu et al. (2022) have demonstrated that the addition of rosemary to animal feed can decrease oxidative spoilage in animal products and enhance the quality of the flesh produced. Consequently, the utilization of herbal antioxidants as feed additives and functional foods for animals is a sustainable and healthful ingredient choice. In addition to enhancing oxidative balance, these interventions will also enable the enhancement of performance, resilience, and overall product quality in both human and animal production systems.

2.6.4 Limitations: Bioavailability and standardization

While curcumin, ashwagandha, and rosemary are herbal compounds with great promise for use in therapy due to their anti-inflammatory and antioxidant benefits, unfortunately, the key issues of bioavailability and standardization render many of them virtually useless for clinical practice. Bioavailability refers to the percentage of a compound that enters the bloodstream and has a biological effect within the body. The therapeutic efficacy of most phytochemicals, such as curcumin and rosmarinic acid, is limited by their low absorption due to high rates of metabolism and low plasma concentration. An example is curcumin's poor oral bioavailability because of its low solubility and rapid liver metabolism. Thus curcumin is often combined with other agents to achieve its bioavailability (Cobley et al., 2018). The use of piperine as an adjuvant or in formulations such as nanoparticles, liposomes, and/or phospholipid mixtures has been used to improve curcumin's bioavailability. On the other hand, the bioavailability of the withanolides in ashwagandha varies depending on the formulation and extraction method.

It is expected that standardized dosing and clinical outcomes should occur with standardized extracts, but the content of phytochemicals can also change depending on the sources of the plant, environmental conditions, and processing (Karthikeyan et al., 2025). For instance, two of the active ingredients found in rosemary, rosmarinic acid and carnosic acid, are also subject to the amount extracted and degrade with cooking or storage (Zhang et al., 2012). All of this underscores the importance of proper analysis and sufficient quality control to ensure not only safety but also efficacy, especially for commercial and research purposes. The complexity of herbal supplements, with the majority containing a number of bioactive compounds, also makes matters more complicated, especially with antagonistic or synergistic interactions between them. They have complex pharmacokinetic and dose-response relationships, especially with interactions between them. In addition, the regulations for herbal supplements also differ from one country to another, and this affects the reliability of the product offered (Chaachouay, 2025). The lack of extensive placebo-controlled human clinical trials to prove the efficacy of using herbal antioxidants is another disadvantage.

Although the data are reliable, inconsistencies may occur because of differences in dosages, formulations, and research designs. Moreover, the long-term safety profiles of these herbal antioxidants, as well as their interactions with conventional medications, are not adequately elucidated (Goodman et al., 2011). Last but not least, the therapeutic benefits of herbal antioxidants are promising, especially for mitigating OS. However, their therapeutic benefits are hindered by issues of bioavailability, formulation, and standardization. To effectively integrate these herbal antioxidants into conventional therapeutic regimens and nutritional practices, these limitations need to be addressed through advanced delivery systems, quality control, and well-planned research trials. A list of the major herbal antioxidants, along with their botanical sources, molecular targets, physiological effects, and additional references, is presented in Table 2.1. These herbal antioxidants play an important role in neuroprotection and antiinflammatory responses in the CNS through the regulation of redox-sensitive pathways, such as Nrf2, NF-κB, PI3K/Akt, and MAPK.

2.7 Future prospects

Using nutrition for healing purposes in both body and mind could transform the approach to helping clients achieve mental well-being from a holistic perspective. In "Nutrition & Psychiatric Disorders" (Accorroni et al., 2025), Baklola discusses nutrition's role in psychiatric disorders, stating that dietary factors can be changed and directly affect one's mental health

TABLE 2.1 Herbal compounds and their molecular targets in antioxidant and antiinflammatory pathways.

Herbal compound	Source plant (Common Name)	Primary targets	Influence	References
Curcumin	Turmeric (*Curcuma longa*)	↑ Nrf2, ↓ NF-κB, ↓ p38/JNK	Antioxidant, antiinflammatory	Shahcheraghi et al. (2022)
Resveratrol	Grapes, Japanese knotweed	↑ SIRT1, ↑ Nrf2, ↓ NF-κB	Activates longevity pathways	Al-Sammak and Al-Hassawi (2024)
Epigallocatechin gallate (EGCG)	Green tea (*Camellia sinensis*)	↑ Nrf2, ↓ NF-κB, ↑ PI3K/Akt	Neuroprotective	Mokra et al. (2023)
Rosmarinic acid	Rosemary, Sage, Lemon balm	↑ Nrf2, ↓ COX-2, ↓ IL-1β	Antioxidant, antineuroinflammatory	Biswkarma and Wadhawan (2025)
Ginsenosides	Ginseng (*Panax ginseng*)	↑ PI3K/Akt, ↑ SIRT1, ↓ JNK	Enhances resilience to stress	He et al. (2022)
Salvianolic acid B	Danshen (*Salvia miltiorrhiza*)	↑ Nrf2, ↓ NF-κB, ↑ HO-1	Cardio- and neuroprotective	Wu et al. (2020)
Quercetin	Onions, apples, capers	↑ Nrf2, ↓ NF-κB, ↓ ROS	Flavonoid antioxidant	Fuentes et al. (2021)
Luteolin	Celery, peppers, chamomile	↓ NF-κB, ↓ microglial activation	Antineuroinflammatory	Singh et al. (2025b)

status. Although the growing field of nutritional neuroscience is still in its infant stages, it has the potential to complement existing pharmacological and psychotherapy techniques by addressing essential biological processes such as inflammation, OS, and neuroplasticity, as well as modulating the gut–brain axis. Marx et al. (2017) assert that incorporating dietary interventions into psychiatric practice may enhance long-term outcomes, mitigate adverse effects, and improve treatment response as the global prevalence of psychological disorders rises.

8-hydroxy-2′-deoxyguanosine (8-OHdG), MDA, and protein carbonyls can be used as biomarkers of damage from OS as they are products of lipid peroxidation, DNA degradation, and protein modification, respectively (Goriuc et al., 2024). Clinicians can monitor OS levels and adapt their therapies over time using quantitative techniques. The development of biosensors and high-throughput assays has made it possible to use these biomarkers in a clinical setting because the cycle between ingestion of nutrients and the physiological effects has been completed (Seyedsadjadi and Grant, 2021). Researchers are currently examining the potential antiinflammatory effects of antioxidant therapies that will specifically target the oxidative pathways involved in mental disorders. Therapeutics that are currently being researched include polyphenol pharmaceuticals, Nrf2 activators, and antioxidants targeted to mitochondria to ameliorate neuroinflammation and help modulate redox state (Deus et al., 2022). These drugs aim to improve on generalized supplementation by focusing on the molecular mechanisms involved in disturbances and by enhancing the specificity of the treatments employed. Certain antioxidants, such as curcumin and resveratrol, are involved in modulating pathways such as NF-kB and MAPK, which are critical for regulating OS and inflammation in the brain (Moosavi et al., 2015; Joshi et al., 2022). The investigation of genetically tailored diets represents one of the most pioneering trends in nutritional psychiatry today. Nutrigenomics and nutrigenetics are revealing gene-diet interactions that affect how a person reacts to nutrients.

Genetic variations in the genes that make inflammatory cytokines, neurotransmitter metabolizing genes, and antioxidant enzyme function genes can make a person more likely to have OS and mental health problems. Using genetics to inform personalized nutrition has the potential to be one of the most efficacious dietary interventions for mental health with minimal side effects (Dixit et al., 2025). The diet-mood model needs to be further studied in order for research related to mechanistic understanding and future therapeutic innovations to develop. Currently, research is being conducted through the use of rodent models of cognitive dysfunction, anxiety, and depression to investigate the effects of high-fat, high-sugar, and antioxidant diets on brain biochemistry and behavior. These models are used to assess oxidative and neuroinflammatory biomarkers, as well as controlled dietary changes. In addition, they will lay the foundation for translational research and allow researchers to evaluate sexual dimorphism, timing of development, and their relationship to the gastrointestinal microbiota (Marx et al., 2017).

2.8 Conclusion

Diet has a significant impact on controlling the inflammatory and redox states of the brain, which in turn affects how emotions are controlled and mental health. Neuronal circuits related to mood, cognition, and stress response are directly impacted by ROS and proinflammatory cytokines. The degree of oxidative burden and neuroinflammation is determined by the type of food consumed, including processing and additives, as well as whether the food contains proinflammatory or antioxidant-rich compounds. While whole foods and natural diets are protective, additives and processed foods like artificial sweeteners and preservatives can increase OS and excitotoxicity.

By controlling important signal pathways, Nrf2, NF-κB, and MAPK, herbal remedies such as extracts of curcumin, ashwagandha, and rosemary exhibit strong antioxidant and antiinflammatory properties. Bioavailability and standardization are limitations of the current state of research in the field of clinical nutrition and applications. Genetic signatures/biomarkers will provide us with a framework for personalized nutrition as well as incorporate nutritional neuroscience into psychiatry. The mechanistic relationship between diet and mood will be elucidated with the use of other animal models and will result in validated dietary-based interventions for mental disorders.

References

Abd El-Hack, M.E., Rudayni, H.A., Aldhalmi, A.K., Youssef, I.M., Arif, M., Alawam, A.S., Allam, A.A., Khafaga, A.F., Ashour, E.A., Khan, M.M.H., 2025. Beyond traditional uses: unveiling the epigenetic, microbiome-modulating, metabolic, and nutraceutical benefits of Salvia officinalis in human and livestock nutrition. J. Funct. Foods 128. doi:10.1016/j.jff.2025.106843.

Abd El-Hakim, Y.M., El-Houseiny, W., EL-Murr, A.E., Ebraheim, L.L.M., Moustafa, A.A., Rahman Mohamed, A.A., 2020. Melamine and curcumin enriched diets modulate the haemato-immune response, growth performance, oxidative stress, disease resistance, and cytokine production in oreochromis niloticus. Aquat. Toxicol. 220. doi:10.1016/j.aquatox.2020.105406.

Accorroni, A., Nencha, U., Bègue, I., 2025. The interdisciplinary synergy between neurology and psychiatry: advancing brain health. Clin. Transl. Neurosci. 9 (1), 18. doi:10.3390/ctn9010018.

Adzic, M., Brkic, Z., Mitic, M., Francija, E., Jovicic, M.J., Radulovic, J., Maric, N.P., 2018. Therapeutic strategies for treatment of inflammation-related depression. Curr. Neuropharmacol. 16 (2), 176–209. doi:10.2174/1570159 ž 15666170828163048.

Agnihotri, A., Aruoma, O.I., 2020. Alzheimer's disease and Parkinson's disease: a nutritional toxicology perspective of the impact of oxidative stress, mitochondrial dysfunction, nutrigenomics and environmental chemicals. J. Am. Coll. Nutr. 39 (1), 16–27. doi:10.1080/07315724.2019.1683379.

Albrakati, A., 2023. Monosodium glutamate induces cortical oxidative, apoptotic, and inflammatory challenges in rats: the potential neuroprotective role of apigenin. Environ. Sci. Pollut. Res. 30 (9), 24143–24153. doi:10.1007/s11356-022-23954-1.

Alessandrello, C., Sanfilippo, S., Minciullo, P.L., Gangemi, S., 2024. An overview on atopic dermatitis, oxidative stress, and psychological stress: possible role of nutraceuticals as an additional therapeutic strategy. Int. J. Mol. Sci. 25 (9), 5020. doi:10.3390/ijms25095020.

Allowitz, K., Taylor, J., Harames, K., Yoo, J., Baloch, O., Ramana, K.V., 2025. Oxidative stress-mediated lipid peroxidation-derived lipid aldehydes in the pathophysiology of neurodegenerative diseases. Curr. Neuropharmacol. 23 (6), 671–685. doi:10.2174/011570159 ž 342720241014164650.

Al-Sammak, M.A.J., Al-Hassawi, W.W., 2024. Resveratrol as an antiaging drug: a review of articles. Iraqi J. Pharm. Sci. 33 (2), 1–13. doi:10.31351/vol33iss3pp1-13.

Angelova, P.R., Abramov, A.Y., 2024. Interplay of mitochondrial calcium signalling and reactive oxygen species production in the brain. Biochem. Soc. Trans. 52 (4), 1939–1946. doi:10.1042/BST20240261.

Ankul, S.S., Chandran, L., Anuragh, S., Kaliappan, I., Rushendran, R., Vellapandian, C., 2023. A systematic review of the neuropathology and memory decline induced by monosodium glutamate in the Alzheimer's disease-like animal model. Front. Pharmacol. 14. doi:10.3389/fphar.2023.1283440.

Anwar, S., Alrumaihi, F., Sarwar, T., Babiker, A.Y., Khan, A.A., Prabhu, S.V., Rahmani, A.H., 2024. Exploring therapeutic potential of catalase: strategies in disease prevention and management. Biomolecules 14 (6). doi:10.3390/biom14060697.

Arushi, N.A., Reena, V.S., Subhanshi Vishwas, P.T., 2025. Mechanism of action of PD: antioxidant, anti-inflammatory and neurorestorative pathways. From Ancient Roots to Nano Routes: Targeted Delivery of Ginkgo and Gastrodia for Parkinson's Disease (PD) Treatment, 100 doi:10.70593/978-93-7185-508-2_2.

Asghar, M., Khan, N., Fatima, M., Arslan, M., Davies, S.J, Haque, N., Ranjan, A., 2024. Feasibility of replacing fish oil with sunflower oil on the growth, body composition, fatty acid profile, antioxidant activity, stress response, and blood biomarkers of Labeo rohita. PLoS One 19 (3), e0299195. doi:10.1371/journal.pone.0299195.

Aswar, U., Patil, R., Akotkar, L., Pawar, A., 2024. Molecular mechanism of action of phytoconstituents in neuropsychiatric disorders. In: NeuroPhytomedicine. CRC Press, India, pp. 83–113. doi:10.1201/9781003389781-5.

Azarfarin, M., Shahla, M.M., Mohaddes, G., Dadkhah, M., 2025. Non-pharmacological therapeutic paradigms in stress-induced depression: from novel therapeutic perspective with focus on cell-based strategies. Acta Neuropsychiatr. 37. doi:10.1017/neu.2024.39.

Bajželj, B., Laguzzi, F., Röös, E., 2021. The role of fats in the transition to sustainable diets. Lancet Planet. Health 5 (9), e644. doi:10.1016/s2542-5196(21)00194-7.

Ballaz, S., Bourin, M., 2023. Anti-inflammatory therapy as a promising target in neuropsychiatric disorders. In: Bourin, M., Costa e Silva, J.A., Halaris, A. (Eds.), Neuroinflammation, Gut-Brain Axis and Immunity in Neuropsychiatric Disorders. Academic Press, Cambridge, MA, USA, pp. 459–486. https://doi.org/10.1016/B978-0-323-90581-1.00015-5.

Bandookwala, M., Sengupta, P., 2020. 3-Nitrotyrosine: a versatile oxidative stress biomarker for major neurodegenerative diseases. Int. J. Neurosci. 130 (10), 1047–1062. doi:10.1080/00207454.2020.1713776.

Barbosa, M.L., de Meneses, A.A.P.M., de Aguiar, R.P.S., de Castro e Sousa, J.M., de Carvalho Melo Cavalcante, A.A., Maluf, S.W., 2020. Oxidative stress, antioxidant defense and depressive disorders: a systematic review of biochemical and molecular markers. Neurol. Psychiatry Brain Res. 36, 65–72. doi:10.1016/j.npbr.2020.02.006.

Basudkar, V., Gujrati, G., Ajgaonkar, S., Gandhi, M., Mehta, D., Nair, S., 2024. Emerging vistas for the nutraceutical Withania somnifera in inflammaging. Pharmaceuticals 17 (5), 597. doi:10.3390/ph17050597.

Batool, R., Umer, M.J., Hussain, B., Anees, M., Wang, Z., 2022. Molecular mechanisms of superoxide dismutase (SODs)-mediated defense in controlling oxidative stress in plants. In: Antioxidant Defense in Plants: Molecular Basis of Regulation. Springer Nature, China, pp. 157–179. doi:10.1007/978-981-16-7981-0_8.

Baufeld, C., Osterloh, A., Prokop, S., Miller, K.R., Heppner, F.L., 2016. High-fat diet-induced brain region-specific phenotypic spectrum of CNS resident microglia. Acta Neuropathol. 132 (3), 361–375. doi:10.1007/s00401-016-1595-4.

Bensid, A., El Abed, N., Houicher, A., Regenstein, J.M., Özogul, F., 2022. Antioxidant and antimicrobial preservatives: properties, mechanism of action and applications in food–a review. Crit. Rev. Food Sci. Nutr. 62 (11), 2985–3001. doi:10.1080/10408398.2020.1862046.

Bhargavi, S., Madhan Shankar, S.R., 2021. Dual herbal combination of Withania somnifera and five Rasayana herbs: a phytochemical, antioxidant, and chemometric profiling. J. Ayurveda Integr. Med. 12 (2), 283–293. doi:10.1016/j.jaim.2020.10.001.

Bhatt, S., Nagappa, A.N., Patil, C.R., 2020. Role of oxidative stress in depression. Drug Discov. Today 25 (7), 1270–1276. doi:10.1016/j.drudis.2020.05.001.

Biswkarma, V.K., Wadhawan, S., 2025. Phytoconstituents ameliorates Alzheimer's disease and cognitive impairments: a review of preclinical studies. Curr. Bioact. Compd. 21 (5). doi:10.2174/0115734072309710240603065813.

Bradlow, R.C.J., Berk, M., Kalivas, P.W., Back, S.E., Kanaan, R.A., 2022. The potential of N-acetyl-L-cysteine (NAC) in the treatment of psychiatric disorders. CNS Drugs 36 (5), 451–482. doi:10.1007/s40263-022-00907-3.

Brymer, K.J., Romay-Tallon, R., Allen, J., Caruncho, H.J., Kalynchuk, L.E., 2019. Exploring the potential antidepressant mechanisms of TNF-α antagonists. In: Quevedo, J., Carvalho, A.F., Zarate, C.A. Jr. (Eds.), Novel Approaches to the Neuropharmacology of Mood Disorders. Academic Press, Cambridge, MA, USA, pp. 283–304. https://doi.org/10.1016/B978-0-12-814732-5.00013-8

Cacciatore, I., Baldassarre, L., Fornasari, E., Mollica, A., Pinnen, F., 2012. Recent advances in the treatment of neurodegenerative diseases based on GSH delivery systems. Oxid. Med. Cell. Long. 2012, 1–12. doi:10.1155/2012/240146.

Cartier, A., Côté, M., Lemieux, I., Pérusse, L., Tremblay, A., Bouchard, C., Després, J.P., 2009. Sex differences in inflammatory markers: what is the contribution of visceral adiposity? Am. J. Clin. Nutr. 89 (5), 1307–1314. doi:10.3945/ajcn.2008.27030Canada.

Cătălina, G.R, Gheorman, V., Gheorman, V., Forţofoiu, M.-C., 2025. The role of neuroinflammation in the comorbidity of psychiatric disorders and internal diseases. Healthcare 13 (7), 837. doi:10.3390/healthcare13070837.

Chaachouay, N., 2025. Synergy, additive effects, and antagonism of drugs with plant bioactive compounds. Drugs Drug Candidates 4 (1), 4. doi:10.3390/ddc4010004.

Cheiran Pereira, G., Piton, E., Moreira dos Santos, B., Ramanzini, L.G., Muniz Camargo, L.F., Menezes da Silva, R., Bochi, G.V., 2022. Microglia and HPA axis in depression: an overview of participation and relationship. World J. Biol. Psychiatry. 23 (3), 165–182. doi:10.1080/15622975.2021.1939154.

Choi, T.G., Kim, S.S., 2019. Physiological functions of mitochondrial reactive oxygen species. In: Das, K.K., Das, S., Biradar, M.S., Bobbarala, V., Tata, S.S. (Eds.), Free Radical Medicine and Biology. IntechOpen, London, UK doi:10.5772/intechopen.88386.

Cobley, J.N., Fiorello, M.L., Bailey, D.M., 2018. 13 reasons why the brain is susceptible to oxidative stress. Redox. Biol. 15, 490–503. doi:10.1016/j.redox.2018.01.008.

Corkey, B.E., 2023. Reactive oxygen species: role in obesity and mitochondrial energy efficiency. Philos. Trans. R. Soc. B, Biol. Sci. 378 (1885). doi:10.1098/rstb.2022.0210.

Csiszár, J., Horváth, E., Bela, K., Gallé, 2016. Glutathione-related enzyme system: Glutathione reductase (GR), Glutathione transferases (GSTs) and Glutathione Peroxidases (GPXs). In: Redox State as a Central Regulator of Plant-Cell Stress Responses. Springer International Publishing, Hungary, pp. 137–158. doi:10.1007/978-3-319-44081-1_7.

Dama, A., Shpati, K., Daliu, P., Dumur, S., Gorica, E., Santini, A., 2024. Targeting metabolic diseases: the role of nutraceuticals in modulating oxidative stress and inflammation. Nutrients 16 (4), 507. doi:10.3390/nu16040507.

Damotharan, K., Sudhakaran, G., Ramu, M., Krishnan, M., Namasivayam S, K.R, Arockiaraj, J., 2024. Biochemical processes mediating neurotoxicity induced by synthetic food dyes: a review of current evidence. Chemosphere 364, 143295. doi:10.1016/j.chemosphere.2024.143295.

Dantzer, R., Konsman, J.-P., Bluthé, R.-M., Kelley, K.W, 2000. Neural and humoral pathways of communication from the immune system to the brain: parallel or convergent? Auton. Neurosci. 85 (1-3), 60–65. doi:10.1016/s1566-0702(00)00220-4.

Deeb, S.E., Ashour, E.A., Youssef, I.M., Alshehry, G., Abuljadayel, D.A., Aljahdali, N., Albaqami, N.M., Bahnas, M.M., Abd El-Hack, M.E., 2025. Enhancing broiler growth, carcass quality, blood parameters and intestinal microbial load: the role of dietary garlic powder as a natural growth promoter. Ann. Anim. Sci. 25 (4), 1467–1478. doi:10.2478/aoas-2025-0024.

De Paula, G.C., Simões, R.F., Garcia-Serrano, A.M., Duarte, J.M.N., 2024. High-fat and high-sucrose diet-induced hypothalamic inflammation shows sex specific features in mice. Neurochem. Res. 49 (12), 3356–3366. doi:10.1007/s11064-024-04243-4.

Deus, C.M., Teixeira, J., Raimundo, N., Tucci, P., Borges, F., Saso, L., Oliveira, P.J., 2022. Modulation of cellular redox environment as a novel therapeutic strategy for Parkinson's disease. Eur. J. Clin. Invest. 52 (10). doi:10.1111/eci.13820.

Dixit, A., Gupta, M., Zaidi, A., Chauhan, K.K., Gupta, R.S., Kumar, C.S., Shukla, N., 2025. ADHD as a system disorder: a narrative review of genetic, nutritional, psychological and environmental interactions. Vasc. Endovasc. Rev. 8 (1s), 145–158.

Dosoky, W.M., Farag, S.A., Almuraee, A.A., Youssef, I.M., Awlya, O.F.A., Abusudah, W.F., Qadhi, A., Arbaeen, A.F., Moustafa, M., Hassan, H., Tellez-Isaias, G., 2024. Vitamin C and/or garlic can antagonize the toxic effects of cadmium on growth performance, hematological, and immunological parameters of growing Japanese quail. Poult. Sci. 103 (3). doi:10.1016/j.psj.2024.103457.

Drevets, W.C., Wittenberg, G.M., Bullmore, E.T., Manji, H.K., 2022. Immune targets for therapeutic development in depression: towards precision medicine. Nat. Rev. Drug Discov. 21 (3), 224–244. doi:10.1038/s41573-021-00368-1.

Farooq, R.K., Asghar, K., Kanwal, S., Zulqernain, A., 2017. Role of inflammatory cytokines in depression: focus on interleukin-1β. Biomed. Rep. 6 (1), 15–20. doi:10.3892/br.2016.807.

Ferreiro, E., Baldeiras, I., Ferreira, I.L., Costa, R.O., Rego, A.C., Pereira, C.F., Oliveira, C.R., 2012. Mitochondrial-and endoplasmic reticulum-associated oxidative stress in Alzheimer's disease: from pathogenesis to biomarkers. Int. J. Cell Biol. 2012 (1), 735206.

Foulkes, L., Blakemore, S.J., 2018. Studying individual differences in human adolescent brain development. Nat. Neurosci. 21 (3), 315–323. doi:10.1038/s41593-018-0078-4.

Fuentes, J., Speisky, H., De Camargo, A.C., Atala, E., Gotteland, M., Olea-Azar, C., 2021. Quercetin oxidation metabolite present in onion peel protects caco-2 cells against the oxidative stress, nf-kb activation, and loss of epithelial barrier function induced by nsaids. J. Agric. Food Chem. 69 (7), 2157–2167. doi:10.1021/acs.jafc.0c07085.

Garay-Sevilla, M.a.E., Rojas, A., Portero-Otin, M., Uribarri, J., 2021. Dietary AGEs as exogenous boosters of inflammation. Nutrients 13 (8), 2802. doi:10.3390/nu13082802.

Garrido-Mesa, N., Zarzuelo, A., Gálvez, J., 2013. Minocycline: far beyond an antibiotic. Br. J. Pharmacol. 169 (2), 337–352. doi:10.1111/bph.12139.

Ghosh, A., Gao, L., Thakur, A., Siu, P.M., Lai, C.W.K., 2017. Role of free fatty acids in endothelial dysfunction. J. Biomed. Sci. 24 (1). doi:10.1186/s12929-017-0357-5.

Giacobbe, J., Benoiton, B., Zunszain, P., Pariante, C.M., Borsini, A., 2020. The anti-inflammatory role of omega-3 polyunsaturated fatty acids metabolites in pre-clinical models of psychiatric, neurodegenerative, and neurological disorders. Front. Psychiatry 11. doi:10.3389/fpsyt.2020.00122.

Goodman, M., Bostick, R.M., Kucuk, O., Jones, D.P., 2011. Clinical trials of antioxidants as cancer prevention agents: past, present, and future. Free Radical Biol. Med. 51 (5), 1068–1084. doi:10.1016/j.freeradbiomed.2011.05.018.

Goriuc, A., Cojocaru, K.A., Luchian, I., Ursu, R.G., Butnaru, O., Foia, L., 2024. Using 8-hydroxy-2′-deoxiguanosine (8-OHdG) as a reliable biomarker for assessing periodontal disease associated with diabetes. Int. J. Mol. Sci. 25 (3). doi:10.3390/ijms25031425.

Guarneri, F., Custurone, P., Papaianni, V., Gangemi, S., 2021. Involvement of RAGE and oxidative stress in inflammatory and infectious skin diseases. Antioxidants 10 (1), 82. doi:10.3390/antiox10010082.

Gugliucci, A., 2017. Formation of fructose-mediated advanced glycation end products and their roles in metabolic and inflammatory diseases. Adv. Nutr. 8 (1), 54–62. doi:10.3945/an.116.013912.

Hatziagapiou, K., Kakouri, E., Lambrou, G.I., Bethanis, K., Tarantilis, P.A., 2019. Antioxidant properties of crocus Sativus L. and its constituents and relevance to neurodegenerative diseases: focus on Alzheimer's and Parkinson's disease. Curr. Neuropharmacol. 17 (4), 377–402. doi:10.2174/1570159 ž 16666180321095705.

He, B.o, Chen, D., Zhang, X., Yang, R., Yang, Y., Chen, P., Shen, Z., Sun, C.De, 2022. Oxidative stress and ginsenosides: an update on the molecular mechanisms. Oxid. Med. Cell. Long. 2022 (1). doi:10.1155/2022/9299574.

He, L., He, T., Farrar, S., Ji, L., Liu, T., Ma, X.i, 2017. Antioxidants maintain cellular redox homeostasis by elimination of reactive oxygen species. Cell. Physiol. Biochem. 44 (2), 532–553. doi:10.1159/000485089.

Hegazy, M.A., Ahmed, S.M., Sultan, S.M., Afifi, O.F., Mohamed, M.A., Azab, A.E., Hassanen, M.A., Zaben, R.K., 2025. Metabolic dysfunction-associated steatotic liver disease and omega-6 polyunsaturated fatty acids: friends or foes. World J. Hepatol. 17 (3). doi:10.4254/wjh.v17.i3.102286.

Houldsworth, A., 2024. Role of oxidative stress in neurodegenerative disorders: a review of reactive oxygen species and prevention by antioxidants. Brain Commun. 6 (1). doi:10.1093/braincomms/fcad356.

Hussain, S.M., Syeda, A.F., Alshammari, M., Alnasser, S., Alenzi, N.D., Alanazi, S.T., Nandakumar, K., 2022. Cognition enhancing effect of rosemary (Rosmarinus officinalis L.) in lab animal studies: a systematic review and meta-analysis. Braz. J. Med. Biol. Res. 55. doi:10.1590/1414-431 ž 2021e11593.

Ighodaro, O.M., Akinloye, O.A., 2018. First line defence antioxidants-superoxide dismutase (SOD), catalase (CAT) and glutathione peroxidase (GPX): their fundamental role in the entire antioxidant defence grid. Alex. J. Med. 54 (4), 287–293. doi:10.1016/j.ajme.2017.09.001.

Iob, E., Kirschbaum, C., Steptoe, A., 2020. Persistent depressive symptoms, HPA-axis hyperactivity, and inflammation: the role of cognitive-affective and somatic symptoms. Mol. Psychiatry 25 (5), 1130–1140. doi:10.1038/s41380-019-0501-6.

Jayaraj, S., Kothandan, L., Murugan, V.B., Prakash, S.K., Vijayakumar, V., Rani, E., 2025. Food toxicology and safety: evaluating dietary risks and functional ingredients. Food Bioeng. 4 (2), 182–198. doi:10.1002/fbe2.70014.

Jiang, S., Liu, H., Li, C., 2021. Dietary regulation of oxidative stress in chronic metabolic diseases. Foods 10 (8), 1854. doi:10.3390/foods10081854.

Jiao, W., Lin, J., Deng, Y., Ji, Y., Liang, C., Wei, S., Jing, X.i, Yan, F., 2025. The immunological perspective of major depressive disorder: unveiling the interactions between central and peripheral immune mechanisms. J. Neuroinflam. 22 (1). doi:10.1186/s12974-024-03312-3.

Jomova, K., Alomar, S.Y., Valko, R., Nepovimova, E., Kuca, K., Valko, M., 2025. The role of redox-active iron, copper, manganese, and redox-inactive zinc in toxicity, oxidative stress, and human diseases. EXCLI J. 24, 880–954. doi:10.17179/excli2025-8449.

Joshi, P., Bisht, A., Joshi, S., Semwal, D., Nema, N.K., Dwivedi, J., Sharma, S., 2022. Ameliorating potential of curcumin and its analogue in central nervous system disorders and related conditions: a review of molecular pathways. Phytother. Res. 36 (8), 3143–3180. doi:10.1002/ptr.7522.

Kaplan, B.J., Rucklidge, J.J., Romijn, A., McLeod, K., 2015. The emerging field of nutritional mental health: inflammation, the microbiome, oxidative stress, and mitochondrial function. Clin. Psychol. Sci. 3 (6), 964–980. doi:10.1177/2167702614555413.

Karthikeyan, M., Mehta, A., Kumar, S.S., Usha, B., Mohan, H., 2025. Safety, dosage, and regulatory aspects in the use of phytochemicals. In: Medicinal Plants and Their Bioactives in Human Diseases. Springer Science+Business Media, India, pp. 207–233. doi:10.1007/978-3-032-01356-9_10.

Ketchem, J.M., Bowman, E.J., Isales, C.M., 2023. Male sex hormones, aging, and inflammation. Biogerontology 24 (1), 1–25. doi:10.1007/s10522-022-10002-1.

Khelfi, A., 2024. Oxidative stress in inflammation. In: Biomarkers of Oxidative Stress: Clinical Aspects of Oxidative Stress. Springer Nature, Algeria, pp. 13–43. doi:10.1007/978-3-031-69962-7_2.

Kim, H., Choi, H.S., Han, K., Sim, W., Suh, H.J., Ahn, Y., 2025. Ashwagandha (Withania somnifera (L.) dunal) root extract containing withanolide a alleviates depression-like behavior in mice by enhancing the brain-derived neurotrophic factor pathway under unexpected chronic mild stress. J. Ethnopharmacol. 340. doi:10.1016/j.jep.2024.119224.

Korte, S.M., Straub, R.H., 2019. Fatigue in inflammatory rheumatic disorders: pathophysiological mechanisms. Rheumatology (Oxford) 58 (Supplement_5), v35. doi:10.1093/rheumatology/kez413.

Kosmopoulou, D., Lafara, M.-P., Adamantidi, T., Ofrydopoulou, A., Grabrucker, A.M., Tsoupras, A., 2024. Neuroprotective benefits of Rosmarinus officinalis and its bioactives against Alzheimer's and Parkinson's diseases. Appl. Sci. 14 (15), 6417. doi:10.3390/app14156417.

Kubera, M., Obuchowicz, E., Goehler, L., Brzeszcz, J., Maes, M., 2011. In animal models, psychosocial stress-induced (neuro)inflammation, apoptosis and reduced neurogenesis are associated to the onset of depression. Prog. Neuropsychopharmacol. Biol. Psychiatry 35 (3), 744–759. doi:10.1016/j.pnpbp.2010.08.026.

Kurowska, A., Ziemichód, W., Herbet, M., Piątkowska-Chmiel, I., 2023. The role of diet as a modulator of the inflammatory process in the neurological diseases. Nutrients 15 (6), 1436. doi:10.3390/nu15061436.

Kuruvilla, B.T., Kaushik, S., Benny, I.J., 2025. Ashwagandha: mechanism and therapeutic benefits. In: Ashwagandha: Potential Drug Candidate from Ancient Ayurvedic Remedy. CRC Press, India, pp. 234–257. doi:10.1201/9781032675961-20.

Lei, A.A., Phang, V.W.X., Lee, Y.Z., Kow, A.S.F., Tham, C.L., Ho, Y.C., Lee, M.T., 2025. Chronic stress-associated depressive disorders: the impact of HPA axis dysregulation and neuroinflammation on the hippocampus—a mini review. Int. J. Mol. Sci. 26 (7). doi:10.3390/ijms26072940.

Li, F., Luo, Q., Guo, T., Zhou, S., Cheng, Z., Pan, H., Tu, J., 2025. The impact of high-sugar diets on central nervous system disorders: mechanisms, pathogenesis, and dietary implication. Ann. Med. 57 (1). doi:10.1080/07853890.2025.2561789.

Li, J., Huang, M., Shen, X., 2014. The association of oxidative stress and pro-inflammatory cytokines in diabetic patients with hyperglycemic crisis. J. Diabetes Complications 28 (5), 662–666. doi:10.1016/j.jdiacomp.2014.06.008.

Liu, D., Wang, Z., Gao, Z.e, Xie, K., Zhang, Q., Jiang, H., Pang, Q.i, 2014. Effects of curcumin on learning and memory deficits, BDNF, and ERK protein expression in rats exposed to chronic unpredictable stress. Behav. Brain Res. 271, 116–121. doi:10.1016/j.bbr.2014.05.068.

Liu, Y., Li, C., Huang, X., Zhang, X.u, Deng, P., Jiang, G., Dai, Q., 2022. Dietary rosemary extract modulated gut microbiota and influenced the growth, meat quality, serum biochemistry, antioxidant, and immune capacities of broilers. Front. Microbiol. 13. doi:10.3389/fmicb.2022.1024682.

Liu, Y., Weng, W., Gao, R., Liu, Y., 2019. New insights for cellular and molecular mechanisms of aging and aging-related diseases: herbal medicine as potential therapeutic approach. Oxid. Med. Cell. Long. 2019, 1–25. doi:10.1155/2019/4598167.

Long, S., Wang, Y., Cheng, R., Deng, L., Chen, L., Dong, Y., 2025. Different IL-1β levels differentially mediate neuroprotection or neurodegeneration and may be related to BDNF. Cytokine 188, 156877. doi:10.1016/j.cyto.2025.156877.

Lugrin, J., Rosenblatt-Velin, N., Parapanov, R., Liaudet, L., 2014. The role of oxidative stress during inflammatory processes. Biol. Chem. 395 (2), 203–230. doi:10.1515/hsz-2013-0241.

Ma, L., Nelson, E.R., 2019. Oxysterols and nuclear receptors. Mol. Cell. Endocrinol. 484, 42–51. doi:10.1016/j.mce.2019.01.016.

Machado, M., Rodriguez-Alcalá, L.M., Gomes, A.M., Pintado, M., 2023. Vegetable oils oxidation: mechanisms, consequences and protective strategies. Food Rev. Int. 39 (7), 4180–4197. doi:10.1080/87559129.2022.2026378.

Majmasanaye, M., Mehrpooya, M., Amiri, H., Eshraghi, A., 2024. Discovering the potential value of coenzyme Q10 as an adjuvant treatment in patients with depression. J. Clin. Psychopharmacol. 44 (3), 232–239. doi:10.1097/jcp.0000000000001845.

Mancini, A., Bruno, C., Vergani, E., d'Abate, C., Giacchi, E., Silvestrini, A., 2021. Oxidative stress and low-grade inflammation in polycystic ovary syndrome: controversies and new insights. Int. J. Mol. Sci. 22 (4), 1667. doi:10.3390/ijms22041667.

Manosso, L.M., Camargo, A., Dafre, A.L., Rodrigues, A.L.S., 2022. Vitamin E for the management of major depressive disorder: possible role of the anti-inflammatory and antioxidant systems. Nutr. Neurosci. 25 (6), 1310–1324. doi:10.1080/1028415X.2020.1853417.

Marx, W., Moseley, G., Berk, M., Jacka, F., 2017. Nutritional psychiatry: the present state of the evidence. Proc. Nutr. Soc. 76 (4), 427–436. doi:10.1017/s0029665117002026.

Mendes, N.F., Kim, Y.B., Velloso, L.A., Araújo, E.P., 2018. Hypothalamic microglial activation in obesity: a mini-review. Front. Neurosci. 12. doi:10.3389/fnins.2018.00846.

Mokra, D., Joskova, M., Mokry, J., 2023. Therapeutic effects of green tea polyphenol (−)-epigallocatechin-3-gallate (EGCG) in relation to molecular pathways controlling inflammation, oxidative stress, and apoptosis. Int. J. Mol. Sci. 24 (1), 340. doi:10.3390/ijms24010340.

Moosavi, F., Hosseini, R., Saso, L., Firuzi, O., 2015. Modulation of neurotrophic signaling pathways by polyphenols. Drug Des. Dev. Ther. 10, 23–42. doi:10.2147/DDDT.S96936.

Morkovin, E., Litvinov, R., Koushner, A., Babkov, D., 2024. Resveratrol and extra virgin olive oil: protective agents against age-related disease. Nutrients 16 (24), 4258. doi:10.3390/nu16244258.

Mottis, G., Kandasamey, P., Peleg-Raibstein, D., 2025. The consequences of ultra-processed foods on brain development during prenatal, adolescent and adult stages. Front. Public Health 13. doi:10.3389/fpubh.2025.1590083.

Müller, L., Di Benedetto, S., 2025. Neuroimmune crosstalk in chronic neuroinflammation: microglial interactions and immune modulation. Front. Cell. Neurosci. 19. doi:10.3389/fncel.2025.1575022.

Negm, M.H., Aldhalmi, A.K., Ashour, E.A., Mohamed, L.A., Youssef, I.M., Moustafa, M., Abdelglil, M.I., Abdel-Shafi, S., Mahgoub, S.A., Abd El-Hack, M.E., Swelum, A.A., 2025. The effect of celery (Apium graveolens) seed powder in the diet on growth, immune function, antioxidant status and gut microbial balance in broiler chickens. Trop. Anim. Health Prod. 57 (8). doi:10.1007/s11250-025-04625-y.

Olmo, B.M.G., Bettes, M.N., DeMarsh, J.W., Zhao, F., Askwith, C., Barrientos, R.M., 2023. Short-term high-fat diet consumption impairs synaptic plasticity in the aged hippocampus via IL-1 signaling. npj Science of Food 7, 35. https://doi.org/10.1038/s41538-023-00211-4.

Orth, M., Bellosta, S., 2012. Cholesterol: its regulation and role in central nervous system disorders. Cholesterol 2012, 1–19. doi:10.1155/2012/292598.

Ozcan, M.E., Gulec, M., Ozerol, E., Polat, R., Akyol, O., 2004. Antioxidant enzyme activities and oxidative stress in affective disorders. Int. Clin. Psychopharmacol. 19 (2), 89–95. doi:10.1097/00004850-200403000-00006.

Papas, A.M., 2019. Diet and antioxidant status. Antioxidant status, diet, nutrition, and health 89–106.

Pateiro, M., Domínguez, R., Varzakas, T., Munekata, P.E.S., Fierro, E.M, Lorenzo, J.M., 2021. Omega-3-rich oils from marine side streams and their potential application in food. Mar. Drugs 19 (5), 233. doi:10.3390/md19050233.

Patel, S.S., Acharya, A., Ray, R.S., Agrawal, R., Raghuwanshi, R., Jain, P., 2020. Cellular and molecular mechanisms of curcumin in prevention and treatment of disease. Crit. Rev. Food Sci. Nutr. 60 (6), 887–939. doi:10.1080/10408398.2018.1552244.

Patterson, S.L., 2015. Immune dysregulation and cognitive vulnerability in the aging brain: interactions of microglia, IL-1β, BDNF and synaptic plasticity. Neuropharmacology 96, 11–18. doi:10.1016/j.neuropharm.2014.12.020.

Peng, Z., Zhang, C., Yan, L., Zhang, Y., Yang, Z., Wang, J., Song, C., 2020. EPA is more effective than DHA to improve depression-like behavior, glia cell dysfunction and hippocampal apoptosis signaling in a chronic stress-induced rat model of depression. Int. J. Mol. Sci. 21 (5), 1769. doi:10.3390/ijms21051769.

Pereira, T.M.C., Côco, L.Z., Ton, A.M.M., Meyrelles, S.S., Campos-Toimil, M., Campagnaro, B.P., Vasquez, E.C., 2021. The emerging scenario of the gut–brain axis: the therapeutic actions of the new actor kefir against neurodegenerative diseases. Antioxidants 10 (11). doi:10.3390/antiox10111845.

Pizza, V., Agresta, A., D'Acunto, C.W., Festa, M., Capasso, A., 2011. Neuroinflamm-aging and neurodegenerative diseases: an overview. CNS Neurol. Disord. Drug Targets 10 (5), 621–634. doi:10.2174/187152711796235014.

Poli, G., Iaia, N., Leoni, V., Biasi, F., 2022. High cholesterol diet, oxysterols and their impact on the gut–brain axis. Redox Exp. Med. 2022 (1), R15. doi:10.1530/rem-22-0003.

Quan, S., Fu, X., Cai, H., Ren, Z., Xu, Y., Jia, L., 2025. The neuroimmune nexus: unraveling the role of the mtDNA-cGAS-STING signal pathway in Alzheimer's disease. Mol. Neurodegener. 20 (1). doi:10.1186/s13024-025-00815-2.

Radbakhsh, S., Butler, A.E., Moallem, S.A., Sahebkar, A., 2024. The effects of curcumin on brain-derived neurotrophic factor expression in neurodegenerative disorders. Curr. Med. Chem. 31 (36), 5937–5952. doi:10.2174/0929867330666230602145817.

Ramesh, M.N., Al-Khusaibi, M., 2020. Cooking and frying of foods. In: Rahman, M.S. (Ed.), Handbook of Food Preservation. CRC Press, Boca Raton, FL, USA, pp. 637–646.

Rapti, E., Adamantidi, T., Efthymiopoulos, P., Kyzas, G.Z., Tsoupras, A., 2024. Potential applications of the anti-inflammatory, antithrombotic and antioxidant health-promoting properties of curcumin: a critical review. Nutraceuticals 4 (4), 562–595. doi:10.3390/nutraceuticals4040031.

Ruiz-Ramírez, A., Chávez-Salgado, M., Peñeda-Flores, J.A., Zapata, E., Masso, F., El-Hafidi, M., 2011. High-sucrose diet increases ROS generation, FFA accumulation, UCP2 level, and proton leak in liver mitochondria. Am. J. Physiol. Endocrinol. Metab. 301 (6), E1198. doi:10.1152/ajpendo.00631.2010.

Samadi, A., Sabuncuoglu, S., Samadi, M., Isikhan, S.Y., Chirumbolo, S., Peana, M., Lay, I., Yalcinkaya, A., Bjørklund, G., 2020. A comprehensive review on oxysterols and related diseases. Curr. Med. Chem. 28 (1), 110–136. doi:10.2174/0929867327666200316142659.

Santos, L.S., Cordeiro, G.S., Matos, R.J.B., Perez, G.S., Silva, R.T., Boaventura, G.T., Barreto-Medeiros, J.M., 2022. High-fat diet promotes hypothalamic inflammation in animal models: a systematic review. Nutr. Rev. 80 (3), 392–399. doi:10.1093/nutrit/nuab033.

Schade, D.S., Shey, L., Eaton, R.P., 2020. Cholesterol review: a metabolically important molecule. Endocr. Pract. 26 (12), 1514–1523. doi:10.4158/EP-2020-0347.

Seyedsadjadi, N., Grant, R., 2021. The potential benefit of monitoring oxidative stress and inflammation in the prevention of non-communicable diseases (NCDs). Antioxidants 10 (1), 15. doi:10.3390/antiox10010015.

Shadel, G.S., Horvath, T.L., 2015. Mitochondrial ROS signaling in organismal homeostasis. Cell 163 (3), 560–569. doi:10.1016/j.cell.2015.10.001.

Shahcheraghi, S.H., Salemi, F., Peirovi, N., Ayatollahi, J., Alam, W., Khan, H., Saso, L., 2022. Nrf2 regulation by curcumin: molecular aspects for therapeutic prospects. Molecules 27 (1). doi:10.3390/molecules27010167.

Shobeiri, P., Kalantari, A., Teixeira, A.L., Rezaei, N., 2022. Shedding light on biological sex differences and microbiota–gut–brain axis: a comprehensive review of its roles in neuropsychiatric disorders. Biol. Sex Differ. 13 (1). doi:10.1186/s13293-022-00422-6.

Singh, A., Tiwari, V., Roy, S., 2025a. Multifaceted role of oxidative stress in neurological disorders. Mol. Biol. Rep. 52 (1). doi:10.1007/s11033-025-10760-4.

Singh, N.K., Bhushan, B., Singh, P., Sahu, K.K., 2025b. Therapeutic expedition of luteolin against brain-related disorders: an updated review. Comb. Chem. High Throughput Screen. 28 (3), 371–391. doi:10.2174/0113862073303342240409060918.

Sivamaruthi, B.S., Raghani, N., Chorawala, M., Bhattacharya, S., Prajapati, B.G., Elossaily, G.M., Chaiyasut, C., 2023. NF-κB pathway and its inhibitors: a promising frontier in the management of Alzheimer's disease. Biomedicines 11 (9). doi:10.3390/biomedicines11092587.

Sobral, M.M.C., Cunha, S.C., Faria, M.A., Ferreira, I.M.P.L.V.O., 2018. Domestic cooking of muscle foods: impact on composition of nutrients and contaminants. Compr. Rev. Food Sci. Food Saf. 17 (2), 309–333. doi:10.1111/1541-4337.12327.

Spinelli, M., Fusco, S., Grassi, C., 2019. Brain insulin resistance and hippocampal plasticity: mechanisms and biomarkers of cognitive decline. Front. Neurosci. 13. doi:10.3389/fnins.2019.00788.

Stadler, R.H., Lineback, D.R., 2008. Maillard reaction of proteins and advanced glycation end products (AGEs). In: Process-Induced Food Toxicants: Occurrence, Formation, Mitigation, and Health Risks. Wiley, pp. 215–242. doi:10.1002/9780470430101.ch2g.

Stathori, G., Vlahos, N.F., Charmandari, E., Valsamakis, G., 2025. Obesity-and high-fat-diet-induced neuroinflammation: implications for autonomic nervous system dysfunction and endothelial disorders. Int. J. Mol. Sci. 26 (9), 4047.

Straub, R.H., Cutolo, M., 2016. Glucocorticoids and chronic inflammation. Rheumatology (Oxford) 55 (suppl 2), ii6. doi:10.1093/rheumatology/kew348.

Sun, Y., Koyama, Y., Shimada, S., 2022. Inflammation from peripheral organs to the brain: how does systemic inflammation cause neuroinflammation? Front. Aging Neurosci. 14. doi:10.3389/fnagi.2022.903455.

Surh, Y.J., Na, H.K., 2008. NF-κB and Nrf2 as prime molecular targets for chemoprevention and cytoprotection with anti-inflammatory and antioxidant phytochemicals. Genes Nutr. 2 (4), 313–317. doi:10.1007/s12263-007-0063-0.

Tamer, F., Ulug, E., Akyol, A., Nergiz-Unal, R., 2020. The potential efficacy of dietary fatty acids and fructose induced inflammation and oxidative stress on the insulin signaling and fat accumulation in mice. Food Chem. Toxicol. 135, 110914. doi:10.1016/j.fct.2019.110914.

Tan, B.L, Norhaizan, M.E., 2024. Nutrients and oxidative stress: biochemistry aspects and pharmacological insights. SpringerBriefs in Food, Health, and Nutrition. Springer Nature, Switzerland doi:10.1007/978-3-031-75319-0.

Teshome, E., Forsido, S.F., Rupasinghe, H.P.V., Keyata, E.O., 2022. Potentials of natural preservatives to enhance food safety and shelf life: a review. Sci. World J. 2022. doi:10.1155/2022/9901018.

Tolosa, L., Caraballo-Miralles, V., Olmos, G., Lladó, J., 2011. TNF-α potentiates glutamate-induced spinal cord motoneuron death via NF-κB. Mol. Cell. Neurosci. 46 (1), 176–186. doi:10.1016/j.mcn.2010.09.001.

Tran, D.M.D., Westbrook, R.F., 2017. A high-fat high-sugar diet-induced impairment in place-recognition memory is reversible and training-dependent. Appetite 110, 61–71. doi:10.1016/j.appet.2016.12.010.

Usturoi, M.G., Rațu, R.N., Crivei, I.C., Veleșcu, I.D., Usturoi, A., Stoica, F., Radu Rusu, R.M., 2025. Unlocking the power of eggs: nutritional insights, bioactive compounds, and the advantages of omega-3 and omega-6 enriched varieties. Agriculture (Switzerland) 15 (3). doi:10.3390/agriculture15030242.

Vaváková, M., Ďuračková, Z., Trebatická, J., 2015. Markers of oxidative stress and neuroprogression in depression disorder. Oxid. Med. Cell. Long. 2015, 1–12. doi:10.1155/2015/898393.

Videla, L.A., Valenzuela, R., Del Campo, A., Zúñiga-Hernández, J., 2023. Omega-3 lipid mediators: modulation of the M1/M2 macrophage phenotype and its protective role in chronic liver diseases. Int. J. Mol. Sci. 24 (21), 15528. doi:10.3390/ijms242115528.

Villa, A., Vegeto, E., Poletti, A., Maggi, A., 2016. Estrogens, neuroinflammation, and neurodegeneration. Endocr. Rev. 37 (4), 372–402. doi:10.1210/er.2016-1007.

Vlaicu, P.A., Untea, A.E., Varzaru, I., Saracila, M., Oancea, A.G., 2023. Designing nutrition for health—incorporating dietary by-products into poultry feeds to create functional foods with insights into health benefits, risks, bioactive compounds, food component functionality and safety regulations. Foods 12 (21). doi:10.3390/foods12214001.

Vollbracht, C., Werner, M., 2024. The role of inflammation and oxidative stress in the pathophysiology of depressions: time to consider vitamin C deficiency. Explor. Neurosci. 3 (4), 287–294. doi:10.37349/en.2024.00050.

Wallace, C.W., Fordahl, S.C., 2022. Obesity and dietary fat influence dopamine neurotransmission: exploring the convergence of metabolic state, physiological stress, and inflammation on dopaminergic control of food intake. Nutr. Res. Rev. 35 (2), 236–251. doi:10.1017/S0954422421000196.

Winiarska-Mieczan, A., Kwiecień, M.ał, Jachimowicz-Rogowska, K., Donaldson, J., Tomaszewska, E., Baranowska-Wójcik, E., 2023. Anti-inflammatory, antioxidant, and neuroprotective effects of polyphenols—polyphenols as an element of diet therapy in depressive disorders. Int. J. Mol. Sci. 24 (3), 2258. doi:10.3390/ijms24032258.

Wu, A., Zhang, J., 2023. Neuroinflammation, memory, and depression: new approaches to hippocampal neurogenesis. J. Neuroinflam. 20 (1). doi:10.1186/s12974-023-02964-x.

Wu, S., Yin, Y., Du, L., 2022. Blood–brain barrier dysfunction in the pathogenesis of major depressive disorder. Cell. Mol. Neurobiol. 42 (8), 2571–2591. doi:10.1007/s10571-021-01153-9.

Wu, Y., Xu, S., Tian, X.Y., 2020. The effect of salvianolic acid on vascular protection and possible mechanisms. Oxid. Med. Cell. Long. 2020. doi:10.1155/2020/5472096.

Xu, X., Pang, Y., Fan, X., 2025. Mitochondria in oxidative stress, inflammation and aging: from mechanisms to therapeutic advances. Signal Transduct. Target. Ther. 10 (1). doi:10.1038/s41392-025-02253-4.

Yan, L., Li, J., 2022. The central nervous system control of energy homeostasis: high fat diet induced hypothalamic microinflammation and obesity. Brain Res. Bull. 185, 99–106. doi:10.1016/j.brainresbull.2022.04.015.

Yu, H., Qin, X., Yu, Z., Chen, Y., Tang, L., Shan, W., 2021. Effects of high-fat diet on the formation of depressive-like behavior in mice. Food Funct. 12 (14), 6416–6431. doi:10.1039/d1fo00044f.

Zárate, R., el Jaber-Vazdekis, N., Tejera, N., Pérez, J.A., Rodríguez, C., 2017. Significance of long chain polyunsaturated fatty acids in human health. Clin. Transl. Med. 6 (1). doi:10.1186/s40169-017-0153-6.

Zhang, Y., Smuts, J.P., Dodbiba, E., Rangarajan, R., Lang, J.C., Armstrong, D.W., 2012. Degradation study of carnosic acid, carnosol, rosmarinic acid, and rosemary extract (Rosmarinus officinalis L.) assessed using HPLC. J. Agric. Food Chem. 60 (36), 9305–9314. doi:10.1021/jf302179c.

Chapter 3

Herbal bioactives with neuroactive potential

Islam Mohamed Youssef[a] **and Mohamed E. Abd El-Hack**[b]

[a] *Animal Production Systems Research Department, Animal Production Research Institute, Agricultural Research Center, Giza, Egypt,* [b] *Department of Industrial Pharmacy, College of Pharmaceutical Sciences and Drug Manufacturing, Misr University for Science and Technology (MUST), Giza, Egypt*

3.1 Introduction

Scientists are becoming more interested in plant-derived metabolites as central nervous system (CNS) health regulators as a result of the rising incidence of mood and neurodegenerative diseases. Through antioxidant, antiinflammatory, and neurotrophic pathways, herbal bioactives such as polyphenols, alkaloids, terpenoids, and adaptogens have a variety of effects on the CNC (Kadiri et al., 2025). These substances, which were once used in traditional herbal medicine, are now recognized as nutraceuticals that can improve synaptic plasticity, neuronal resistance, and reverse oxidative and inflammatory insults that make people more prone to mental illness and cognitive decline (Abd El-Hack et al., 2025; Tkaczenko et al., 2025).

Polyphenols, one of the most researched forms of neuroprotection, are found in these classes of phytochemicals. They are present in tea, wine, fruits, and vegetables and include flavonoids, phenolic acids, and stilbenes (Silva and Pogačnik, 2020; Negm et al., 2025). By increasing brain-derived neurotrophic factor (BDNF) and controlling signaling cascades such as ERK-CREB and PI3K/Akt, flavonoids have been shown to improve synaptic plasticity (Amidfar et al., 2024). Studies on humans and animals have shown that supplementing flavonoids enhances learning performance, memory, and attention (Vauzour, 2014). Phenolic acids such as caffeic and ferulic acid are implicated in neuroprotection through the scavenging of reactive oxygen species (ROS) and activation of the Nrf2 pathway (Singh et al., 2022), whereas stilbenes such as resveratrol modulate sirtuin (SIRT1) signaling for enhancing the neuronal lifespan and suppressing neuroinflammation (Moraes et al., 2020).

Plant-Derived Feed Additives and Animal-Based Foods: Linking Animal Health, Product Quality, and Consumer Perceptions. DOI: https://doi.org/10.1016/B978-0-443-45446-2.00009-0

Alkaloids, another important class of plant chemicals, contain biologically active substances such as caffeine, berberine, and nicotine, which affect neurotransmitter systems. Caffeine is an antagonist at adenosine receptors to increase alertness and cognitive function (Sharma et al., 2023). Berberine has an antidepressant and anxiolytic action on serotonin and dopamine, as well as via increasing AMP-Activated Protein Kinase (AMPK) activity (Gao et al., 2024). Although nicotine possesses neuroprotective effects in models of Parkinson's and Alzheimer's disease, its addictive liability means it requires context-dependent therapeutic application (Mouhape et al., 2019).

Some of the well-studied major botanicals, such as curcumin, resveratrol, and ashwagandha, are neuroactive. Curcumin is the most prominent curcuminoid of Curcuma longa and has shown antioxidant, antiinflammatory, and monoamine oxidase (MAO) inhibitory activity that aids in improved mood and memory function (Fuloria et al., 2022). Dose-response and bioavailability are, however, emergent issues in clinic-to-bedside translation of herbal neuroactives.

Curcumin, for instance, is poorly absorbed and metabolized rapidly, necessitating nanoformulations or adjuvants such as piperine to enhance systemic bioavailability (Yakubu et al., 2025). Plant bioactives are a diverse collection of phytochemicals that act by convergent molecular pathways to ensure neuronal integrity, withstand oxidative and inflammatory stressors, and enhance stress resilience. Developing these naturally engineered compounds into effective, evidence-based therapies for maintaining brain health and defending against neurodegenerative diseases requires a thorough understanding of their mechanisms, pharmacokinetics, and clinical significance.

3.2 Phytochemical categories relevant to brain health

3.2.1 Polyphenols

Plant foods high in polyphenols (from fruits, vegetables, tea, cocoa, red wine, and others) are made of a variety of chemical structures and fall into the category of phytochemicals. Their neuroprotective effects are primarily due to their antiinflammatory and antioxidant properties that can reduce oxidative damage, which is known to be one of the foremost contributors to neurodegenerative diseases such as, but not limited to, Parkinson's and Alzheimer's diseases. Flavonoids, phenolic acids, or hydroxycinnamic acids, and stilbenes are some of the subclassifications of polyphenols with differing mechanisms of action. As examples, flavonoids (e.g., quercetin and catechins) scavenge ROS and modify neuronal survival signaling pathways (Kim et al., 2020); phenolic acids (e.g., caffeic acid and ferulic acid) reduce neuroinflammation and lipid peroxidation (Singh et al., 2022); and stilbenes (e.g., resveratrol) offer protection to neurons by enhancing mitochondrial function and activating sirtuin pathways (Moraes et al., 2020).

The neuroprotective effects of a phenolic-enriched maple syrup extract, which is rich in flavonoids, stilbenes, and phenolic acids, were investigated in another interesting study. At high concentrations, the phenolic-enriched maple syrup extract was found to inhibit amyloid-beta (Aβ1–42) fibrils, a hallmark of Alzheimer's disease, by as much as 63.5%. This effect is comparable to the effect of resveratrol (Rose et al., 2021). Furthermore, the phenolic-enriched maple syrup extract was found to inhibit inflammatory markers such as nitric oxide synthase (NOS), IL-6, PGE2, and TNF-α by as much as 87.6%. It also reduced H_2O_2-induced oxidative stress in microglial cells by 16.1%. The phenolic-enriched maple syrup extract also has the potential for use as a dietary neuroprotective agent, which increases cell viability and protects Caenorhabditis elegans from Aβ-induced neurotoxicity (Wei et al., 2022). These phenolic compounds have a high chance of being used as an integrated approach for the prevention, cessation, or slowing down of neurodegenerative diseases, as they have the capacity to protect the brain from neuroinflammation and oxidative stress, which play a crucial role in brain health.

3.2.2 Flavonoids

Flavonoids are a subclass of polyphenolic compounds that can be found in many plants, such as tea, berries, citrus fruits, and cocoa. They have gained substantial interest due to their neuroprotective properties. Flavonoids have been shown to enhance synaptic plasticity and to modulate BDNF—both of which are fundamental to cognitive function and the durability of the brain (Al Amin et al., 2025). The expression of BDNF is closely related to processes involved in learning and memory and is essential for the survival, differentiation, and synaptic remodeling of neurons (Kowiański et al., 2018). In addition, some flavonoids, including epicatechin, hesperidin, and quercetin, have been shown to increase BDNF levels, as well as activate the PI3K/Akt and ERK signaling pathways, both of which are important for synaptic plasticity and strength (Amidfar et al., 2024).

Dietary flavonoids play a collaborative role in these functions. For example, berries contain flavanols and anthocyanins that cross the blood–brain barrier and accumulate in memory centers of the brain, including the hippocampus (Kelly et al., 2018). Epigallocatechin gallate, a compound found in tea, specifically in green tea, which contains a number of compounds known as catechins, increases synaptic connectivity and reduces neuroinflammation (Khalatbary and Khademi, 2020). Other flavonoids, specifically flavanones such as naringenin found in citrus fruits, have been linked to neurogenesis and improved cognitive functions (Pontifex et al., 2021). In the process known as long-term potentiation, which is a cellular mechanism of learning, the functions are supported by compounds found in cocoa, specifically flavanols such as epicatechin, which increase blood flow to the brain and induce the expression of BDNF (Fiore et al., 2025).

It was found that foods rich in flavonoids increase cognitive functions both in humans and animal models. It was observed that, in rodent models, flavonoids increased memory test scores, which resulted in increased BDNF levels in the hippocampus, along with changes in synaptic density and dendritic spine morphology, which were associated with the effects (Stagni et al., 2017). It was also found that flavonoids, as a supplement, increased mood, working memory, and executive functions, especially in older people (Gardener et al., 2021). Flavonoids confer neuroprotection through the instantaneous inhibition of intracellular signaling pathways involved in synaptic function and plasticity, as well as antioxidant effects. Flavonoids are promising dietary candidates to maintain brain health and stop cognitive dysfunction because of their multimodal mechanism (Bellavite, 2023).

3.2.3 Alkaloids

Some plants contain nitrogenous phytochemicals called alkaloids, which are what give them their strong effects on the CNS. Caffeine, berberine, and nicotine are the most important alkaloids for brain function because they affect mood regulation and neurotransmitter systems (Alamgir, 2018). Tea and coffee contain caffeine, a methylxanthine alkaloid that acts as an antagonist on adenosine receptors to promote alertness and cognitive function. In addition, it raises norepinephrine and dopamine levels, which lift the spirits and lessen weariness (Sharma et al., 2023). By altering monoamine neurotransmitters and activating AMPK pathways, which are implicated in energy metabolism and neuroplasticity, berberine, an isoquinoline alkaloid found in plants such as Berberis, demonstrates antidepressant-like properties (Gao et al., 2024). Despite being context-dependent because of its capacity for addiction, nicotine acts on postsynaptic neurons' nicotinic acetylcholine receptors to promote dopamine release, which momentarily elevates mood and attention (Letsinger et al., 2022).

The effects of arecoline, caffeine, and nicotine on gene expression in the nucleus accumbens (NAc), a region of the mouse brain involved in reward and mood control, have recently been investigated and compared. These findings revealed that both nicotine and caffeine have profoundly affected the biological mechanisms involved in synaptic function. Caffeine increased locomotor activity and affected genes involved in neurotransmitter release, while mice treated with nicotine displayed enhanced gene expression in substance dependence and synaptic signaling. These findings provide molecular evidence of the neuromodulatory effects of these alkaloids and their influence on mood and behavioral changes (Huang et al., 2025). The ability of alkaloids to influence mood and neurotransmission makes them important in neuropsychiatric studies and in treating mood and cognitive disorders, even though there are aspects of dosage and context that need to be considered when using these alkaloids therapeutically. The major classes of phytochemicals and their neuroactive effects are presented in Table 3.1 .

TABLE 3.1 Major phytochemical classes and their neuroactive functions.

Phytochemical class	Representative compounds	Primary neuroactive mechanisms	Neuroprotective/ neuromodulatory effects	References
Flavonoids	Quercetin, kaempferol, epigallocatechin gallate (EGCG)	↑ Nrf2/HO-1 signaling, ↓ NF-κB activation, inhibition of MAO-A/B, antioxidant radical scavenging	Protection against oxidative stress and neuroinflammation; improved cognitive function and memory	Kumar et al. (2024); Singh et al. (2025)
Phenolic acids	Caffeic acid, ferulic acid, rosmarinic acid	Antioxidant via ROS scavenging, ↑ BDNF, inhibition of lipid peroxidation	Reduced neuronal apoptosis and improved synaptic plasticity	Szwajgier et al. (2017); Rojas-García et al. (2023)
Curcuminoids	Curcumin	Activation of Nrf2/ARE pathway, inhibition of amyloid-β aggregation, ↓ microglial activation	Protection against Alzheimer's and depression-related oxidative stress	Mandal et al. (2020)
Stilbenes	Resveratrol, pterostilbene	↑ SIRT1/Nrf2 signaling, ↓ Aβ accumulation, antiapoptotic	Enhancement of mitochondrial function and neurogenesis; protection from ischemic injury	Cui et al. (2024)

(*continued on next page*)

TABLE 3.1 Major phytochemical classes and their neuroactive functions—cont'd

Phytochemical class	Representative compounds	Primary neuroactive mechanisms	Neuroprotective/ neuromodulatory effects	References
Alkaloids	Huperzine A, berberine, nicotine	Acetylcholinesterase inhibition, AMPK activation, antiinflammatory	Improvement of learning and memory; modulation of neurotransmission	Singh et al. (2025)
Terpenoids	Ginsenosides, limonene, asiaticoside	Regulation of PI3K/Akt and mitogen-activated protein kinase (MAPK) pathways, antioxidants, and antiinflammatory	Enhanced neuronal survival, protection against excitotoxicity	Xu et al. (2022)
Carotenoids	Lycopene, β-carotene, astaxanthin	Antioxidant defense, membrane stabilization, modulation of neuroinflammation	Protection against cognitive decline and oxidative brain injury	Fakhri et al. (2021)
Sulfur-containing compounds	Sulforaphane, allicin	Nrf2/HO-1 pathway activation, detoxification enzyme induction	Attenuation of oxidative and inflammatory brain damage	Egbujor et al. (2022)
Lignans	Sesamin, schisandrin	Antioxidant enzyme modulation, regulation of mitochondrial function	Neuroprotection via improved energy metabolism and reduced oxidative stress	Das et al. (2019)

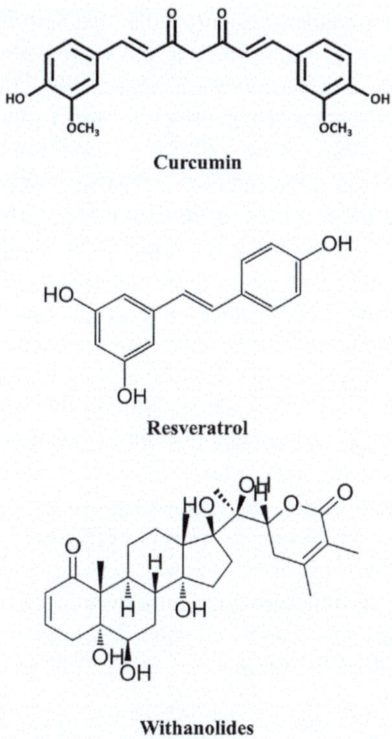

FIGURE 3.1 Chemical structures of curcumin, resveratrol, and withanolides.

3.3 Curcumin, resveratrol, ashwagandha: Key examples

3.3.1 Curcumin

Because of the variety of biochemical characteristics, curcumin, the primary curcuminoid of the rhizome of Curcuma longa, or turmeric, has become a popular compound for its potential as a potent neuroprotective agent (Fig. 3.1). Curcumin, a polyphenol compound, has demonstrated potent antioxidant activity, effectively scavenging ROS and regulating endogenous antioxidant pathways such as glutathione peroxidase, catalase, and SOD (Abbas et al., 2020). Oxidative stress, a major risk factor for the development of neurodegenerative diseases such as Parkinson's and Alzheimer's, is reduced by the antioxidant activity of curcumin. Curcumin also has the potential to inhibit the activity of MAOs, specifically MAO-A and MAO-B, which are responsible for the breakdown of neurotransmitters such as norepinephrine, dopamine, and serotonin. Curcumin has the potential to become an effective antidepressant and antianxiety drug as it would increase the levels of MAO activity, resulting in a positive impact on the improvement of mood and mental health states.

Arguably, the most remarkable neuroprotective activity of curcumin is the anti-amyloidogenic activity, where the curcumin molecule binds to Aβ fibrils, disrupts their structure, and promotes disaggregation, thereby inhibiting the aggregation of amyloid-beta (Aβ) peptides to form plaques associated with Alzheimer's disease (Abbas et al., 2015). In addition, curcumin has been demonstrated to possess antiinflammatory properties by modulating microglial activation and reducing the levels of proinflammatory cytokines such as TNF-α, IL-1β, and IL-6, thereby maintaining the function of synapses and the structure of neurons through these cooperative effects (Cianciulli et al., 2022).

Historically, curcumin's poor absorption and rapid metabolism have hindered its bioavailability, but new innovative formulations, such as liposomal delivery, encapsulation in nanoparticles, and coadministration with piperine, have significantly improved curcumin's systemic bioavailability and therapeutic efficacy (Yakubu and Pandey, 2024). In preclinical studies, curcumin treatment improved memory function by reducing the burden of amyloid plaques, increasing the level of oxidative markers, and improving memory function. Furthermore, curcumin can penetrate the blood–brain barrier, providing further support for the use of curcumin as a therapeutic agent for neurodegenerative diseases (Baghcheghi et al., 2025). Given its antiinflammatory, antiamyloid, MAO inhibitory, and antioxidant properties, curcumin can be considered a potentially beneficial natural compound for the health of the brain.

3.3.2 Resveratrol

Resveratrol, or trans-3,5,4′-trihydroxystilbene, is a polyphenolic compound present in peanuts, red wine, berries, and grapes (Fig. 3.1) (Siddiqui et al., 2025). Because it primarily activates sirtuin 1 (SIRT1), a NAD+-dependent deacetylase involved in cellular stress resilience, metabolism, and longevity, it has neuroprotective and anti-aging properties that make it highly desirable. According to Chen et al. (2020), resveratrol activation of SIRT1 enhances mitochondrial function, permits DNA repair, and modifies transcription factors such as FOXO and PGC-1α, which are essential for neuronal survival and plasticity.

Resveratrol has several antiinflammatory properties by targeting various mechanisms involved in regulating neuroinflammation; one is by suppressing the NF-kappaB signaling pathway and therefore decreasing the level of proinflammatory cytokines. In many neurodegenerative diseases, prolonged or sustained neuroinflammation can worsen neuronal injury; therefore understanding how resveratrol regulates neuroinflammation is very important. Resveratrol also provides protection against oxidative stress through decreased lipid peroxidation and increased expression/activity of antioxidant enzymes (de Sá Coutinho et al., 2018). Also, resveratrol has been shown to inhibit migration and proliferation of hepatocellular carcinoma cells via post-translational modulation of the PI3K/AKT pathway through the activation of SIRT1. Upon resveratrol treatment, proapoptotic proteins such as Bax and cleaved PARP were significantly increased, while the expression of the antiapoptotic protein

Bcl-2 decreased, promoting the process of apoptosis. In addition, resveratrol also enhanced the expression and phosphorylation of the tumor suppressor DLC1, further supporting its regulatory role in cells and longevity (Chai et al., 2017). Collectively, given the ability of resveratrol to activate the SIRT1 pathway in the regulation of antiinflammatory and prosurvival signaling, the findings of these studies provide evidence of a potential role for resveratrol in promoting brain health and longevity and thus serve as a promising drug candidate for neuroprotection through modulation of primary molecular targets in the brain and by crossing the blood–brain barrier.

3.3.3 Ashwagandha

Ashwagandha, the sacred herb in Ayurveda, fits into the category of classic adaptogens—the phytonutrients that assist the body in adapting to biophysiological and chemical stressors (Fig. 3.1). Its neuroprotective action is attributed to bioactive molecules, such as withanolides, which control stress response systems, particularly the hypothalamic–pituitary–adrenal (HPA) axis. Ashwagandha can reduce cortisol levels, the primary stress hormone, thereby reducing chronic stress and its noxious effects on brain function (Kuśmierska et al., 2024).

In a randomized, double-blind, placebo-controlled trial, ashwagandha extract strongly improved immune markers and reduced biomarkers of stress in healthy adults. Healthy adults on 60 mg of WS extract showed greater levels of immunoglobulins (IgA, IgG, IgM), cytokines (IFN-γ, IL-4), and TBNK cells (CD3+, CD4+, CD8+, CD19+, NK cells), indicating increased immune strength. These were demonstrated alongside improved mood and reduced anxiety, suggesting a systemic adaptogenic effect (Tharakan et al., 2021).

Ashwagandha also enhances GABAergic tone, which is the basis of its anxiolytic and sedative properties. The major inhibitory neurotransmitter in the brain is gamma-aminobutyric acid (GABA), and its modulation is essential for the maintenance of emotional balance as well as protection against excitotoxicity (Jayaram et al., 2025). Ashwagandha enhances the sensitivity of GABA receptors as well as the production of GABA, leading to improved sleep, decreased anxiety, and improved cognition (Park et al., 2023). The herb's neuroprotective actions comprise promotion of neurogenesis, reduction of oxidative stress, and modulation of neuroinflammation. Its ability to influence an array of pathways—suppression of cortisol, immunomodulation, and augmentation of GABAergic—makes ashwagandha a system agent for brain health and resistance to stress. Table 3.2 shows a comparative summary of curcumin, resveratrol, and ashwagandha.

3.4 Mechanisms of neuroprotection

3.4.1 Antioxidant activity

Oxidative stress is one of the primary causes of neuronal damage and neurodegenerative diseases through a surplus of ROS and reactive nitrogen species

TABLE 3.2 Comparative summary of curcumin, resveratrol, and ashwagandha.

Phytochemical	Source	Key neuroprotective mechanisms	Cognitive effects	Mood/stress effects	Bioavailability challenges	References
Curcumin	Turmeric (*Curcuma longa*)	Antioxidant (ROS scavenging), MAO inhibition, antiamyloid activity, NF-κB inhibition	Improves memory and attention in older adults	Reduces depressive symptoms, modulates HPA axis	Poor absorption, enhanced by piperine and nanoformulations	Small et al. (2018); Yuan et al. (2025)
Resveratrol	Grapes, red wine, berries	SIRT1 activation, antiinflammatory, mitochondrial enhancement, antioxidant	Enhances executive function and cerebral blood flow	Reduces anxiety and inflammation; potential antidepressant effects	Rapid metabolism (glucuronidation/sulfation); low systemic levels	Martemucci et al. (2024); Ciupei et al. (2024)
Ashwagandha	*Withania somnifera*	Adaptogen: Reduces cortisol, enhances GABAergic tone, antioxidant	Supports cognitive resilience under stress	Reduces anxiety and cortisol; improves sleep and mood	Generally well-absorbed; standardization of withanolide content needed	Kuśmierska et al. (2024)

(RNS) over the brain's antioxidant defense (Houldsworth, 2024). The brain is also particularly vulnerable since it possesses high oxygen demand, high lipid content, and relatively less antioxidant capacity (Lee et al., 2020). Antioxidant mechanisms cause neuroprotection by neutralizing ROS/RNS and activating the endogenous defense mechanism. The most important among them is the nuclear factor erythroid 2-related factor 2 (Nrf2) pathway, which is a key regulator of cellular antioxidant responses (Tonev and Momchilova, 2023). Under oxidative stress, Nrf2 is liberated from its cytoplasmic repressor Keap1 and translocates to the nucleus, where it binds to antioxidant response elements (AREs) in the DNA, thus inducing the expression of genes coding for antioxidant enzymes such as heme oxygenase-1 (HO-1), glutathione S-transferase, and NAD(P)H quinone dehydrogenase 1 (NQO1) (Giudice et al., 2010).

HO-1, in turn, is a major enzyme that cleaves heme into biliverdin, carbon monoxide, and free iron, all of which are cytoprotective (Origassa and Câmara, 2013). After that, biliverdin is reduced to bilirubin, which is a strong antioxidant. In Alzheimer's and Parkinson's disease models, it has been demonstrated that inducing HO-1 through Nrf2 activation shields neurons from oxidative damage (Jazwa and Cuadrado, 2010). Furthermore, it has been shown that a number of phytochemicals, including sulforaphane, resveratrol, and curcumin, alter the Nrf2/HO-1 pathway, increasing the brain's resistance to oxidative stress (Hannan et al., 2020). Fig. 3.2 demonstrates how phytochemical-induced activation of the Keap1–Nrf2-ARE signaling pathway increases the expression of antioxidant enzymes, thereby reducing neurodegenerative events.

Antioxidants such as vitamin E, vitamin C, and polyphenols prevent lipid peroxidation, protein oxidation, and DNA damage by directly scavenging ROS and RNS (Bratovcic, 2020). Furthermore, antioxidants stabilize the structures of cells and support the functioning of neurons by stopping the cascading effect of free radicals. Mitochondria-targeting antioxidants such as MitoQ and CoQ10 have also been shown in studies to effectively reduce mitochondrial oxidative stress, one of the leading causes of neurodegeneration (Fields et al., 2023). In conclusion, antioxidant activity is a cornerstone of neuroprotection, both direct neutralization of toxic species and activation of transcriptional programs such as the Nrf2/HO-1 pathway. These mechanisms collectively reduce oxidative load, preserve neuronal integrity, and support cognitive function and are therefore critical targets for treatment in neurodegenerative disease.

3.4.2 Antiinflammatory activity

Neuroinflammation is characteristic of most neurodegenerative diseases, including Alzheimer's disease, Parkinson's disease, and multiple sclerosis. It is characterized by activation of glial cells, particularly microglia and astrocytes, and the secretion of proinflammatory cytokines, chemokines, and ROS (Kwon and Koh, 2020). Acute inflammation is protective, but chronic neuroinflammation causes

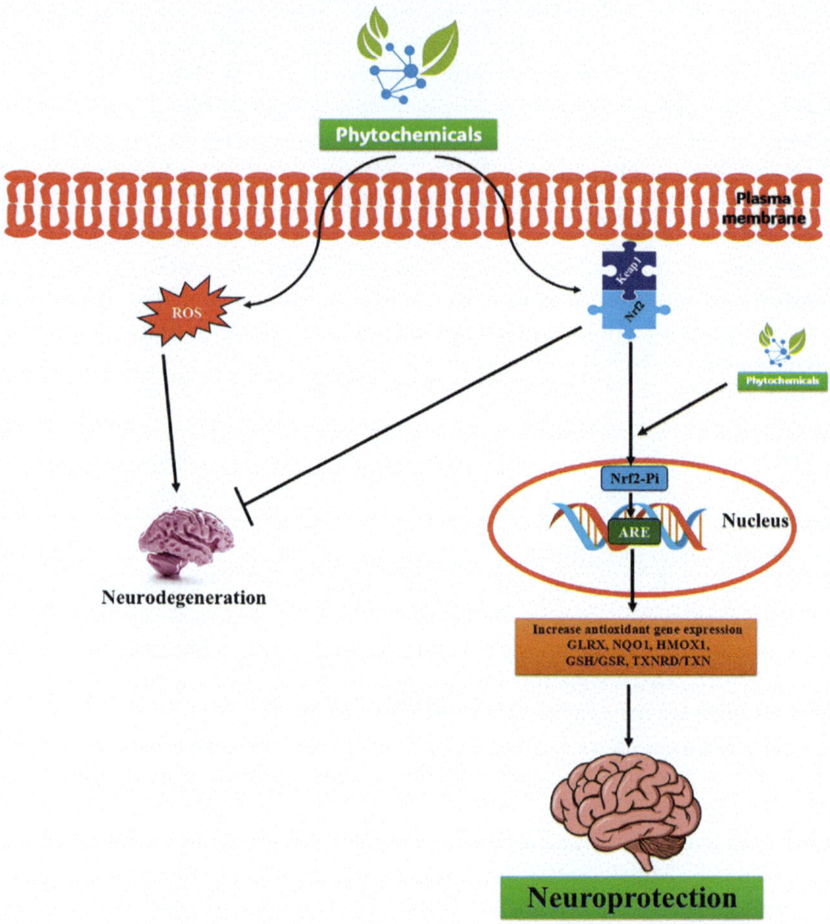

FIGURE 3.2 Activation of the Keap1–Nrf2–ARE signaling pathway by phytochemicals enhances antioxidant enzyme expression, mitigating neurodegenerative processes.

neuronal dysfunction, synaptic loss, and cell death. So, antiinflammatory mechanisms are essential for neuroprotection, and the targeting of proinflammatory mediators has become a promising therapeutic strategy (Wang et al., 2018).

An important regulator of inflammation is the nuclear factor kappa-light-chain-enhancer of activated B cells (NF-κB) signaling pathway. Under basal conditions, NF-κB is sequestered in the cytoplasm by inhibitory proteins (IκBs). When stressors such as oxidative stress, pathogens, or cytokines activate IκBs, they become phosphorylated and degraded. After entering the nucleus, NF-κB initiates the transcription of proinflammatory genes such as cyclooxygenase-2 (COX-2), interleukins (such as IL-1β and IL-6), and tumor necrosis factor-alpha

(TNF-α) (Kannan et al., 2025). Persistent NF-κB activation in the CNS results in chronic inflammation and neuronal damage (Anilkumar and Wright-Jin, 2024).

Numerous experimental studies clearly show that inhibiting the activation of NF-kB results in less overall neuroinflammation and a protective effect against cognitive decline (Liu et al., 2017). Phytochemicals that inhibit the signaling cascade of NF-kB, and consequently inhibit TNF-α and COX-2, include curcumin, resveratrol, and epigallocatechin gallate (EGCG). COX-2 is an inducible enzyme that adds to the production of prostaglandins, and COX-2 is overexpressed in the neuroinflammatory state; this overexpression leads to neuron death and impairment of synaptic function (Mokra et al., 2023). Antiinflammatory medications are capable of preventing neuronal death by blocking the production of COX-2, thus preventing the production of proinflammatory prostaglandins (Famitafreshi and Karimian, 2020). In addition, TNF-α, the main cytokine in the inflammatory cascade, serves two functions in the brain; it can cause synaptic plasticity and neurogenesis under normal conditions, but when it is chronically overexpressed, it can cause excitotoxicity, the breakdown of the blood–brain barrier, and neuron death. Antiinflammatory treatments, which block TNF-α, also improve cognitive functioning and decrease neurodegeneration according to preclinical studies (Soltani Khaboushan et al., 2022).

Moreover, the progression of neuroinflammation is also affected by microglial polarization, whereby microglia are either antiinflammatory (M2) or proinflammatory (M1) phenotypes. This effect can be altered to have a neuroprotective effect through therapeutic means, whereby M2 microglia are encouraged and M1 is inhibited (Guo et al., 2022). A number of pharmacological and phytochemical agents have been identified to have the effect of increasing the release of neurotrophic factors, decreasing the production of proinflammatory cytokines, and altering microglia function (Zahedipour et al., 2022). Lastly, one of the major ways through which neuroprotection is achieved is through antiinflammatory activity, whereby it acts through the downregulation of TNF-α, inhibition of NF-κB, and inhibition of COX-2 expression, leading to the prevention of synaptic loss, maintenance of neuronal function, and slowing down the progression of neurodegenerative diseases. There is great potential for maintaining cognitive function and reducing age-related neurological decline by modifying neuroinflammation through dietary, pharmaceutical, or lifestyle changes.

3.4.3 Adaptogenic properties

A class of natural compounds known as adaptogens helps the body maintain homeostasis, strengthens its resistance to stress, and guards against the harmful effects of long-term stress exposure (Pawar and Shivakumar, 2012). The main way that adaptogens in neuroprotection work is by controlling the body's main stress response system, the HPA axis (Panossian, 2017). Chronic HPA axis activation leads to elevated levels of cortisol, the primary glucocorticoid hormone,

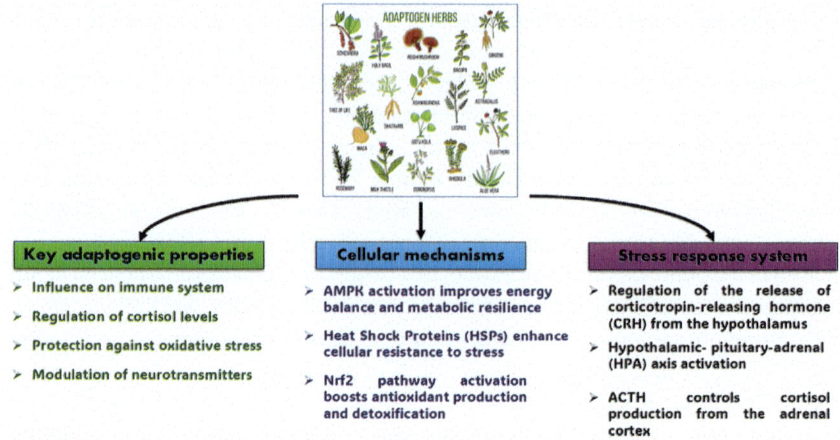

FIGURE 3.3 The main properties of adaptogens.

for an extended period, thereby causing harm to hippocampal neurogenesis, reducing synaptic plasticity, and contributing to mood disorders, as well as cognitive decline. Adaptogens regulate the activity of the HPA axis and therefore reduce cortisol levels as well as prevent stress-induced neuronal damage (Sharan and Vellapandian, 2024). Fig. 3.3 shows the main properties of adaptogens.

One of the best-studied adaptogens is *Withania somnifera* (ashwagandha), which has demonstrated significant effects on resistance to stress and neuroendocrine regulation (Haber et al., 2024). Ashwagandha supplementation has been shown by clinical trials to lower serum cortisol levels, improve sleep quality, and reduce symptoms of anxiety and depression. These actions are mediated by withanolides, the bioactive compounds of ashwagandha, which influence GABAergic transmission and impact gene expression involved in stress (Speers et al., 2021). Ashwagandha acts as a countermeasure for excitotoxicity caused by chronic stress and maintains emotional stability by enhancing inhibitory neurotransmission because of an increase in GABAergic tone (Mallick and Banerjee, 2024). Moreover, adaptogens regulate the expression of heat shock proteins (HSPs), particularly Hsp70, which play a role in the defense mechanism against cellular stress (Panossian, 2017). HSPs maintain neuronal integrity by enhancing protein folding, preventing aggregation, and allowing the degradation of damaged proteins (Albinhassan et al., 2025). Moreover, it has been shown that adaptogens, such as Panax ginseng and Rhodiola rosea, modulate monoaminergic pathways, including the dopamine, norepinephrine, and serotonin systems, which are crucial for emotional and cognitive function (Panossian and Wikman, 2010).

On a molecular basis, adaptogens interact with transcription factors that regulate metabolism, cell survival, and resistance to oxidative stress, such as nuclear factor erythroid 2-related factor 2 (Nrf2) and forkhead box O (FOXO)

(Panossian et al., 2021). Through interaction with the HPA axis, these processes create a complex of protective mechanisms that protect against stress-induced brain injury. In addition, adaptogens have been shown to play a role in neuroprotection by reducing the expression of proinflammatory cytokines such as TNF-α and interleukin-6 (IL-6) (Cai et al., 2022).

Research on nuts reveals that they have a few health support properties, including improving cognitive abilities, stimulating the brain (hippocampus) to generate new nerve cells, and reducing the markers of oxidative stress and inflammation. Self-reported research on people shows that those taking adaptogens can tolerate more stress, feel less anxious, and have improved cognitive abilities (Wróbel-Biedrawa and Podolak, 2024). Adaptogenic effects that help to modify the HPA axis, reduce cortisol levels, enhance GABAergic transmission, and enhance cellular defenses against stress create an exceptional neuroprotective mechanism. Adaptogens support neuron health, elevate cognitive capability, and provide long-term protection from the adverse cardiovascular impacts of chronic stress and neurodegeneration by way of multiple mechanisms of action.

3.5 Dose-response considerations

3.5.1 Bioavailability

The neuroprotective phytochemicals' bioavailability is a significant factor affecting their therapeutic efficacy. Bioavailability is the proportion of a particular chemical that enters the bloodstream in an active state and is bioavailable to carry out its biological effects (Ali and Siddique, 2019). Despite their strong activity in vitro, many phytochemicals have low bioavailability because of their poor solubility, gastrointestinal instability, extensive metabolism, and low permeability through biological membranes (Karaś et al., 2017). Curcumin, the primary curcuminoid in *Curcuma longa*, or turmeric, is one example. Despite curcumin's strong antioxidant, antiinflammatory, and antiamyloid qualities, its limited aqueous solubility, extensive hepatic and intestinal metabolism, and low bioavailability rates limit its clinical utility (Far et al., 2024). When taken orally, curcumin undergoes extensive metabolism in the gastrointestinal tract and liver through sulfation and glucuronidation. These conjugation reactions significantly reduce the plasma concentration of free curcumin, limiting its availability to target tissues such as the brain (Tsai et al., 2011). The hydrophobic characteristics of curcumin limit its solubility in the aqueous phase of the gastrointestinal tract, which further impedes absorption. As a result, even high oral dosages have minimal systemic exposure, necessitating innovative approaches to provide higher bioavailability (Zheng and McClements, 2020).

The most investigated method involves the use of adjuvants as inhibitors of metabolic enzymes. Piperine, a known bioenhancer and a glucuronidation inhibitor of the liver and intestinal tissues, is an alkaloid compound of black pepper (Patil et al., 2022). It has been established that when piperine and curcumin are taken as a combination, the bioavailability of curcumin may increase up to 2000%, significantly increasing the therapeutic potential of curcumin. Piperine could increase the half-life and plasma concentration of curcumin by the reversible inhibition of UDP-glucuronosyltransferase enzymes (Heidari et al., 2023).

The use of formulations, particularly nanoformulations (e.g., liposomes, micelles, solid-lipid nanoparticles, and polymeric nanoparticles), has been demonstrated to be an effective means for overcoming bioavailability obstacles associated with curing diseases, such as the use of curcumin. In addition, these different types of nanoformulations have been demonstrated to enhance solubility, inhibit enzymatic degradation, and facilitate the transport of substances across epithelial barriers (Shi et al., 2025). This capability allows curcumin to be encapsulated in a lipid or polymer matrix via these drug delivery systems, thereby providing improved stability and bioavailability to curcumin. For example, liposomes containing curcumin have exhibited improved pharmacokinetics and delivery to the brain in animal models. Furthermore, curcumin nanoparticles have been shown to increase bioavailability and therapeutic efficacy in preclinical models of neurodegenerative diseases (Palanisamy et al., 2025). The usability of the methods demonstrates that formulation science can be effectively used to maximize the dose-response relationships. If the phytonutrients are not bioavailable, then no matter how potent they are, they will be unable to reach levels necessary to elicit a pharmacological response in the tissue of interest. Therefore overcoming absorption hurdles through adjuvants and nanoformulations is required to access the high potential of promising in vitro findings as clinically valuable interventions (Paul-Chima et al., 2024).

3.5.2 Metabolism

Metabolism is the controlling principle of the pharmacokinetics and pharmacodynamics of neuroprotective phytochemicals and is directly accountable for their dose-response relationships. Following absorption of a compound, it is metabolically transformed primarily in the liver through Phase I and Phase II reactions. These reactions are vital to the metabolism of lipophilic compounds into more hydrophilic compounds that can be excreted, but have a significant effect on phytochemical bioactivity and therapeutic usefulness too (Rathaur and SR, 2020). Phase I metabolism is usually oxidation, reduction, or hydrolysis, catalyzed by cytochrome P450S (CYPs) and other enzymes. These reactions put in or unveil functional groups that prepare the compound for Phase II metabolism. In Phase II, conjugation processes such as glucuronidation, sulfation, and methylation occur, catalyzed by enzymes such as UDP-glucuronosyltransferases (UGTs) and sulfotransferases (SULTs). The process adds water-solubility and facilitates renal or biliary excretion (David Josephy et al., 2005).

Although such metabolic processes are designed to detoxify xenobiotics, they also decrease the therapeutic effect of phytochemicals. For example, resveratrol, a grape and red wine polyphenol, is rapidly metabolized to glucuronide and sulfate conjugates that have the potential to have reduced biological activity compared to the parent compound (Delmas et al., 2011). Similarly, curcumin undergoes extensive Phase II metabolism to have limited plasma levels of free curcumin and thin tissue distribution. These processes decrease the compound's ability to be transported to target sites such as the brain, thereby affecting its neuroprotective activity (Di Meo et al., 2019).

Apart from hepatic metabolism, gut microbiota is also responsible for affecting the bioavailability and pharmacology of phytochemicals. Trillions of microorganisms in the human gut have the capacity to enzymatically metabolize compounds in foods. These microbiologically catalyzed conversions can stimulate or inhibit phytochemicals, producing metabolites with altered pharmacological profiles (Khan et al., 2025). For example, the gut microbiota converts pomegranate ellagitannins into neuroprotective and antiinflammatory urolithins. Flavonoids such as quercetin and catechins can break down into smaller phenolic acids that may help to maintain or enhance their biological activity (García-Villalba et al., 2023). In addition, some phytochemicals will alter the composition of the gut microbiota by promoting the growth of beneficial bacteria and suppressing the growth of pathogenic bacteria. This two-way relationship may also suggest that phytochemicals could mediate their neuroprotective effects indirectly through mechanisms related to the gut microbiota, including modulating systemic inflammation and producing neuroactive compounds (Santhiravel et al., 2022).

3.5.3 Therapeutic windows

An important principle in the use of neuroprotective agents from phytonutrients for the treatment of disease is the therapeutic window. The therapeutic window is the range of doses over which a drug has a beneficial effect and does not produce any significant side effects (Xu et al., 2021). This is a particularly important concept with bioactive compounds that target sensitive areas of the body, such as the CNS, because it helps ensure that the compounds are effective while minimizing toxicity. In many cases, the consumption of most phytonutrients occurs through the food chain or from a naturally occurring source, making it difficult to determine the pharmacological threshold for phytochemicals, as the levels of consumption of a particular phytochemical are usually unknown; therefore establishing standardized dosing guidelines can be challenging (Chaachouay, 2025).

Phytochemicals such as flavonoids, alkaloids, and polyphenols often exhibit a hormetic response curve, which is also known as a biphasic dose-response curve. This type of model indicates that at high doses, phytochemicals can have no effect and could be toxic; however, at lower to moderate doses, phytochemicals can be protective (Jodynis-Liebert and Kujawska, 2020). For example, quercetin and catechins (types of flavonoids) were found to be antioxidants, but when consumed at high amounts, they can act as prooxidants, producing ROS and contributing to oxidative stress. These findings illustrate the necessity for appropriate, thorough preclinical and clinical studies to identify optimal dosing ranges (Tumilaar et al., 2024).

The therapeutic windows of alkaloids such as nicotine, berbine, and caffeine all have narrow therapeutic windows. In moderate doses, caffeine is consumed nearly universally for its stimulating effects and thus enhancing mental clarity and attentiveness, but in excessive doses, it can cause neurotoxicity, tachycardia, anxiety, and insomnia (Cappelletti et al., 2015). Berberine, although utilized for its antiinflammatory effects and metabolic effects in excessive doses, can result in hepatotoxicity and gastrointestinal (GI) side effects (Zhao et al., 2018). While nicotine can promote the release of chemical messengers (neurotransmitters) and could improve attentiveness given certain conditions, it is highly addictive and neurotoxic at excessive doses (Tiwari et al., 2020). Both cases demonstrate that achieving the maximum therapeutic benefit of these alkaloids without the potential for adverse effects is based upon an individual's ability to absorb, metabolize, and develop a response to these compounds. Genetic polymorphisms of metabolic enzymes, the composition of the gut microbiota, and the pharmacokinetics of aging are all factors that can affect an individual's ability to receive a specific dose of an alkaloid (Sera and McPherson, 2012). In addition, the method and formulation used to deliver the therapeutic alkaloid can directly impact the size of the therapeutic window.

First-pass metabolism (Bhandare et al., 2024) may reduce the bioavailability of oral dosage. Novel and innovative dosage forms such as liposomes, nanocarriers, and coadministration with bioenhancers such as piperine can enhance systemic exposure and shift the therapeutic window down to lower, safer doses than the conventional high-dose regimen and provide improved dosing stability (Jacob et al., 2025). Therefore safe and effective use of neuroprotective phytochemicals depends on the identification and characterization of therapeutic windows. Formulation techniques, variability, and dose-response relationships need to maintain the appropriate balance between efficacy and toxicity. Establishing standardized dosage guidelines and individualized treatment protocols will be important for developing candidate compounds toward safe and effective clinical use with increasing recognition of phytochemicals as treatment options.

3.6 Human and animal trials

3.6.1 Cognitive enhancement

Several human studies have shown promising cognitive results of neuroprotective phytochemicals on memory, attention, and executive function. Some of these compounds include curcumin, resveratrol, flavonoids, and omega-3 fatty acids.

Studies have suggested that some of these compounds could promote neuroplasticity and reduce oxidative stress, thereby reducing inflammation and promoting blood flow to the brain (Howes et al., 2020). For example, curcumin has been shown to enhance working memory and attention in healthy older people. In a randomized controlled trial, older adults taking curcumin had significantly less fatigue and less severe mood problems but also showed an increase in verbal memory and sustained attention compared to the placebo group (Cox et al., 2020). These benefits are thought to result from the antioxidant and anti-inflammatory effects of curcumin as well as its effects on neurotransmitter systems.

Resveratrol also improves cognition, particularly in older populations, by promoting synaptic plasticity and neuronal metabolism via activation of SIRT1 and enhancement of mitochondrial function (Moraes et al., 2020). Further, resveratrol supplementation has been demonstrated to increase cerebral perfusion and performance on assessment of cognitive ability and memory in clinical trials (Wightman et al., 2015). A number of other flavonoids found in berries and cocoa have produced similar results in terms of cognition. For instance, cocoa flavanols have been shown to enhance executive function and processing speed, likely because of improved vascular function and increased levels of BDNF (Socci et al., 2017). Phytochemicals appear to enhance cognitive functions through multiple mechanisms such as increased synaptic signaling, anti-inflammatory effects, and modulation of neurotrophic factors. However, as with any supplement, generalization to the evidence-based clinical applications is hampered by the heterogeneity of study design, dosing, and bioavailability (Venkatesan et al., 2015).

3.6.2 Mood and stress outcomes

Multiple compounds have demonstrated anxiolytic and antidepressant-like properties in human clinical studies, and plant-based compounds have also been examined for their ability to help regulate mood and stress (Tkaczenko et al., 2025). Research has shown that ashwagandha, a classical adaptogen, has been extensively studied for its ability to reduce cortisol levels and enhance resilience to stress (Ossowska et al., 2025). Randomized controlled trials have found that supplementation with ashwagandha results in significantly lower serum cortisol levels, improved quality of sleep, and alleviation of symptoms of anxiety and depression. The mechanism through which these effects occur is likely because of ashwagandha's ability to increase GABAergic tone and modulate the HPA axis (Majeed et al., 2023).

Furthermore, curcumin has shown antidepressant effects in clinical settings. Its activity in inhibiting MAO, reducing inflammatory cytokines, and increasing BDNF levels assists in mood stabilization (Zhang et al., 2020). Supplementation of curcumin in clinical trials of major depressive disorder patients revealed a significant reduction in depressive symptoms, which was equivalent to standard

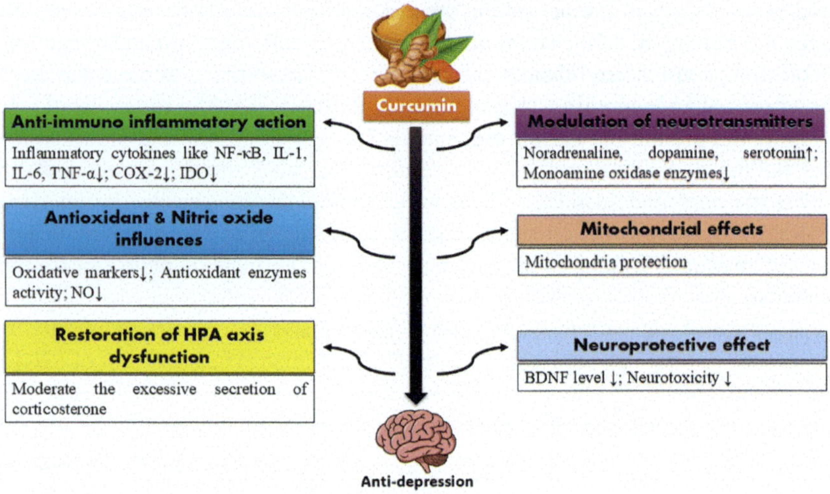

FIGURE 3.4 Possible antidepressant mechanisms of curcumin.

antidepressant treatment (Fig. 3.4) (Lopresti and Drummond, 2017). Resveratrol, through its antiinflammatory and neuroprotective actions, also exhibited anxiolytic and mood-improving actions, although larger studies are needed (Shayganfard, 2020). These results support the application of phytochemicals as adjunctive or alternative therapy to conventional psychotropic medication, particularly in mild to moderate illness. Formulation and dosing variability, as well as differences in study population characteristics, need to be considered in interpreting results. Table 3.3 shows clinical studies on phytochemicals and mental health.

3.6.3 Animal studies

Animal models have also been instrumental in establishing the neuroprotective effects of phytochemicals. Behavioral tests such as the forced swim test (FST), elevated plus maze (EPM), and open field test are common methods used to identify antidepressant and anxiolytic activity (Figueiredo Cerqueira et al., 2023). Reduction in immobility time in the FST is an antidepressant-like effect, while increased time spent in open arms in the EPM indicates reduced anxiety. Phytochemicals such as curcumin, resveratrol, and ashwagandha have consistently shown positive effects in such paradigms (Balogun et al., 2025).

Biochemical studies in such models show alterations in key neurobiological parameters. For example, curcumin-treated animals show increased hippocampal BDNF levels, reduced corticosterone levels, and enhanced antioxidant enzyme activity (Lopresti, 2017). Resveratrol has been shown to enhance SIRT1 and suppress proinflammatory cytokines in the brain (Sarubbo et al., 2018). Ashwagandha increases GABA receptor expression and reduces markers of

TABLE 3.3 Clinical studies on phytochemicals and mental health.

Phytochemical	Design and population	Dose and duration	Primary mental health outcomes measured	Main result	References
Saffron (*Crocus sativus*)	Double-blind randomized controlled trial (RCT); adults with mild–moderate major depression	Saffron 30 mg/day vs fluoxetine 20 mg/day; 6 weeks	HDRS (Hamilton Depression Rating Scale)	Saffron showed similar antidepressant efficacy to fluoxetine with comparable side effects	Noorbala et al. (2005)
Saffron	Double-blind RCT; patients with mild–moderate depression	Saffron 30 mg/day vs fluoxetine 20 mg/day; 6 weeks	HDRS, adverse events	Saffron was as effective as fluoxetine in reducing depressive symptoms	Shahmansouri et al. (2014)
Hypericum perforatum	Systematic review & meta-analysis of RCTs for depression	Various preparations and doses across studies (typical 300–900 mg/day)	Depression severity scales (multiple trials)	Overall efficacy superior to placebo for mild–moderate depression; results vs standard antidepressants mixed	Linde et al. (2005)
Curcumin (bioavailable formula Biocurcumax)	12-month, randomized, placebo-controlled, double-blind; community-dwelling older adults (cognitive outcomes)	Curcumin 1500 mg/day; 12 months	Cognitive test battery (memory, attention)	No clear prevention of global cognitive decline in full cohort; subgroup/secondary analyses suggested some cognitive benefits—evidence mixed and formulation-dependent	Rainey-Smith et al. (2016)

(*continued on next page*)

TABLE 3.3 Clinical studies on phytochemicals and mental health—cont'd

Phytochemical	Design and population	Dose and duration	Primary mental health outcomes measured	Main result	References
Curcumin (CurQfen®)	Randomized, double-blind, placebo-controlled trial (cognitive impairment)	400 mg × 2/day (800 mg/day) formulations; duration per trial arms	Cognitive scales (MoCA/MMSE/other) and safety	Reported improvement in some cognitive measures vs placebo (supports formulation and dose-specific effects)	Das et al. (2023)
Resveratrol	Double-blind, placebo-controlled, crossover acute studies in healthy adults	Single oral doses (e.g., 250–500 mg) in acute crossover designs	Cerebral blood flow (near-infrared spectroscopy), cognitive task performance	Acute doses increased cerebral blood-flow variables; mixed effects on cognitive performance—suggest vascular/acute cognitive modulation	Kennedy et al. (2010)
Ashwagandha (*Withania somnifera*)	Prospective, randomized, double-blind, placebo-controlled trial; stressed but otherwise healthy adults	Standardized root extract, 300 mg twice daily (600 mg/day); 60 days	Perceived Stress Scale (PSS), anxiety scales, well-being measures	Significant reduction in stress and anxiety vs placebo; good tolerability	Chandrasekhar et al. (2012)

(*continued on next page*)

Ashwagandha	Randomized, double-blind, placebo-controlled trial; healthy adults with stress	High-concentration full-spectrum extract; 60 days	PSS (primary), cortisol, sleep, anxiety	Ashwagandha reduced perceived stress and cortisol vs placebo; improvements in secondary outcomes reported	Lopresti et al. (2019); Pandit et al. (2024)
Saffron	Review and meta-analysis of RCTs vs placebo/fluoxetine for depression	Trials typically 30 mg/day; 4–8 weeks	Depression rating scales	Meta-analytic evidence supports saffron's efficacy vs placebo and comparable effects to some selective serotonin reuptake inhibitors (SSRIs) in short-term trials	Khaksarian et al. (2019); Dai et al. (2020)

oxidative stress, as one would predict from its anxiolytic profile (Speers et al., 2021).

3.6.4 Limitations and gaps

Research on neuroprotective phytochemicals has produced promising results; however, there are some similarities between research conducted using animal subjects and human subjects. One major limitation of these studies is that many have very small sample sizes, thus decreasing the statistical power of the analyses and limiting the generalizability of the results. For studies with fewer than 50 participants, it is difficult to detect subtle differences between groups and to account for differences in individuals. Another complication is found in the variability of dosing, formulation, and methods of assessing outcomes used in various studies. For example, the numerous ways to consume curcumin, such as common extract, nanoparticle, and combined with piperine, create different bioavailability profiles (Baker et al., 2021).

The problem of placebo effects is of significant concern. In studies of how plants and plant-sourced compounds affect mood or cognitive functioning, self-rated items (e.g., anxiety and memory) are more susceptible to bias, and because of the difference in taste or color of the plant-based supplements, are less likely to be blinded. The short-term nature of most studies precludes any long-term safety and/or efficacy determination (Knezevic et al., 2025).

The underutilization of ethnically diverse communities is another major issue. Sample sizes are so small across ethnic groups that there are not enough numbers available to establish statistically valid distributions; therefore many studies end up being performed in countries such as Canada and the United Kingdom, where the age and ethnicity of all participants tend to be uniform and homogeneous, therefore affecting both their metabolic reactions as well as all other dose-titration parameters necessary to determine therapy effectiveness (Roman, 2025). Furthermore, research that has focused on elucidating potential interactions between phytochemicals and conventional medicines remains very limited, making this an area of great interest to the community engaged in clinical practice (Roman, 2025). Future studies must be conducted using a sufficiently sized, randomized, well-designed, and longitudinally followed study cohort with a standardized experimental methodology and a representative multi-ethnic sample population to assess whether phytochemicals can be safely added into evidence-based neurotherapeutic care methods.

3.7 Conclusion

Herbal bioactives propose a new option for neuroprotection and psychotropics because of their potential complex actions against the pathophysiological processes involved in many brain disorders. Acting through a variety of antiinflammatory, neurotrophic, and antioxidant mechanisms, many different polyphenols (flavonoids, phenolic acids, stilbenes, and alkaloids), as well as adaptogenic plants, collectively provide an additive effect on neuron functions. These compounds, such as curcumin, resveratrol, and ashwagandha, illustrate this potential since they modulate such signaling pathways as Nrf2/HO-1, SIRT1, NF-κB, and the HPA axis, reducing oxidative injury, neuroinflammation, and stress-related dysfunctions.

Although robust experimental and clinical evidence verifies improved cognitive functioning, mood, and stress resistance, there are difficulties in standardizing the composition, increasing bioavailability, and defining therapeutic doses. Enhanced nanoformulations, phytochemical standardization, and research in the gut–brain axis would drive translational impacts. Ultimately, use of these herbal bioactives in prevention and adjunctive therapy could ensure maintenance of neurocognitive function as well as prevent neurodegenerative disease progression. Iterative mechanistic, pharmacokinetic, and large-scale clinical studies are inevitable to validate their efficacy and safety, leading to evidence-based use of herbal products in modern-day neuropharmacology.

References

Abbas, S., Latif, M.S., Shafie, N.S., Ghazali, M.I., Kormin, F., 2020. Neuroprotective expression of turmeric and curcumin. Food Res. 4 (6), 2366–2381. doi:10.26656/fr.2017.4(6).363.

Abbas Bukhari, S., Jantan, I., 2015. Synthetic curcumin analogs as inhibitors of amyloid peptide aggregation: potential therapeutic and diagnostic agents for Alzheimer's disease. Mini Rev. Med. Chem. 15 (13), 1110–1121. doi:10.2174/138955751513150923101841.

Abd El-Hack, M.E., Rudayni, H.A., Aldhalmi, A.K., Youssef, I.M., Arif, M., Alawam, A.S., Allam, A.A., Khafaga, A.F., Ashour, E.A., Khan, M.M.H., 2025. Beyond traditional uses: unveiling the epigenetic, microbiome-modulating, metabolic, and nutraceutical benefits of Salvia officinalis in human and livestock nutrition. J. Funct. Foods 128. doi:10.1016/j.jff.2025.106843.

Alamgir, A.N.M., 2018. Secondary metabolites: secondary metabolic products consisting of C and H; C, H, and O; N, S, and P elements; and O/N heterocycles. Therapeutic Use of Medicinal Plants and Their Extracts: Volume 2: Phytochemistry and Bioactive Compounds. Springer International Publishing, Cham, pp. 165–309.

Al Amin, M., Dehbia, Z., Nafady, M.H., Zehravi, M., Kumar, K.P., Haque, M.A., Baig, M.S., Farhana, A., Khan, S.L., Afroz, T., Koula, D., Tutone, M., Nainu, F., Ahmad, I., Emran, T.B., 2025. Flavonoids and Alzheimer's disease: reviewing the evidence for neuroprotective potential. Mol. Cell. Biochem. 480 (1), 43–73. doi:10.1007/s11010-023-04922-w.

Albinhassan, T.H., Alharbi, B.M., AlSuhaibani, E.S., Mohammad, S., Malik, S.S., 2025. Small heat shock proteins: protein aggregation amelioration and neuro- and age-protective roles. Int. J. Mol. Sci. 26 (4). doi:10.3390/ijms26041525.

Ali, F., Siddique, Y.H., 2019. Bioavailability and pharmaco-therapeutic potential of luteolin in overcoming Alzheimer's disease. CNS Neurol. Disord. Drug Targets 18 (5), 352–365. doi:10.2174/1871527318666190319141835.

Amidfar, M., Garcez, M.L., Askari, G., Bagherniya, M., Khorvash, F., Golpour-Hamedani, S., de Oliveira, J., 2024. Role of BDNF signaling in the neuroprotective and memory-enhancing effects of flavonoids in Alzheimer's disease. CNS Neurol. Disord. Drug Targets 23 (8), 984–995. doi:10.2174/1871527323666230912090856.

Anilkumar, S., Wright-Jin, E., 2024. NF-κB as an inducible regulator of inflammation in the central nervous system. Cells 13 (6), 485. doi:10.3390/cells13060485.

Baghcheghi, Y., Razazpour, F., Mirzaee, F., Dalfardi, M., Pourfridoni, M., Hedayati-Moghadam, M., 2025. Exploring the molecular mechanisms of curcumin in modulating memory impairment in neurodegenerative disorders. Mol. Biol. Rep. 52 (1), 301–4851. doi:10.1007/s11033-024-10115-5.

Baker, D.H., Vilidaite, G., Lygo, F.A., Smith, A.K., Flack, T.R., Gouws, A.D., Andrews, T.J., 2021. Power contours: optimising sample size and precision in experimental psychology and human neuroscience. Psychol. Methods 26 (3), 295–314. doi:10.1037/met0000337.

Balogun, S., Ajen, M., Igboke, S., Alugbuo, S., Asogwa, F., Agboola, A.R., 2025. Exploring the role of phytochemicals in managing depression: mechanisms, therapeutic potential, and future directions. AROC Pharm. Biotechnol. 05 (02), 01–17. doi:10.53858/arocpb05020117.

Bellavite, P., 2023. Neuroprotective potentials of flavonoids: experimental studies and mechanisms of action. Antioxidants 12 (2), 280. doi:10.3390/antiox12020280.

Bhandare, A., Nannor, K.M., 2024. Bioavailability in drug design and development: a comprehensive review. World J. Pharm. Res. 13 (17), 145–168.

Bratovcic, A., 2020. Antioxidant enzymes and their role in preventing cell damage. Acta Sci. Nutr. Health 4 (3), 01–07. doi:10.31080/asnh.2020.04.0659.

Cai, M., Park, H.R., Yang, E.J., 2022. Nutraceutical interventions for post-traumatic stress disorder in animal models: a focus on the hypothalamic–pituitary–adrenal axis. Pharmaceuticals 15 (7). doi:10.3390/ph15070898.

Cappelletti, S., Daria, P., Sani, G., Aromatario, M., 2015. Caffeine: cognitive and physical performance enhancer or psychoactive drug? Curr. Neuropharmacol. 13 (1), 71–88. doi:10.2174/1570159ž13666141210215655.

Chaachouay, N., 2025. Synergy, additive effects, and antagonism of drugs with plant bioactive compounds. Drugs Drug Candidates 4 (1), 4. doi:10.3390/ddc4010004.

Chai, R., Fu, H., Zheng, Z., Liu, T., Ji, S., Li, G., 2017. Resveratrol inhibits proliferation and migration through SIRT1 mediated post-translational modification of PI3K/AKT signaling in hepatocellular carcinoma cells. Mol. Med. Rep. 16 (6), 8037–8044. doi:10.3892/mmr.2017.7612.

Chandrasekhar, K., Kapoor, J., Anishetty, S., 2012. A prospective, randomized double-blind, placebo-controlled study of safety and efficacy of a high-concentration full-spectrum extract of ashwagandha root in reducing stress and anxiety in adults. Indian J. Psychol. Med. 34 (3), 255–262. doi:10.4103/0253-7176.106022.

Chen, C., Zhou, M., Ge, Y., Wang, X., 2020. SIRT1 and aging related signaling pathways. Mech. Ageing Dev. 187, 111215. doi:10.1016/j.mad.2020.111215.

Cianciulli, A., Calvello, R., Ruggiero, M., Panaro, M.A., 2022. Inflammaging and brain: curcumin and its beneficial potential as regulator of microglia activation. Molecules 27 (2), 341. doi:10.3390/molecules27020341.

Ciupei, D., Colişar, A., Leopold, L., Stănilă, A., Diaconeasa, Z.M., 2024. Polyphenols: from classification to therapeutic potential and bioavailability. Foods 13 (24). doi:10.3390/foods13244131.

Cox, K.H.M., White, D.J., Pipingas, A., Poorun, K., Scholey, A., 2020. Further evidence of benefits to mood and working memory from lipidated curcumin in healthy older people: a 12-week, double-blind, placebo-controlled, partial replication study. Nutrients 12 (6), 1678. doi:10.3390/nu12061678.

Cui, C., Zheng, J., Zhang, H., Xing, Z., 2024. Pterostilbene ameliorates oxidative stress and neuronal apoptosis after intracerebral hemorrhage via the sirtuin 1-mediated Nrf2 pathway in vivo and in vitro. J. Stroke Cerebrovasc. Dis. 33 (11), 107950. doi:10.1016/j.jstrokecerebrovasdis.2024.107950.

Dai, L., Chen, L., Wang, W., 2020. Safety and efficacy of saffron (Crocus sativus L.) for treating mild to moderate depression: a systematic review and meta-analysis. J. Nerv. Ment. Dis. 208 (4), 269–276. doi:10.1097/NMD.0000000000001118.

Das, M., Devi, K.P., 2019. A mini review on the protective effect of lignans for the treatment of neurodegenerative disorders. J. Nutr. Food Lipid Sci. 2019, 40–53.

Das, S.S., Gopal, P.M., Thomas, J.V., Mohan, M.C., Thomas, S.C., Maliakel, B.P., Krishnakumar, I.M., Sasidharan, B.C.P., 2023. Influence of CurQfen®-curcumin on cognitive impairment: a randomized, double-blinded, placebo-controlled, 3-arm, 3-sequence comparative study. Front. Dement. 2. doi:10.3389/frdem.2023.1222708.

David Josephy, P., Peter Guengerich, F., Miners, 2005. "Phase I and phase II" drug metabolism: terminology that we should phase out? Drug Metab. Rev. 37 (4), 575–580.

Delmas, D., Aires, V., Limagne, E., Dutartre, P., Mazué, F., Ghiringhelli, F., Latruffe, N., 2011. Transport, stability, and biological activity of resveratrol. Acad. Sci. 1215 (1), 48–59. doi:10.1111/j.1749-6632.2010.05871.x.

de Sá Coutinho, D., Pacheco, M.T., Frozza, R.L., Bernardi, A., 2018. Anti-inflammatory effects of resveratrol: mechanistic insights. Int. J. Mol. Sci. 19 (6). doi:10.3390/ijms19061812.

Di Meo, F., Margarucci, S., Galderisi, U., Crispi, S., Peluso, G., 2019. Curcumin, gut microbiota, and neuroprotection. Nutrients 11 (10), 2426. doi:10.3390/nu11102426.

Egbujor, M.C., Petrosino, M., Zuhra, K., Saso, L., 2022. The role of organosulfur compounds as Nrf2 activators and their antioxidant effects. Antioxidants 11 (7), 1255. doi:10.3390/antiox11071255.

Fakhri, S., Piri, S., Farzaei, M.H., Sobarzo-Sánchez, E., 2021. Recent updates on the neuroprotective role of carotenoids: astaxanthin and beyond. Global Perspectives on Astaxanthin: From Industrial Production to Food, Health, and Pharmaceutical Applications. Elsevier, Iran, pp. 719–740. doi:10.1016/B978-0-12-823304-7.00035-0.

Famitafreshi, H., Karimian, M., 2020. Prostaglandins as the agents that modulate the course of brain disorders. Degener. Neurol. Neuromuscul. Dis. 10, 1–13. doi:10.2147/dnnd.s240800.

Far, B.F., Safaei, M., Pourmolaei, A., Adibamini, S., Shirdel, S., Shirdel, S., Emadi, R., Kaushik, A.K., 2024. Exploring curcumin-loaded lipid-based nanomedicine as efficient targeted therapy for Alzheimer's diseases. ACS Appl. Bio Mater. 7 (6), 3535–3555. doi:10.1021/acsabm.4c00112.

Fields, M., Marcuzzi, A., Gonelli, A., Celeghini, C., Maximova, N., Rimondi, E., 2023. Mitochondria-targeted antioxidants, an innovative class of antioxidant compounds for neurodegenerative diseases: perspectives and limitations. Int. J. Mol. Sci. 24 (4), 3739. doi:10.3390/ijms24043739.

Figueiredo Cerqueira, M.M., Castro, M.M.L., Vieira, A.A., Kurosawa, J.A.A., Junior, F.L.A., Mendes, F.C.C.S., Sosthenes, M.C.K., 2023. Comparative analysis between open field and elevated plus maze tests as a method for evaluating anxiety-like behavior in mice. Heliyon 9 (4). doi:10.1016/j.heliyon.2023.e14522.

Fiore, M., Terracina, S., Ferraguti, G., 2025. Brain neurotrophins and plant polyphenols: a powerful connection. Molecules 30 (12), 2657. doi:10.3390/molecules30122657.

Fuloria, S., Mehta, J., Chandel, A., Sekar, M., Rani, N.N.I.M., Begum, M.Y., Subramaniyan, V., Chidambaram, K., Thangavelu, L., Nordin, R., Wu, Y.S., Sathasivam, K.V., Lum, P.T., Meenakshi, D.U., Kumarasamy, V., Azad, A.K., Fuloria, N.K., 2022. A comprehensive review on the therapeutic potential of Curcuma longa Linn. in relation to its major active constituent curcumin. Front. Pharmacol. 13. doi:10.3389/fphar.2022.820806.

Gao, Y., Nie, K., Wang, H., Dong, H., Tang, Y., 2024. Research progress on antidepressant effects and mechanisms of berberine. Front. Pharmacol. 15. doi:10.3389/fphar.2024.1331440.

García-Villalba, R., Tomás-Barberán, F.A., Iglesias-Aguirre, C.E., Giménez-Bastida, J.A., González-Sarrías, A., Selma, M.V., Espín, J.C., 2023. Ellagitannins, urolithins, and neuroprotection: human evidence and the possible link to the gut microbiota. Mol. Aspects Med. 89. doi:10.1016/j.mam.2022.101109.

Gardener, S.L., Rainey-Smith, S.R., Weinborn, M., Bondonno, C.P., Martins, R.N., 2021. Intake of products containing anthocyanins, flavanols, and flavanones, and cognitive function: a narrative review. Front. Aging Neurosci. 13. doi:10.3389/fnagi.2021.640381.

Giudice, A., Arra, C., Turco, M.C., 2010. Review of molecular mechanisms involved in the activation of the Nrf2-ARE signaling pathway by chemopreventive agents. Methods Mol. Biol. 647, 37–74. doi:10.1007/978-1-60761-738-9_3.

Guo, S., Wang, H., Yin, Y., 2022. Microglia polarization from M1 to M2 in neurodegenerative diseases. Front. Aging Neurosci. 14. doi:10.3389/fnagi.2022.815347.

Haber, M., Czachor, A., Kula, P., Juśkiewicz, A., Grelewicz, O., Kucy, N., Servaas, E., Kotula, A., Siemiątkowski, R., 2024. Ashwagandha as an adaptogen: its influence on sleep patterns, stress response, and anxiety in modern life. J. Educ. Health Sport 68, 55327. doi:10.12775/jehs.2024.68.55327.

Hannan, M.A., Dash, R., Sohag, A.A.M., Haque, M.N., Moon, I.S., 2020. Neuroprotection against oxidative stress: phytochemicals targeting TrkB signaling and the Nrf2-ARE antioxidant system. Front. Mol. Neurosci. 13. doi:10.3389/fnmol.2020.00116.

Heidari, H., Bagherniya, M., Majeed, M., Sathyapalan, T., Jamialahmadi, T., Sahebkar, A., 2023. Curcumin-piperine co-supplementation and human health: A comprehensive review of preclinical and clinical studies. Phytother. Res. 37 (4), 1462–1487. doi:10.1002/ptr.7737.

Houldsworth, A., 2024. Role of oxidative stress in neurodegenerative disorders: a review of reactive oxygen species and prevention by antioxidants. Brain Commun. 6 (1). doi:10.1093/braincomms/fcad356.

Howes, M.J.R., Perry, N.S.L., Vásquez-Londoño, C., Perry, E.K., 2020. Role of phytochemicals as nutraceuticals for cognitive functions affected in ageing. Br. J. Pharmacol. 177 (6), 1294–1315. doi:10.1111/bph.14898.

Huang, S., Wang, X., Zhou, F., 2025. Comparative effects of arecoline, caffeine, and nicotine on transcription level in the nucleus accumbens of mice. Genomics 117 (2), 110986. doi:10.1016/j.ygeno.2025.110986.

Jacob, S., Kather, F.S., Boddu, S.H.S., Rao, R., Nair, A.B., 2025. Vesicular carriers for phytochemical delivery: a comprehensive review of techniques and applications. Pharmaceutics 17 (4). doi:10.3390/pharmaceutics17040464.

Jayaram, L., Puttegowda, D., Pushpa, V.H., Patil, S.M., Ramu, R., 2025. GABA and plant-derived therapeutics. GABA in Plants: Biosynthesis, Plant Development, and Food Security. Wiley, India, pp. 321–341. doi:10.1002/9781394217786.ch18.

Jazwa, A., Cuadrado, A., 2010. Targeting heme oxygenase-1 for neuroprotection and neuroinflammation in neurodegenerative diseases. Curr. Drug Targets 11 (12), 1517–1531. doi:10.2174/138945011109011517.

Jodynis-Liebert, J., Kujawska, M., 2020. Biphasic dose-response induced by phytochemicals: experimental evidence. J. Clin. Med. 9 (3), 718. doi:10.3390/jcm9030718.

Kadiri, S.K., Tiwari, P., Roy, S.P., Bhowmik, S., 2025. Herbal medicine and neurological diseases. Herbal Pharmacopeia: Nanotechnology and Advancing Drug Discovery. CRC Press, India, pp. 287–304. doi:10.1201/9781003544418-13.

Kannan, G., Paul, B.M., Thangaraj, P., 2025. Stimulation, regulation, and inflammaging interventions of natural compounds on nuclear factor kappa B (NF-kB) pathway: a comprehensive review. Inflammopharmacology 33 (1), 145–162. doi:10.1007/s10787-024-01635-4.

Karaś, M., Jakubczyk, A., Szymanowska, U., Złotek, U., Zielińska, E., 2017. Digestion and bioavailability of bioactive phytochemicals. Int. J. Food Sci. Technol. 52 (2), 291–305. doi:10.1111/ijfs.13323.

Kelly, E., Vyas, P., Weber, J., 2018. Biochemical properties and neuroprotective effects of compounds in various species of berries. Molecules 23 (1), 26. doi:10.3390/molecules23010026.

Kennedy, D.O., Wightman, E.L., Reay, J.L., Lietz, G., Okello, E.J., Wilde, A., Haskell, C.F., 2010. Effects of resveratrol on cerebral blood flow variables and cognitive performance in humans: a double-blind, placebo-controlled, crossover investigation. Am. J. Clin. Nutr. 91 (6), 1590–1597. doi:10.3945/ajcn.2009.28641United Kingdom.

Khaksarian, M., Behzadifar, M., Behzadifar, M., Alipour, M., Jahanpanah, F., Re, T.S., Firenzuoli, F., Zerbetto, R., Bragazzi, N.L., 2019. The efficacy of Crocus sativus (Saffron) versus placebo and Fluoxetine in treating depression: a systematic review and meta-analysis. Psychol. Res. Behav. Manag. 12, 297–305. doi:10.2147/PRBM.S199343.

Khalatbary, A.R., Khademi, E., 2020. The green tea polyphenolic catechin epigallocatechin gallate and neuroprotection. Nutr. Neurosci. 23 (4), 281–294. doi:10.1080/1028415X.2018.1500124.

Khan, A., Abass, S., Nizami, S.A.I., Shariq, M., Zahiruddin, S., Parveen, B., Parveen, R., 2025. The gut health revolution: herbs and dietary phytochemicals in balancing gut microbiota for optimal human health. Curr. Pharm. Biotechnol. 26 (14), 2339–2360. doi:10.2174/0113892010313921240923125946.

Kim, T.Y., Leem, E., Lee, J.M., Kim, S.R., 2020. Control of reactive oxygen species for the prevention of parkinson's disease: the possible application of flavonoids. Antioxidants 9 (7), 1–28. doi:10.3390/antiox9070583.

Knezevic, N.N., Sic, A., Worobey, S., Knezevic, E., 2025. Justice for placebo: placebo effect in clinical trials and everyday practice. Medicines 12 (1), 5. doi:10.3390/medicines12010005.

Kowiański, P., Lietzau, G., Czuba, E., Waśkow, M., Steliga, A., Moryś, J., 2018. BDNF: a key factor with multipotent impact on brain signaling and synaptic plasticity. Cell. Mol. Neurobiol. 38 (3), 579–593. doi:10.1007/s10571-017-0510-4.

Kumar, A., Angelopoulou, E., Pyrgelis, E.-S., Piperi, C., Mishra, A., 2024. Harnessing therapeutic potentials of biochanin A in neurological disorders: pharmacokinetic and pharmacodynamic overview. Chem. Biodiversity 21 (8). doi:10.1002/cbdv.202400709.

Kuśmierska, M., Kuśmierski, J., Kwaśniewska, O., 2024. Exploring the therapeutic potential of ashwagandha (*Withania somnifera*) supplementation in alleviating stress and stress-related disorders. Qual. Sport 15, 51854. doi:10.12775/qs.2024.15.51854.

Kwon, H.S., Koh, S.H., 2020. Neuroinflammation in neurodegenerative disorders: the roles of microglia and astrocytes. Transl. Neurodegener. 9 (1). doi:10.1186/s40035-020-00221-2.

Lee, K.H., Cha, M., Lee, B.H., 2020. Neuroprotective effect of antioxidants in the brain. Int. J. Mol. Sci. 21 (19), 1–29. doi:10.3390/ijms21197152.

Letsinger, A.C., Gu, Z., Yakel, J.L., 2022. α7 nicotinic acetylcholine receptors in the hippocampal circuit: taming complexity. Trends Neurosci. 45 (2), 145–157. doi:10.1016/j.tins.2021.11.006.

Linde, K., Berner, M., Egger, M., Mulrow, C., 2005. St John's wort for depression: meta-analysis of randomised controlled trials. Br. J. Psychiatry 186, 99–107. doi:10.1192/bjp.186.2.99.

Liu, Q., Chen, Y., Shen, C., Xiao, Y., Wang, Y., Liu, Z., Liu, X., 2017. Chicoric acid supplementation prevents systemic inflammation-induced memory impairment and amyloidogenesis via inhibition of NF-κB. FASEB J. 31 (4), 1494–1507. doi:10.1096/fj.201601071r.

Lopresti, A.L., 2017. Curcumin for neuropsychiatric disorders: A review of in vitro, animal and human studies. J. Psychopharmacol. 31 (3), 287–302. doi:10.1177/0269881116686883.

Lopresti, A.L., Drummond, P.D., 2017. Efficacy of curcumin, and a saffron/curcumin combination for the treatment of major depression: A randomised, double-blind, placebo-controlled study. J. Affect. Disord. 207, 188–196. doi:10.1016/j.jad.2016.09.047.

Lopresti, A.L., Smith, S.J., Malvi, H., Kodgule, R., Wane, D., 2019. An investigation into the stress-relieving and pharmacological actions of an ashwagandha (Withania somnifera) extract: a randomized, double-blind, placebo-controlled study. Medicine (United States) 98 (37). doi:10.1097/MD.0000000000017186.

Majeed, M., Nagabhushanam, K., Mundkur, L., 2023. A standardized ashwagandha root extract alleviates stress, anxiety, and improves quality of life in healthy adults by modulating stress hormones: results from a randomized, double-blind, placebo-controlled study. Medicine 102 (41), e35521. doi:10.1097/md.0000000000035521.

Mallick, K., Banerjee, S., 2024. Plants affecting serotonergic neurotransmission. Role of Herbal Medicines: Management of Lifestyle Diseases. Springer Nature, India, pp. 211–229. doi:10.1007/978-981-99-7703-1_11.

Mandal, M., Jaiswal, P., Mishra, A., 2020. Role of curcumin and its nanoformulations in neurotherapeutics: a comprehensive review. J. Biochem. Mol. Toxicol. 34 (6). doi:10.1002/jbt.22478.

Martemucci, G., Khalil, M., Luca, A.D., Abdallah, H., D'Alessandro, A.G, 2024. Comprehensive strategies for metabolic syndrome: how nutrition, dietary polyphenols, physical activity, and lifestyle modifications address diabesity, cardiovascular diseases, and neurodegenerative conditions. Metabolites 14 (6), 327. doi:10.3390/metabo14060327.

Mokra, D., Joskova, M., Mokry, J., 2023. Therapeutic effects of green tea polyphenol (–)-epigallocatechin-3-gallate (EGCG) in relation to molecular pathways controlling inflammation, oxidative stress, and apoptosis. Int. J. Mol. Sci. 24 (1), 340. doi:10.3390/ijms24010340.

Moraes, D.S., Moreira, D.C., Andrade, J.M.O., Santos, S.H.S., 2020. Sirtuins, brain and cognition: a review of resveratrol effects. IBRO Rep. 9, 46–51. doi:10.1016/j.ibror.2020.06.004.

Mouhape, C., Costa, G., Ferreira, M., Abin-Carriquiry, J.A., Dajas, F., Prunell, G., 2019. Nicotine-induced neuroprotection in rotenone in vivo and in vitro models of Parkinson's disease: evidences for the involvement of the labile iron pool level as the underlying mechanism. Neurotox. Res. 35 (1), 71–82. doi:10.1007/s12640-018-9931-1.

Negm, M.H., Aldhalmi, A.K., Ashour, E.A., Mohamed, L.A., Youssef, I.M., Moustafa, M., Abdelglil, M.I., Abdel-Shafi, S., Mahgoub, S.A., Abd El-Hack, M.E., Swelum, A.A., 2025. The effect of celery (Apium graveolens) seed powder in the diet on growth, immune function, antioxidant status and gut microbial balance in broiler chickens. Trop. Anim. Health Prod. 57 (8). doi:10.1007/s11250-025-04625-y.

Noorbala, A.A., Akhondzadeh, S., Tahmacebi-Pour, N., Jamshidi, A.H., 2005. Hydro-alcoholic extract of Crocus sativus L. versus fluoxetine in the treatment of mild to moderate depression: a double-blind, randomized pilot trial. J. Ethnopharmacol. 97 (2), 281–284. doi:10.1016/j.jep.2004.11.004.

Origassa, C.S.T., Câmara, N.O.S., 2013. Cytoprotective role of heme oxygenase-1 and heme degradation derived end products in liver injury. World J. Hepatol. 5 (10), 541–549. doi:10.4254/wjh.v5.i10.541Brazil.

Ossowska, W., Czarnota, M.,Ösztreicher, W., 2025. Ashwagandha and its role in stress reduction, hormonal balance, and athletic recovery: an evidence-based review. J. Educ. Health and Sport 83, 61874. doi:10.12775/jehs.2025.83.61874.

Palanisamy, S., Balasubramanian, A., Kanchana, M., Sisubalan, N., Selvakesavan, R.K., 2025. Curcumin-loaded nanoparticles in neurodegenerative diseases: Alzheimer's, Parkinson's, and amyotrophic lateral sclerosis. Nanotechnol. Life Sci. 98, 225–237. doi:10.1007/978-3-031-80185-3_8.

Pandit, S., Srivastav, A.K., Sur, T.K., Chaudhuri, S., Wang, Y., Biswas, T.K., 2024. Effects of Withania somnifera extract in chronically stressed adults: a randomized controlled trial. Nutrients 16 (9). doi:10.3390/nu16091293.

Panossian, A., 2017. Understanding adaptogenic activity: specificity of the pharmacological action of adaptogens and other phytochemicals. Acad. Sci. 1401 (1), 49–64. doi:10.1111/nyas.13399.

Panossian, A., Wikman, G., 2010. Effects of adaptogens on the central nervous system and the molecular mechanisms associated with their stress—protective activity. Pharmaceuticals 3 (1), 188–224. doi:10.3390/ph3010188.

Panossian, A.G., Efferth, T., Shikov, A.N., Pozharitskaya, O.N., Kuchta, K., Mukherjee, P.K., Banerjee, S., Heinrich, M., Wu, W., Guo, D., Wagner, H., 2021. Evolution of the adaptogenic concept from traditional use to medical systems: pharmacology of stress- and aging-related diseases. Med. Res. Rev. 41 (1), 630–703. doi:10.1002/med.21743.

Park, C.W., Hong, K.B., Suh, H.J., Ahn, Y., 2023. Sleep-promoting activity of amylase-treated ashwagandha (Withania somnifera L. Dunal) root extract via GABA receptors. J. Food Drug Anal. 31 (2), 278–288. doi:10.38212/2224-6614.3456.

Patil, R., Aher, P., Bagad, P., Ekhande, S., 2022. Herbal bioenhancers in veterinary phytomedicine. Drug Deliv Technol Herb Bioenhancers Pharm 325. doi:10.1515/9783110746808-012.

Paul-Chima, U.O., Ugwu, C.N., Alum, E.U., 2024. Integrated approaches in nutraceutical delivery systems: optimizing ADME dynamics for enhanced therapeutic potency and clinical impact. RPS Pharmacol. Pharm. Rep. 3 (4). doi:10.1093/rpsppr/rqae024.

Pawar, V.S., Shivakumar, H., 2012. A current status of adaptogens: natural remedy to stress. Asia Pac. J. Trop. Dis. 2 (1), S480. doi:10.1016/S2222-1808(12)60207-2.

Pontifex, M.G., Malik, M.M.A.H., Connell, E., Müller, M., Vauzour, D., 2021. Citrus polyphenols in brain health and disease: current perspectives. Front. Neurosci. 15. doi:10.3389/fnins.2021.640648.

Rainey-Smith, S.R., Brown, B.M., Sohrabi, H.R., Shah, T., Goozee, K.G., Gupta, V.B., Martins, R.N., 2016. Curcumin and cognition: a randomised, placebo-controlled, double-blind study of community-dwelling older adults. J. Nutr. 115 (12), 2106–2113. doi:10.1017/S0007114516001203.

Rathaur, P., SR, K.J., 2020. Metabolism and Pharmacokinetics of phytochemicals in the human body. Curr. Drug Metab. 20 (14), 1085–1102. doi:10.2174/1389200221666200103090757.

Rojas-García, A., Fernández-Ochoa, Á., Cádiz-Gurrea, M.d.l.L., Arráez-Román, D., Segura-Carretero, A., 2023. Neuroprotective effects of agri-food by-products rich in phenolic compounds. Nutrients 15 (2). doi:10.3390/nu15020449.

Roman, Y., 2025. Bridging the United States population diversity gaps in clinical research: roadmap to precision health and reducing health disparities. Pers. Med. 22 (3), 193–203. doi:10.1080/17410541.2025.2504329.

Rose, K.N., Barlock, B.J., DaSilva, N.A., Johnson, S.L., Liu, C., Ma, H., Nelson, R., Akhlaghi, F., Seeram, N.P., 2021. Anti-neuroinflammatory effects of a food-grade phenolic-enriched maple syrup extract in a mouse model of Alzheimer's disease. Nutr. Neurosci. 24 (9), 710–719. doi:10.1080/1028415X.2019.1672009.

Santhiravel, S., Bekhit, A.E.D.A., Mendis, E., Jacobs, J.L., Dunshea, F.R., Rajapakse, N., Ponnampalam, E.N., 2022. The impact of plant phytochemicals on the gut microbiota of humans for a balanced life. Int. J. Mol. Sci. 23 (15). doi:10.3390/ijms23158124.

Sarubbo, F., Esteban, S., Miralles, A., Moranta, D., 2018. Effects of resveratrol and other polyphenols on SIRT1: relevance to brain function during aging. Curr. Neuropharmacol. 16 (2), 126–136. doi:10.2174/1570159ž15666170703113212.

Sera, L.C., McPherson, M.L., 2012. Pharmacokinetics and pharmacodynamic changes associated with aging and implications for drug therapy. Clin. Geriatr. Med. 28 (2), 273–286. doi:10.1016/j.cger.2012.01.007.

Shahmansouri, N., Farokhnia, M., Abbasi, S.H., Kassaian, S.E., Noorbala Tafti, A.A., Gougol, A., Yekehtaz, H., Forghani, S., Mahmoodian, M., Saroukhani, S., Arjmandi-Beglar, A., Akhondzadeh, S., 2014. A randomized, double-blind, clinical trial comparing the efficacy and safety of Crocus sativus L. with fluoxetine for improving mild to moderate depression in post percutaneous coronary intervention patients. J. Affect. Disord. 155 (1), 216–222. doi:10.1016/j.jad.2013.11.003.

Sharan, P., Vellapandian, C., 2024. Hypothalamic-pituitary-adrenal (HPA) axis: unveiling the potential mechanisms involved in stress-induced Alzheimer's disease and depression. Cureus 16 (8). doi:10.7759/cureus.67595.

Sharma, V.K., Sharma, A., Verma, K.K., Kumar Gaur, P., Kaushik, R., Abdali, B., 2023. A comprehensive review on pharmacological potentials of caffeine. J. Appl. Pharm. Sci. Res. 6 (3), 16–26. doi:10.31069/japsr.v6i3.04.

Shayganfard, M., 2020. Molecular and biological functions of resveratrol in psychiatric disorders: a review of recent evidence. Cell Biosci. 10 (1), 2045–3701. doi:10.1186/s13578-020-00491-3.

Shi, Y.Y, Li, X., Li, Z., Sun, J., Gao, T., Wei, G., 2025. Qie Guo, Nano-formulations in disease therapy: designs, advances, challenges, and future directions. J. Nanobiotechnol. 23 (1), 1477–3155. doi:10.1186/s12951-025-03442-7.

Siddiqui, A., Timalwar, M., Bhuruk, S., Rahul Tandale, S.R., Chaudhari, R., 2025. Reactive oxygen species, and the science of aging: a mini review. Eur. J. Med. Plants 36 (3), 52–61. doi:10.9734/ejmp/2025/v36i31263.

Silva, R.F.M., Pogačnik, L., 2020. Polyphenols from Food and natural products: neuroprotection and safety. Antioxidants 9 (1), 61. doi:10.3390/antiox9010061.

Singh, D.K, Kumari, B., Sinha, S., Fatima, K., 2025. Phytochemicals for the improvement of cognitive function through cholinergic anti-inflammatory pathway. Curr. Indian Sci. 03. doi:10.2174/012210299ž309791240914125655.

Singh, S., Arthur, R., Upadhayay, S., Kumar, P., 2022. Ferulic acid ameliorates neurodegeneration via the Nrf2/ARE signalling pathway: a review. Pharmacol. Res. Mod. Chin. Med. 5, 100190. doi:10.1016/j.prmcm.2022.100190.

Singh, T., Singh, H., Singh, B., 2025. Protective effects of Fraxinus Micrantha against thioacetamide-induced hepatic encephalopathy: a focus on antioxidant and neuroprotective mechanisms. Metab. Brain Dis. 40 (7), 1–14.

Small, G.W., Siddarth, P., Li, Z., Miller, K.J., Ercoli, L., Emerson, N.D., Martinez, J., Wong, K.P., Liu, J., Merrill, D.A., Chen, S.T., Henning, S.M., Satyamurthy, N., Huang, S.C., Heber, D., Barrio, J.R., 2018. Memory and brain amyloid and tau effects of a bioavailable form of curcumin in non-demented adults: a double-blind, placebo-controlled 18-month trial. Am. J. Geriatr. Psychiatry 26 (3), 266–277. doi:10.1016/j.jagp.2017.10.010.

Socci, V., Tempesta, D., Desideri, G., De Gennaro, L., Ferrara, M., 2017. Enhancing human cognition with cocoa flavonoids. Front. Nutr. 4. doi:10.3389/fnut.2017.00019.

Soltani Khaboushan, A., Yazdanpanah, N., Rezaei, N., 2022. Neuroinflammation and proinflammatory cytokines in epileptogenesis. Mol. Neurobiol. 59 (3), 1724–1743. doi:10.1007/s12035-022-02725-6.

Speers, A.B., Cabey, K.A., Soumyanath, A., Wright, K.M., 2021. Effects of withania somnifera (Ashwagandha) on stress and the stress related neuropsychiatric disorders anxiety, depression, and insomnia. Curr. Neuropharmacol. 19 (9), 1468–1495. doi:10.2174/1570159ž19666210712151556.

Stagni, F., Giacomini, A., Guidi, S., Emili, M., Uguagliati, B., Salvalai, M.E, Bortolotto, V., Grilli, M., Rimondini, R., Bartesaghi, R., 2017. A flavonoid agonist of the TrkB receptor for BDNF improves hippocampal neurogenesis and hippocampus-dependent memory in the Ts65Dn mouse model of DS. Exp. Neurol. 298 (Pt A), 79–96. doi:10.1016/j.expneurol.2017.08.018.

Szwajgier, D., Borowiec, K., Pustelniak, K., 2017. The neuroprotective effects of phenolic acids: molecular mechanism of action. Nutrients 9 (5), 477. doi:10.3390/nu9050477.

Tharakan, A., Shukla, H., Benny, I.R, Tharakan, M., George, L., Koshy, S., 2021. Immunomodulatory effect of Withania somnifera (Ashwagandha) extract—a randomized, double-blind, placebo controlled trial with an open label extension on healthy participants. J. Clin. Med. 10 (16), 3644. doi:10.3390/jcm10163644.

Tiwari, R.K., Sharma, V., Pandey, R.K., Shukla, S.S., 2020. Nicotine addiction: neurobiology and mechanism. J. Pharmacopuncture 23 (1), 1–7. doi:10.3831/KPI.2020.23.001.

Tkaczenko, H., Buyun, L., Kołodziejska, R., Kamiński, P., Kurhaluk, N., 2025. Neuroactive phytochemicals as multi-target modulators of mental health and cognitive function: an integrative review. Int. J. Mol. Sci. 26 (18), 8907. doi:10.3390/ijms26188907.

Tonev, D., Momchilova, A., 2023. Oxidative stress and the nuclear factor erythroid 2-related factor 2 (Nrf2) pathway in multiple sclerosis: focus on certain exogenous and endogenous Nrf2 activators and therapeutic plasma exchange modulation. Int. J. Mol. Sci. 24 (24), 17223. doi:10.3390/ijms242417223.

Tsai, Y.M., Chien, C.F., Lin, L.C., Tsai, T.H., 2011. Curcumin and its nano-formulation: the kinetics of tissue distribution and blood-brain barrier penetration. Int. J. Pharm. 416 (1), 331–338. doi:10.1016/j.ijpharm.2011.06.030.

Tumilaar, S.G., Hardianto, A., Dohi, H., Kurnia, D., Ahmed, M., 2024. A comprehensive review of free radicals, oxidative stress, and antioxidants: overview, clinical applications, global perspectives, future directions, and mechanisms of antioxidant activity of flavonoid compounds. J. Chem. 2024, 1–21. doi:10.1155/2024/5594386.

Vauzour, D., 2014. Effect of flavonoids on learning, memory and neurocognitive performance: relevance and potential implications for Alzheimer's disease pathophysiology. J. Sci. Food Agric. 94 (6), 1042–1056. doi:10.1002/jsfa.6473.

Venkatesan, R., Ji, E., Kim, S.Y., 2015. Phytochemicals that regulate neurodegenerative disease by targeting neurotrophins: a comprehensive review. BioMed. Res. Int. 2015. doi:10.1155/2015/814068.

Wang, J., Song, Y., Chen, Z., Leng, S.X., 2018. Connection between systemic inflammation and neuroinflammation underlies neuroprotective mechanism of several phytochemicals in neurodegenerative diseases. Oxid. Med. Cell. Long. 2018. doi:10.1155/2018/1972714.

Wei, C.C., Li, S.W., Wu, C.T., How, C.M., Pan, M.H., 2022. Dietary methylglyoxal exposure induces Alzheimer's disease by promoting amyloid β accumulation and disrupting autophagy in Caenorhabditis elegans. J. Agric. Food Chem. 70 (32), 10011–10021. doi:10.1021/acs.jafc.2c03411.

Wightman, E.L., Haskell-Ramsay, C.F., Reay, J.L., Williamson, G., Dew, T., Zhang, W., Kennedy, D.O., 2015. The effects of chronic trans-resveratrol supplementation on aspects of cognitive function, mood, sleep, health and cerebral blood flow in healthy, young humans. Br. J. Nutr. 114 (9), 1427–1437. doi:10.1017/S0007114515003037.

Wróbel-Biedrawa, D., Podolak, I., 2024. Anti-neuroinflammatory effects of adaptogens: a mini-review. Molecules 29 (4), 866. doi:10.3390/molecules29040866.

Xu, B., Bai, L., Chen, L.u, Tong, R., Feng, Y., Shi, J., 2022. Terpenoid natural products exert neuroprotection via the PI3K/Akt pathway. Front. Pharmacol. 13. doi:10.3389/fphar.2022.1036506.

Xu, H., Wang, E., Chen, F., Xiao, J., Wang, M., Vergara, D., 2021. Neuroprotective phytochemicals in experimental ischemic stroke: mechanisms and potential clinical applications. Oxid. Med. Cell. Long. 2021 (1). doi:10.1155/2021/6687386.

Yakubu, J., Natsaridis, E., du Toit, T., Barata, I.S., Tagit, O., Pandey, A.V., 2025. Nanoparticles with curcumin and piperine modulate steroid biosynthesis in prostate cancer. Sci. Rep. 15 (1). doi:10.1038/s41598-025-98102-z.

Yakubu, J., Pandey, A.V., 2024. Innovative delivery systems for curcumin: exploring nanosized and conventional formulations. Pharmaceutics 16 (5), 637. doi:10.3390/pharmaceutics16050637.

Yuan, J., Pi, C., Shen, H., Zhou, B.i, Wei, Y., Dechsupa, N., Zhao, L., 2025. Potential therapeutic benefits of curcumin in depression or anxiety induced by chronic diseases: a systematic review of mechanistic and clinical evidence. Front. Pharmacol. 16. doi:10.3389/fphar.2025.1638645.

Zahedipour, F., Hosseini, S., Henney, N., Barreto, G., Sahebkar, A., 2022. Phytochemicals as inhibitors of tumor necrosis factor alpha and neuroinflammatory responses in neurodegenerative diseases. Neural Regener. Res. 17 (8), 1675–1684. doi:10.4103/1673-5374.332128.

Zhang, Y., Li, L.i, Zhang, J., 2020. Curcumin in antidepressant treatments: an overview of potential mechanisms, pre-clinical/clinical trials and ongoing challenges. Basic Clin. Pharmacol. Toxicol. 127 (4), 243–253. doi:10.1111/bcpt.13455.

Zhao, Z., Wei, Q., Hua, W., Liu, Y., Liu, X., Zhu, Y., 2018. Hepatoprotective effects of berberine on acetaminophen-induced hepatotoxicity in mice. Biomed. Pharmacother. 103, 1319–1326. doi:10.1016/j.biopha.2018.04.175.

Zheng, B., McClements, D.J., 2020. Formulation of more efficacious curcumin delivery systems using colloid science: enhanced solubility, stability, and bioavailability. Molecules 25 (12), 2791. doi:10.3390/molecules25122791.

Chapter 4

Mechanisms, applications and prospects of functional phytofeed additives: Insights for livestock production

Muhammad Muneeb[a], **Arooj Fatima**[b], **Chaudhry Ahsan Akram**[c], **Ehsaan Ullah Khan**[a], **Mohamed E. Abd El-Hack**[d] and **Sohail Ahmad**[e]

[a] *Department of Animal Nutrition, Faculty of Animal Production and Technology, University of Veterinary and Animal Sciences, Lahore, Pakistan,* [b] *Department of Veterinary Medicine, Faculty of Veterinary Sciences, University of Veterinary and Animal Sciences, Lahore, Pakistan,* [c] *Department of Livestock and Poultry Production, Faculty of Veterinary Sciences, Bahauddin Zakariya University, Multan, Pakistan,* [d] *Department of Industrial Pharmacy, College of Pharmaceutical Sciences and Drug Manufacturing, Misr University for Science and Technology (MUST), Giza, Egypt,* [e] *Department of Poultry Production, Faculty of Animal Production and Technology, University of Veterinary and Animal Sciences, Lahore, Pakistan*

4.1 Introduction

Functional feed additives are the nonnutritive compounds added to animal feeds not only to enhance their feed efficiency and growth but also physiological functions such as immunity, metabolism, and gut integrity. Functional feed additives are appreciated in comparison to traditional additives that only provide essential nutrients (e.g., vitamins and amino acids) to animals. These additives have attracted considerable interest in additives based on herbs, spices, and plant extracts that have multimodal biological effects, such as antiinflammatory, antimicrobial, antioxidant, and immunomodulatory (Pandey et al., 2019). As the livestock and poultry industry is moving toward more sustainable methods, the function of such additives in reconnecting productivity with welfare has become more critical.

Rapid growth of animal agriculture before the start of the 21st century, with the massive consumption of antibiotic growth promoters (AGPs), was aimed at maximizing feed efficacy and reducing subclinical infections. Nevertheless, there is increasing evidence of AGPs in relation to antimicrobial resistance

Plant-Derived Feed Additives and Animal-Based Foods: Linking Animal Health, Product Quality, and Consumer Perceptions. DOI: https://doi.org/10.1016/B978-0-443-45446-2.00012-0

and the presence of drug residues in animal products, which has led to several regulatory interventions. One of the milestones was a policy of the European Union's regulation (1831/2003) that prohibited the use of antibiotics as growth promoters in animals (Pandey et al., 2019). This was a shift in paradigm, and it caused the world to seek natural, safe, and effective alternatives to AGPs. Functional feed additives have become a candidate of choice because they can mimic most of the positive effects of AGPs without the development of resistance (Mandey and Sompie, 2021). National governments, international trade policies, and consumer-led market reforms have vindicated their adoption.

The use of phytofeed additives has also increased due to consumers' awareness and their demand for clean-labeled animal products, i.e., products without synthetic and chemical additives as well as those without antibiotics. Consumers in the modern world are directly relating food safety to animals and environmental agriculture. This has compelled producers to redesign animal diets to satisfy market and regulatory requirements. Phytofeed additives provide a comprehensive solution in this area because they not only increase growth but also regulate the quality of meat, egg composition, and milk characteristics (Adedokun and Olojede, 2019). Furthermore, the aspect of these additives that enables them to lower the disease burden by promoting natural immune boosting justifies the inclusion of these additives in holistic animal health practices.

The need to develop climate-smart livestock nutrition has further increased the desire to use functional phytofeed additives. Livestock has been a major source of greenhouse gas emissions and environmental degradation by producing methane. Nutrition-based interventions are greatly required for managing manure and the land use associated with the production of feed. Phytofeed additives have been found to decrease enteric methane generation, improve the use of nitrogen, and alleviate stress related to metabolism (Fushai et al., 2025). For example, tannins and saponins can inhibit the process of methanogenesis in ruminants, whereas essential oils may enhance the rate of digestion in monogastric animals. These interventions can ensure that animal production systems meet the goal of climate resilience and remain profitable and productive (Godde et al., 2021).

The gastrointestinal tract is one of the most vital physiological targets of functional feed additives, and it represents a central point of nutrient absorption, immunity, and homeostasis of microbes. Proper gut health is directly translated to high feed efficiency, resistance to disease, and growth potential. Phytoadditives have a variety of effects on gut health, which include improvements in the mucosal barrier, changes in inflammatory cytokines and oxidative stress biomarkers, as well as changes in microbial populations (Adedokun and Olojede, 2019). These additives make the gut microbiota stable and diverse for better digestion, as well as lower the pathogenic load and reduce the reliance on medicines. In swine and poultry, this has been linked to a considerable improvement in the feed conversion ratio, villus height, and immune markers (Burden, 2020).

Economically, better gut health associated with the use of these functional additives will reduce production costs due to decreased morbidity, mortality, and therapeutic interventions. Feed represents the greatest investment in animal production, and any intervention to improve its usage has a huge financial payback. Moreover, healthier animals are also able to gain faster and achieve market weight more effectively, which saves time-to-slaughter and resource input per unit of output. With an increase in protein demand and animal production efficiency, phyto-feed additives provide a scientific, economic, and environmentally friendly solution (Steiner and Syed, 2015). Their multifactorial advantages have placed them not only as emerging feed additives but also as the potential means for future-oriented animal production.

4.2 Classification, sources, and forms of phytofeed additives

Functional phytofeed additives are natural plant materials that have some benefit rather than simple nutrition. These additives possess bioactive agents/compounds like polyphenols, flavonoids, essential oils, and saponins. They possess abilities to perform physiological functions, enhance gut health, or have antimicrobial and antiinflammatory properties (Wang et al., 2024). These additives affect biological processes, enhance feed efficiency, and limit the use of synthetic antibiotics, which is important for sustainable animal production (Placha et al., 2022).

4.2.1 Classification by botanical source

Functional phytofeed additives could be botanically categorized. Oregano, thyme, cinnamon, and other herbs and spices are high in essential oils and phytochemicals, which have a high antimicrobial and antioxidant effect (Abd El-Hack and Alagawany, 2022). Carvacrol and thymol, which are present in oregano, regulate gut microbiota and prevent pathogens (Feye et al., 2020). Cinnamon contains cinnamaldehyde, which is an antimicrobial and antiinflammatory compound, and thyme as a supplement to the enzymes of the digestive tract and improves nutrient uptake in broilers (Hafeez et al., 2016).

Another category is the plant extracts and essential oils that include concentrated bioactives such as rosemary extract, clove oil, and eucalyptus oil produced through solvent extraction or steam distillation. They contain a lot of terpenes and phenolics that destabilize bacterial membranes and prevent pathogens (Placha et al., 2022). Eugenol is found in clove oil, and it has shown antibacterial, antifungal, and analgesic properties in livestock. The antioxidant effect of rosemary extract is commonly employed to stabilize the quality of feed and develop animal immunity (Tchoursinov et al., 2020).

In addition to the whole plant sources and extracts, individual secondary metabolites have been isolated and also used as functional feed additives. These secondary metabolites are polyphenols, flavonoids, tannins, saponins, and alkaloids. Polyphenols and flavonoids are potent antioxidants that safeguard the cells against oxidative injury and inflammation (Wang et al., 2024). Tannins have antimicrobial effects and regulate protein digestion but can also inhibit nutrient absorption in large doses. Saponins stimulate immunity and lower cholesterol, supporting animal health and productivity (Abd El-Hack and Alagawany, 2022).

4.2.2 Processing forms of phytofeed additives

Several processing forms of phytofeed additives exist that affect their stability, bioavailability, and effectiveness. The most basic one is raw plant dried herbs, which are simply included in the diet of animals. The crude forms are less expensive, but can contain unpredictable concentrations of bioactive compounds and have shorter shelf-life (Tchoursinov et al., 2020). Extracts made with either alcohol or water contain more standardized phytochemical levels and biological activity.

Sophisticated processing techniques, such as encapsulation and nanoformulation, have been developed to overcome the shortcomings of the traditional forms. Encapsulation methods ensure the stability of the sensitive bioactive compounds against degradation in feed processing and in the gut to achieve a desirable concentration of the additives at target sites, so they can interact with the gut microbiota more effectively (Feye et al., 2020). The nanoformulations also contribute to increased absorption by reducing the size of particles, leading to their enhanced bioavailability at the cellular level. Such technologies result in the next generation of additives that have increased functional performance in animal diets.

4.2.3 Source variability and standardization

Biological activity of the phytofeed additives depends on the type of plants used, geography, time of harvesting, and processing. Indeed, the essential oils of oregano or thyme vary based on the soil and weather (Wang et al., 2024). Picking the plant at the optimum time of maturity yields bioactives in the highest possible quantity, whereas extraction processes, such as infusion, maceration, or distillation, can change the composition and strength of the final product to a great extent (Tchoursinov et al., 2020). Recent studies highlight the need to thoroughly characterize the chemicals as well as test their functionality in vivo to justify their regulation and general usage (Placha et al., 2022). With the increasing demand for natural and sustainable solutions for feeds, addressing the variability and standardization of sources will be critical in maximizing the advantages of functional phytofeed additives (Table 4.1).

4.3 Bioactive components and their functional roles

The phytogenic feed additives (PFAs) offer a good source of a wide array of bioactive substances, which are important to animal health, livestock

TABLE 4.1 Phytofeed additives and their health benefits in animals (Shahbakht et al., 2024; Wang et al., 2024).

Phytofeed additives	Botanical source	Main active compounds	Benefits in animals
Oregano	*Origanum vulgare*	Carvacrol, thymol	Antibacterial, antioxidant, improved nutrient absorption
Cinnamon	*Cinnamomum verum*	Cinnamaldehyde	Antimicrobial, improved intestinal morphology, enhanced growth
Green tea	*Camellia sinensis*	Catechins, theaflavins	Antioxidant, antimicrobial, improved meat quality
Clove	*Syzygium aromaticum*	Eugenol	Antimicrobial, antioxidant, increased digestive enzyme activity
Neem	*Azadirachta indica*	Azadirachtin, nimbin	Antiparasitic, immunostimulant, supports gut health
Thyme	*Thymus vulgaris*	Thymol, carvacrol	Antimicrobial, growth promotion, improved gut morphology
Fenugreek	*Trigonella foenum-graecum*	Saponins, flavonoids	Enhanced digestion, antiinflammatory, improved milk yield in dairy animals

productivity, and feed efficiency. These are naturally occurring compounds of a broad range of functional activity, such as antioxidant, antiinflammatory, antimicrobial, antiprotozoal, and immunomodulatory (Kholif and Olafadehan, 2021; Singh et al., 2024).

The presence of these phytochemicals in the diet of livestock and poultry has become a potential alternative to the synthetic growth promoters, particularly in the face of rising international regulations on antibiotic use (Alghirani et al., 2021; Ayalew et al., 2022). It has been reported in many studies that phytogenic additives have the potential to enhance rumen fermentation rates, minimize methane production, enhance immune stimulation, and contribute to overall animal welfare (Ramdani et al., 2023; Tsiplakou et al., 2023). Moreover, the fact that phytochemicals can be used to regulate gut microbiota and prevent the existence of pathogenic microorganisms renders them viable candidates in sustainable animal production (Tchoursinov et al., 2020). The following text discusses the most important categories of these bioactive components, including polyphenols, essential oils, saponins, tannins, and flavonoids, and their modes of action and applicability in contemporary animal nutrition (Bagno et al., 2018; Miękus et al., 2020).

4.3.1 Polyphenols

Polyphenols make a large group of plant compounds that are famous because of their strong antioxidant and antiinflammatory effects. These molecules remove reactive oxygen species (ROS) and make cellular structures resistant to damage caused by oxidative stress in animals. Some of the polyphenols that have been the most extensively studied in terms of their health-promoting properties include green tea catechins and grape seed proanthocyanidins (Singh et al., 2024). Dietary polyphenols have been attributed to increasing rumen fermentation, nutrient uptake, and decreasing inflammation in ruminants. Their contribution to immune functioning and the modulation of gut microbiota is also well recognized (Ramdani et al., 2023).

4.3.2 Essential oils and terpenoids

Volatile plant compounds include essential oils and terpenoids that have a strong antimicrobial effect, which is mainly driven by the destabilization of the membrane of the microbe. The compounds such as thymol and carvacrol enter the cell walls of bacteria and change the permeability, resulting in cell lysis (Kholif and Olafadehan, 2021). These characteristics make essential oils effective against pathogenic bacteria in the gastrointestinal tract of livestock. Moreover, the essential oils also help in making the feed palatable and have the capacity to adjust the ruminal fermentation, thereby increasing the feed consumption. They also play the role of holistic health modulators in the diets of animals because of their anti-inflammatory and antioxidative properties (Kholif and Olafadehan, 2021).

4.3.3 Saponins

Saponins, belonging to the glycoside group, possess a typical foaming property and are predominantly obtained from legumes such as yucca and quillaja. Saponins have been shown to have antiprotozoal activity in ruminants, resulting in a decrease in methane emissions and enhancing energy efficiency (Ramdani et al., 2023). In addition, it has been observed that saponins enhance nitrogen source and lessen ammonia emissions from manure, which leads to environmental sustainability. Their bioactivity is further expanded to regulating the immune system and promoting general gut health, especially in ruminants (Ramdani et al., 2023).

4.3.4 Tannins

Polyphenolic compounds known as tannins are classified as condensed or hydrolyzable depending on their structure. These compounds have high protein-binding powers and hence affect the protein degradation in the rumen. Specifically, condensed tannins lower ruminal protein degradation and increase the proportion/supply of undegraded dietary protein to the intestine, therefore increasing the efficiency of nitrogen utilization (Tsiplakou et al., 2023). However, at high inclusion levels, they can interfere with palatability and even nutrient uptake. In functional feed additives, tannins have also been identified as antimicrobial, antioxidant, and antiparasitic agents (Bagno et al., 2018).

4.3.5 Flavonoids

Polyphenols, and especially flavonoids, have various biological activities such as antimicrobial, antioxidant and immunomodulatory activities. The typical flavonoids in feed additives to improve animal health are quercetin and luteolin, which alleviate inflammatory and oxidative stress (Miękus et al., 2020). These substances affect the composition of the gut microbiome, favoring useful microbes and inhibiting pathogens. Flavonoid supplementation has also resulted in better growth performance and immune status in poultry (Alghirani et al., 2021). They are available and useful in the formulation of functional feeds due to their large distribution in plants.

4.3.6 Other compounds: Alkaloids, glucosinolates, and organosulfur compounds

Other bioactive classes such as alkaloids, glucosinolates, and organosulfur compounds are also important, in addition to the major classes. Plant-based secondary metabolites, which are nitrogen-containing heterocyclic compounds and are formed by transformation of amino acids (lysine, tyrosine, and tryptophan), are called alkaloids. Alkaloids in functional feed additives exert positive physiological functions, including antimicrobial, antioxidant, and antiinflammatory properties at low doses (Dell'Anno, 2023).

4.3.6.1 Glucosinolates

Sulfuric-containing secondary metabolites that are important to plant defense and flavor are the glucosinolates (GSLs), which are mainly present in *Brassica* species (Blažević et al., 2020; Bischoff, 2021). *Brassica* crops are identified to have around 30 types of GSLs (Blažević et al., 2020). Their breakdown products

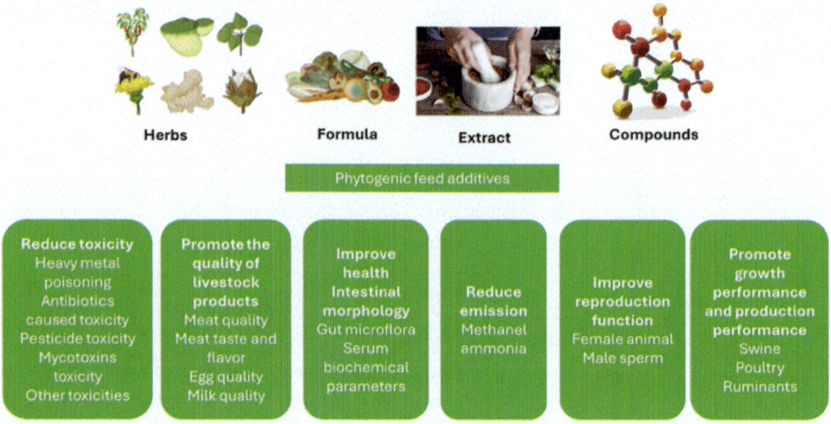

FIGURE 4.1 Overview of phytogenic feed additives, including main classifications and functions (Wang et al., 2024).

not only shield plants against pests and pathogens, but they also have anticarcinogenic properties (Bischoff, 2021). Phytochemicals in cruciferous vegetables regulate immunity by inhibiting inflammation, enhancing the detox enzyme, eliminating the ROS, and promoting natural defense (Nair et al., 2018).

4.3.6.2 Organosulfur compounds

Garlic organosulfur compounds, specifically alliin and allicin, the metabolites, have significant anticancer activities because they induce apoptosis, enhance detoxification enzymes, and regulate DNA repair processes (Nuutila et al., 2003; Nicastro et al., 2015). These compounds have a significant effect on tumor cell growth and also increase the efficacy of chemotherapy agents of paclitaxel and 5-fluorouracil (Zou et al., 2016). They have therapeutic potential that is supported by in vitro and in vivo studies in the models of breast, liver, and cervical cancer. It is clinically proven that there is less risk of cancer in a population that consumes a high amount of garlic (Torres-Sanchez et al., 2000) (Fig. 4.1).

4.4 Mechanisms of action of phytogenic feed additives

PFAs are known to have various biological functions that lead to animal health, high production, and feed efficiency. Such mechanisms include gut regulation, immune regulation, and metabolism. This part presents the key pathways through which these natural compounds play their positive roles (Fig. 4.2).

4.4.1 Modulation of gut microbiota

The most visible effect of PFAs is their involvement in regulating the gut microbiota to mimic the action of prebiotics. Phytogenic compounds, such as

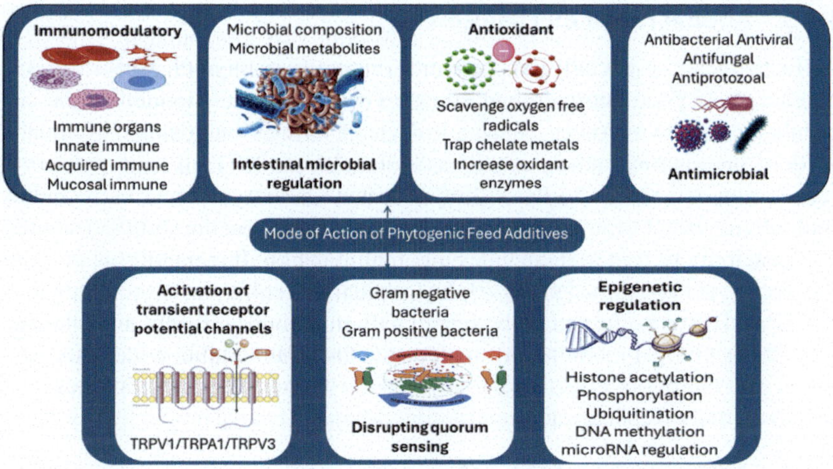

FIGURE 4.2 Potential modes of action of phytogenic feed additives (Wang et al., 2024).

polyphenols and oligosaccharides, promote the growth of beneficial microorganisms, such as *Lactobacillus* and *Bifidobacterium*, and at the same time, prevent the growth of pathogenic microorganisms, such as *Salmonella* and *Clostridium* (Yaqoob et al., 2021). The selective activation of intestinal commensal organisms creates a competitive exclusion setting, in which there is a failure of colonization by the pathogens (Kaur et al., 2021). Plant-based prebiotics such as mannan oligosaccharides are substrates of helpful microbes, which ultimately support gut health and the immune system (Shashidhara and Devegowda, 2003). Such intestinal microbiota balance finally enhances digestion, nutrient intake, and performance (Mazanko et al., 2018).

4.4.2 Enhancement of gut barrier function

The other important mechanism is the increase in intestinal barrier integrity, which is achieved by the upregulation of tight junction proteins, i.e., occludin, claudins, and zonula occludens. The tight junctions play a critical role in the preservation of intestinal permeability and the prevention of the translocation of harmful bacteria and toxins (Latek et al., 2022). These junctions have been demonstrated to be reinforced with phytogenic compounds such as flavonoids and sulfur-containing bioactives, which preserve the mucosal structure. This thing is necessary in the period of inflammatory conditions when the barrier is generally damaged. Also, vitamin supplementation of PFAs has been linked to high levels of villus height and depth of the crypt, which are measures of nutrient absorption and gut condition (Kaschubek et al., 2018). Gut integrity enhances the survival and productivity of animals.

4.4.3 Antimicrobial mechanisms

Phytochemicals, especially the essential oils such as thymol, carvacrol, and eugenol, have good antimicrobial effects. These chemicals destabilize the cell membranes of the microbes, causing leakage of cellular material and eventually death of the microorganism (Kholif and Olafadehan, 2021). It is an indiscriminate action and results in the killing of a wide variety of bacteria, including antibiotic-resistant strains. Other phytogenic additives have the ability to neutralize the activity of bacterial enzymes that inhibit the production of vital proteins and nucleic acids (Ahmed et al., 2024). Carvacrol interferes with the production of ATP and disturbs the metabolic functioning of microbes (Kholif and Olafadehan, 2021). Gut pH modulation supplements the antimicrobial effect and has a synergistic inhibitory effect on the growth of pathogens. These properties have great potential to replace traditional antibiotics in the animal production systems.

4.4.4 Antioxidant and antiinflammatory effects

PFAs are also reported to have strong antioxidant and antiinflammatory properties. Polyphenols, flavonoids, and terpenes are compounds that scavenge ROS and prevent cells from undergoing oxidative stress (Miękus et al., 2020). These substances also inhibit the cycling of nuclear factor kappa B (NF-kB), which is a major mediator of cytokines. The inhibition of NF-kB results in the reduced synthesis of proinflammatory cytokines (TNF-α and IL-6) that can otherwise cause chronic inflammation (Kaschubek et al., 2018). In addition, these bioactives also increase the action of endogenous antioxidant enzymes such as glutathione peroxidase and superoxide dismutase. A combination of these measures prevents cellular redox imbalance and reduces the damage caused by inflammation, which will enhance the well-being of animals.

4.4.5 Rumen fermentation modulation

PFAs have shown great influence on the ruminal fermentation process derived by rumen microbes in ruminants. They change the rumen microbial population, decreasing protozoa and methanogenic archaea and hence reducing methane production (Ramdani et al., 2023). Saponins diminish the level of rumen protozoa, linked to the methanogen populations, which results in a changed fermentation profile with more propionate and a reduced amount of methane (Tsiplakou et al., 2023). This transition will make the energy more efficient, and fewer greenhouse gases will be emitted. Moreover, the control of volatile fatty acids (VFAs) profile optimizes ruminant energy supply, enhancing growth performance and the ratio of feed/gain.

4.4.6 Metabolic regulation

PFAs also regulate the systemic metabolism, such as lipid metabolism and immune modulation. Some phytogenic chemicals control lipid synthesis and degradation pathways through changes in gene regulation to optimize the ratio between lipogenesis and lipolysis (Flees et al., 2021). For example, the bioactives in garlic and turmeric have the ability to regulate the major metabolic enzymes, consequently lowering the fat storage and enhancing the carcass quality. Another essential benefit is immune modulation, in which phytogenics activate innate and adaptive immunity by cytokines and leukocytes (Yaqoob et al., 2021). To a large extent, these regulatory functions are responsible not only for promoting disease resistance but also for enhancing productive performance and animal welfare.

4.5 Applications in livestock and poultry production

4.5.1 Growth performance improvements

PFAs have been shown to be effective in enhancing feed efficiency and growth performance of livestock through stimulation of digestion and increased nutrient absorption. These substances stimulate the release of digestive enzymes and bile acids and help the species (poultry, pigs, and ruminants) to optimize the use of the feed (Abdelli et al., 2021). A number of studies have reported that phytogenic compounds supplementation can positively affect body weight gain and feed conversion ratios in animals, which means that they lead to increased metabolic efficiency (Bartoš et al., 2016). These results highlight that phytogenics can be used as a viable natural alternative growth promoter in the postantibiotic era.

4.5.2 Gut health and immunity enhancement

Phytogenic additives can be critical to maintain the gut health by enhancing barrier integrity, immunity of the mucosal area, and altering microbial communities (Shehata et al., 2022). The bioactive plant compounds stimulate tight junction proteins such as occludin and claudin that decrease the intestinal permeability and restrict the translocation of pathogens (Abdelli et al., 2021). Higher gut integrity is associated with effective nutrient absorption and prevents gastrointestinal infections in animals (Torres-Rodriguez et al., 2007). Moreover, phytogenics can regulate the action of cytokines, increase the level of antiinflammatory mediators, and decrease the level of proinflammatory cytokines, including TNF-a and IL-6, which contributes to the stability of the immune system (Coles et al., 2021). PFAs also improve the immunity of poultry and livestock by boosting both natural and adaptive immunity (Tellez-Isaias et al., 2021).

4.5.3 Reduction of pathogen load

Phytogenics play an important role in controlling pathogens in livestock production through antimicrobial activity in various modes (Karásková et al., 2015). Thymol, carvacrol, and cinnamaldehyde compounds interfere with bacterial cell membranes and suppress the expression of virulence factors, which lowers the gut colonization of *Salmonella, Escherichia coli*, and *Clostridium perfringens* (Shehata et al., 2022). In birds, these attributes can provide a good microbial

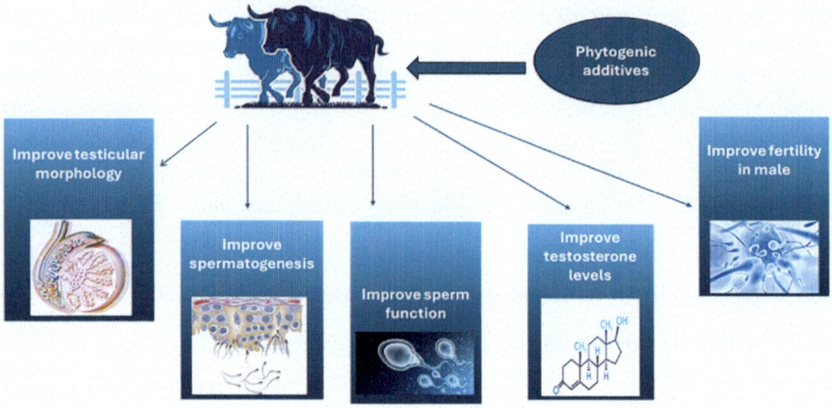

FIGURE 4.3 Effects of phytogenic feed additives on animal reproduction (Swelum et al., 2021).

balance by promoting beneficial bacteria (e.g., *Lactobacillus* and *Bifidobacterium*) that compete with harmful bacteria (Hernandez-Patlan et al., 2019). Further, these changes lead to improved food safety and product quality (Tellez-Isaias et al., 2021). Such a twofold effect (antimicrobial action and probiotic-like benefits) renders phytogenics very useful in the reduction of pathogenic loads and in lowering the risks of the spread of foodborne zoonoses (Ahmed et al., 2024).

4.5.4 Role in animal reproduction

Phytogenics have also proved to have significant effects in improving reproductive performance in livestock species (Hashem et al., 2013). For example, supplementation of propolis enhanced the sexual activity, spermatogenesis, and antioxidant status in rabbit bucks, especially during stressful climatic conditions (Hashem et al., 2013). The extracts of the *Moringa oleifera* leaf enhanced the semen quality, sperm motility, and antioxidant enzyme activities in the rams, which is an indication of the reproductive-promoting effect of polyphenol-rich plants (El-Desoky et al., 2017). Bulls supplemented with herbs and boars supplemented with ginseng root showed increased sperm viability, oxidative stress, and seminal plasma antioxidant enzyme levels (Yun et al., 2016). Likewise, ginger extract was found to stimulate Leydig cells to produce more testosterone, whereas cinnamon oil enhanced libido and semen quality in rams, showing a variety of reproductive benefits (El-Azrak et al., 2017) (Fig. 4.3).

4.5.5 Environmental benefits

The environmental use of PFAs has been increasingly popular because of their contribution to the reduction of the environmental footprint of livestock production (Bartoš et al., 2016). In ruminants, tannin and saponins have an effect on rumen fermentation, inhibiting the growth of methanogenic archaea and protozoa and thus leading to the production of lower levels of methane (Ahmed et al., 2024). The result of this fermentation conversion is the production of more propionate, which is more efficient in energy production and reducing the emission of greenhouse gases (Kil and Stein, 2010). In monogastric livestock such as pigs and poultry, phytogenic supplementation has lowered the nitrogen excretion and release of ammonia, a factor that enhances the quality of air within intensive production systems (Gomez-Osorio et al., 2021). These results show that phytogenics can be used to ensure that livestock is productive and sustainable, through a notable reduction in environmental pollutants (Yirga, 2015).

4.5.6 Product quality

Phytogenic feed supplements improve the quality and the functional value of animal-derived foods, including meat, milk, and eggs. The oxidative stability of meat, tenderness, and fatty acid profiles have also improved (such as increased percentage of polyunsaturated fatty acids) in meat produced by animals (Abdelli et al., 2021). Phytogenic supplementation in dairy production was found to raise the antioxidants in milk and improve milk fat composition with a higher level of conjugated linoleic acids and omega-3 fatty acids (Ahmed et al., 2024). Plant extract supplementation increased the yolk pigmentation of layer hens because of natural carotenoids and fortified omega-3 fatty acids, which are the attributes desired by consumers (Reyes-Camacho et al., 2020). Less lipid peroxidation in the animal products also increases the shelf life and marketability, which gives producers an added economic advantage (Ma et al., 2021) (Table 4.2).

4.6 Regulatory considerations

4.6.1 Toxicological limits and safety margins

The use of PFAs would be subject to the consideration of their toxicology profile so that they are safe to the animal as well as to consumers. Although the plant-derived compounds are natural, their effects are dose-dependent, where low levels are healthy and large doses may be toxic. The reason is that alkaloids and saponins, among others, can lead to a decrease in palatability, hepatic toxicity, and homolysis when consumed in unacceptable concentrations (Rao et al., 2013; Pandey et al., 2019). The safety levels are based on the calculation of no-observed-adverse-effect levels (NOAEL) and species difference uncertainty factors. Practically, additives are applied below the toxicity levels, but phytochemical composition may vary thus it is necessary to monitor it continuously. Toxicological evaluations must combine acute and chronic testing in target animals and incorporate tissue residue testing that would be of importance to human health (Steiner and Syed, 2015).

TABLE 4.2 Some selective studies showing outcomes of phytofeed additives in various animal models.

Species	Additive and dose	Main outcomes
Broiler Chicken	Essential-oil blend, 100–300 mg/kg feed	Improved FCR and BW gain; ↓ *E. coli*/*Salmonella* counts in cecal (Abdelli et al., 2021; Ayalew et al., 2022)
Pig (Grow-Finish)	Phytogenic premix, label dose	↑ Growth rate; ↓ NH3 in barn and GHG emissions (Karásková et al., 2015; Bartoš et al., 2016)
Dairy Cow	Tannin-saponin blend, ~10–30 g/d	↓ Methane proxy and NH_3-N; stable milk yield with improved N use efficiency (Kholif and Olafadehan, 2021; Ahmed et al., 2024)
Laying Hen	Plant-derived phytogenic, 100–200 mg/kg	↑ Egg quality metrics and shell traits; improved gut morphology (Abdelli et al., 2021; Alghirani et al., 2021)
Aquaculture (*Tilapia*)	Herbal extract 0.5%–1.0%	↑ Growth and feed efficiency; ↓ Opportunistic pathogens (Karásková et al., 2015)

4.6.2 Risks of antinutritional factors at high dosages

Tannins, phytases, and glucosinolates are examples of secondary plant metabolites that could hinder nutrient absorption or metabolic efficiency at high levels, but are beneficial at moderate levels (Rao et al., 2013). High concentrations of tannins form complexes with amino acids and decrease the rate of digestion of proteins. The glucosinolates contained in *Brassica* species could influence thyroid activity (Tchoursinov et al., 2020). Higher consumption of saponin-containing supplements in poultry decreases the feed intake and growth performance (Abdelli et al., 2021).

In addition, volatile essential oils, including thymol or eugenol, could be effective against pathogens; however, they might be irritative to the mucosa or cause oxidative stress when their level is in excess of the recommended level (Ayalew et al., 2022). Therefore it is essential to maintain the optimum level of these additives in the diet. The emergence of encapsulation and controlled-release technology can reduce these risks without decreasing efficiency (Abigarl et al., 2023).

4.6.3 Interactions with other feed components

Phytogenic interactions on the feed components determine its safety and efficacy. Bioactive phytochemicals may have a synergistic or antagonistic effect with nutrients, minerals, or other additives in the feed formulations. Polyphenols help increase antioxidant levels, but can chelate zinc and iron, decreasing the bioavailability of these minerals (Pandey et al., 2019). Similarly, essential oils cause changes in rumen fermentation and may disrupt the digestion of fibrous diets (Soni, 2021). Combination with other additives, such as probiotics and organic acids, can either enhance the positive effect or lead to the diminution of the efficacy, which depends on the dosage and the interactions between the additives and the matrix (Steiner and Syed, 2015). These interactions show the necessity of holistic approaches and more precise nutrition that compares both individual and combined effects (Abigarl et al., 2023).

4.6.4 Global regulatory overview

PFAs are subject to the same classification as sensory additives or zootechnical additives. Depending on the purpose of use, they are regulated by the European Food Safety Authority (EFSA). According to the EFSA guidelines, a comprehensive dose considering the safety of animals, users, consumers, and the environment has to be presented (Steiner and Syed, 2015). Before being authorized, critical parameters are tested, including toxicological levels, content of residues in edible tissues, effectiveness, and safety to the environment (Abdelli et al., 2021). PFAs in the United States are considered Generally Recognized as Safe (GRAS; in which a substance is classified as safe due to its historical usage or scientific validation) by the Food and Drug Administration (FDA) (Pandey et al., 2019). The regulatory framework in the United States focuses more on safety assurance, where product claims of PFAs are substantiated, and new bioactive components may require a separate approval procedure (Abigarl et al., 2023). Regulation varies greatly in Asia, where safety and efficacy demonstrations have to be made by the Ministry of Agriculture, whereas systems in South Asian countries are less standardized (Rao et al., 2013; Ayalew et al., 2022). Adoption of PFAs is increasing due to increasing bans on the use of AGPs, although inconsistent approval procedures exist (Abigarl et al., 2023).

4.6.5 Importance of standardization for reproducibility

Phytogenic composition varies based on species, cultivation, extraction, and formulation, which has an impact on adoption, efficacy, and safety. Reproducibility requires active compounds to be standardized using established analytical techniques, including HPLC (Steiner and Syed, 2015). A combination of omics methods and accurate farming data can also help predictability and result uniformity to an even greater extent (Abigarl et al., 2023).

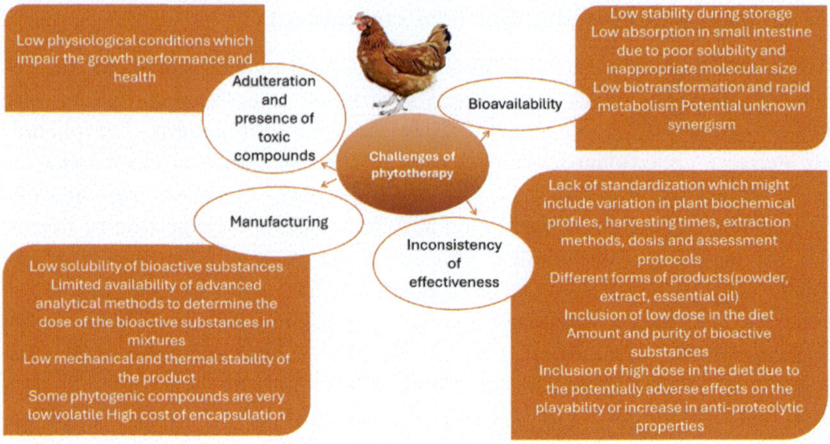

FIGURE 4.4 Challenges related to the application of phytogenic substances in poultry (Shehata et al., 2022).

4.7 Challenges, probable solutions, and future prospects

4.7.1 Variability in efficacy due to plant source and processing

One of the major challenges of PFAs is their variation in efficacy because of differences in the source of plants, their cultivation conditions, and processing (Tchoursinov et al., 2020). Active compounds like flavonoids, essential oils, and alkaloids can also be different depending on the type of soil, climate, harvest time, and storage (Ayalew et al., 2022). The method of processing also affects bioactivity since cold pressing can preserve antioxidants and high-temperature drying can destroy volatile components, resulting in worse functional results (Alghirani et al., 2021). This inconsistency leads to unstable outcomes of livestock trials, as some can demonstrate significant impact on growth and gut health, whereas others show only minimal effects (Pandey et al., 2019). The future development will need standardized cultivation and processing procedures to minimize variability and guarantee consistent efficacy (Fig. 4.4).

4.7.2 Need for standardization and identification of active compounds

The intricacy of phytogenic additives illustrates the necessity of quick standardization and identification of their active substances (Latek et al., 2022). Phytogenics have more bioactive compounds in them, which have not been characterized completely and are not well understood compared to synthetic additives, which have a known structure. Therefore it is not easy to determine dose-response relationships using such additives (Saeed et al., 2022). Promoting standardization would enhance consistency in trials and increase regulatory approval with a reliable data set on safety and efficacy (Steiner and Syed, 2015). More sophisticated analytical instruments, LC-MS and NMR, may be useful in detecting the marker compounds, such as carvacrol (used as an antimicrobial agent or saponins used to reduce methane) (Soni, 2021). Creating the standard formulas with certified profiles will increase the reproducibility and facilitate their wider implementation in commercial livestock systems.

4.7.3 Development of synergistic blends

Forming phytogenic compound synergies is an effective strategy to maximize the use of the phytogenic compounds. Most bioactive molecules work in a complementary way, and combinations of them could be more effective in antimicrobial and antioxidant action than individual molecules (Alghirani et al., 2021). On the same note, the tannins/saponins have been associated with less emission of methane and better use of nutrients by ruminants (Soni, 2021). Nevertheless, not all interactions can be friendly; they can even decrease nutrient uptake or act as neutral, and the increased costs of such synergistic formulations are also an issue. The next step in this field should be a systematic screening of phytochemical combinations through in vitro and in vivo models to determine the formulation with the best synergies to reduce its limitations (Morshedy et al., 2021; Abigarl et al., 2023).

4.7.4 Use of encapsulation and nanodelivery for stability and targeted release

The unstable nature of phytogenic compounds during storage, feed processing and digestion restricts their activity in animals (Ayalew et al., 2022). Various polyphenols and essential oils are volatile or oxidizable, which causes a decrease in bioactivity in the gut. Encapsulation of these compounds in lipids or polymers helps in enhancing their stability and allows them to be released at the target sites in a controlled manner (Saeed et al., 2022). The nanodelivery strategies, e.g., nanoemulsions or nanoparticle vectors, also change bioavailability by increasing solubility and improving transport through biological membranes. The technologies can also be used to release these antimicrobials or antioxidants in the gut areas where they are required the most. Price and scale are obstacles to adoption, although these technologies are very promising. So, the future of PFAs can be determined by designing affordable encapsulation systems.

Another key opportunity in the future is the incorporation of phytogenics into the precision nutritional approaches and climate-resilient livestock production. Precision nutrition aims at designing feed formulations to meet precise requirements of animals (in terms of species, breed, age, and health status) to ensure efficient consumption of resources and minimization of waste (Roy and George, 2020). In this regard, phytogenics may be important as they are able to regulate gut microbiota, improve the absorption of nutrients, and improve immunity in a manner that is complementary to customized feeding programs. Furthermore, their contribution to mitigating methane emissions, nitrogen excretion, and oxidative stress is consistent with the global sustainability aim and climate-smart farming approaches (Soni, 2021). With the escalation of climate obstacles, incorporating phytogenic feed supplements in precision and sustainable agricultural methods would be necessary to attain productivity and environmental sustainability (Morshedy et al., 2021).

4.8 Conclusion

PFAs have become a very promising natural alternative to synthetic growth promoters and antibiotics in livestock production. The combination of multiple functions, i.e., antimicrobial, antioxidant, and immunomodulatory activities, makes their bioactive compounds (including essential oils, polyphenols, alkaloids, tannins, and saponins) of wide interest in animal nutrition and reaffirms their value as feed additives. The application of PFAs is also in line with the world trend of sustainable and responsible animal farming. Rising limitations on the use of antibiotics in feed have led to an immediate need to find suitable and natural substitutes that will ensure the productivity of animals without interfering with food safety. Phytogenic additives serve this purpose by decreasing the number of pathogenic bacteria, such as *Salmonella* and *Escherichia coli*, as well as decreasing the environmental impact of livestock production. They are capable of enhancing feed conversion efficiency, which lowers the feed input demands, thus maximizing resource utilization in a time where there is growing demand for animal proteins. Moreover, the additives reduce greenhouse gas emissions and limit the release of nitrogen, which directly translates to climate-smart livestock production and a sustainable food system. Therefore strategic use of phytogenics in the feed of animals not only is a solution to production issues but also a backbone of the environmentally responsible production of animals. Besides, phytogenics increase the quality of the products: they increase the tenderness of meat, stabilize oxidation, enrich the milk with useful fatty acids, and elevate the pigmentation and nutritional value of egg yolks. Such improvements lead to the development of functional foods that respond to the demand of the consumers who seek healthier and more natural foods. To conclude, PFAs have the most versatile benefits, which go beyond increasing the performance of animals and are also associated with food safety and environmental sustainability. Even though some obstacles to the use of these natural products, such as variability in efficacy, standardization, and ways of delivery, are still present, the constant research and technological development present a pathway to a more uniform and useful application of these natural products. Finally, PFAs are holistic therapeutic agents to enhance the well-being of consumers, animal health, and decrease the environmental footprint of livestock farming.

References

Abd El-Hack, M.E., Alagawany, M., 2022. Antibiotic Alternatives in Poultry and Fish Feed. Bentham Science Publishers, Sharjah, UAE doi:10.2174/97898150490151220101.

Abdelli, N., Solà-Oriol, D., Francisco Pérez, J., 2021. Phytogenic feed additives in poultry: achievements, prospective and challenges. Animals 11 (12), 3471. doi:10.3390/ani11123471.

Abigarl, N., Jessica, P., Tapiwanashe, M., Sakhile, N., Mongameli, M., Samantha, S., Jeffrey, C., Alban, M., 2023. Incorporation of functional feed ingredients to substitute antimicrobials in animal nutrition: opportunities for livestock production in developing countries. Int. J. Livest. Prod. 14 (2), 44–57. doi:10.5897/ijlp2023.0820.

Adedokun, S.A., Olojede, O.C., 2019. Optimizing gastrointestinal integrity in poultry: the role of nutrients and feed additives. Front. Vet. Sci. 5. doi:10.3389/fvets.2018.00348.

Ahmed, M.G., Elwakeel, E.A., El-Zarkouny, S.Z., Al-Sagheer, A.A., 2024. Environmental impact of phytobiotic additives on greenhouse gas emission reduction, rumen fermentation manipulation, and performance in ruminants: an updated review. Environ. Sci. Pollut. Res. 31 (26), 37943–37962. doi:10.1007/s11356-024-33664-5.

Alghirani, M.M., Chung, E.L.T., Jesse, F.F.A., Sazili, A.Q., Loh, T.C., 2021. Could phytobiotics replace antibiotics as feed additives to stimulate production performance and health status in poultry? J. Adv. Vet. Res. 11 (4), 254–265.

Ayalew, H., Zhang, H., Wang, J., Wu, S., Qiu, K., Qi, G., Tekeste, A., Wassie, T., Chanie, D., 2022. Potential feed additives as antibiotic alternatives in broiler production. Front. Vet. Sci. 9. doi:10.3389/fvets.2022.916473.

Bagno, O.A., Prokhorov, O.N., Shevchenko, S.A., Shevchenko, A.I., Dyadichkina, T.V., 2018. Use of phytobiotücts in farm animal feeding. Sel'skokhozyaistvennaya Biol. 53 (4), 687–697. doi:10.15389/agrobiology.2018.4.687eng.

Bartoš, P., Dolan, A., Smutný, L., Šístková, M., Celjak, I., Šoch, M., Havelka, Z., 2016. Effects of phytogenic feed additives on growth performance and on ammonia and greenhouse gases emissions in growing-finishing pigs. Anim. Feed Sci. Technol. 212, 143–148. doi:10.1016/j.anifeedsci.2015.11.003.

Bischoff, K.L., 2021. Glucosinolates Nutraceuticals: Efficacy, Safety and Toxicity. Elsevier, United States, pp. 903–909. doi:10.1016/B978-0-12-821038-3.00053-7.

Blažević, I., Montaut, S., Burčul, F., Olsen, C.E, Burow, M., Rollin, P., Agerbirk, N., 2020. Glucosinolate structural diversity, identification, chemical synthesis and metabolism in plants. Phytochemistry 169, 112100. doi:10.1016/j.phytochem.2019.112100.

Burden, Y. O., 2020. The role of feed additives in mitigating the effect of stressors on growth, digestibility, intestinal morphology, permeability, and immune response in poultry. Theses and Dissertations–Animal andFood Sciences, 121. University of Kentucky https://uknowledge.uky.edu/animalsci_etds/121.

Coles, M.E., Forga, A.J., Señas-Cuesta, R., Graham, B.D., Selby, C.M., Uribe, Á.J., Martínez, B.C., Angel-Isaza, J.A., Vuong, C.N., Hernandez-Velasco, X., Hargis, B.M., Tellez-Isaias, G., 2021. Assessment of lippia origanoides essential oils in a salmonella typhimurium, eimeria maxima, and clostridium perfringens challenge model to induce necrotic enteritis in broiler chickens. Animals 11 (4). doi:10.3390/ani11041111.

Dell'Anno, M., 2023. Functional feed additives for animal health. Doctoral dissertation, Department of Veterinary Medicine and Animal Sciences. University of Milan, Italy https://hdl.handle.net/20.500.14242/73962.

El-Azrak, K., Hashem, N., Sallam, S., 2017. Effect of cinnamon oil administration on semen characteristics and sexual libido of Barki rams. 2nd International Conference on Veterinary and Animal Science. Cairo, Egypt.

El-Desoky, N.I., Hashem, N.M., Elkomy, A., Abo-elezz, Z.R., 2017. Physiological response and semen quality of rabbit bucks supplemented with Moringa leaves ethanolic extract during summer season. Animal 11 (9), 1549–1557. doi:10.1017/s1751731117000088.

Feye, K.M., Swaggerty, C.L., Kogut, M.H., Ricke, S.C., Piva, A., Grilli, E., 2020. The biological effects of microencapsulated organic acids and botanicals induces tissue-specific and dose-dependent changes to the Gallus gallus microbiota. BMC Microbiol. 20 (1). doi:10.1186/s12866-020-02001-4.

Flees, J.J., Ganguly, B., Dridi, S., 2021. Phytogenic feed additives improve broiler feed efficiency via modulation of intermediary lipid and protein metabolism–related signaling pathways. Poult. Sci. 100 (3), 100963. doi:10.1016/j.psj.2020.12.060.

Fushai, F., Chitura, T., Oke, O.E., 2025. Climate-smart livestock nutrition in semi-arid Southern African agricultural systems. Front. Vet. Sci. 12. doi:10.3389/fvets.2025.1507152.

Godde, C.M., Mason-D'Croz, D., Mayberry, D.E., Thornton, P.K., Herrero, M., 2021. Impacts of climate change on the livestock food supply chain: a review of the evidence. Glob. Food Secur. 28, 100488. doi:10.1016/j.gfs.2020.100488.

Gomez-Osorio, L.M., Yepes-Medina, V., Ballou, A., Parini, M., Angel, R., 2021. Short and medium chain fatty acids and their derivatives as a natural strategy in the control of necrotic enteritis and microbial homeostasis in broiler chickens. Front. Vet. Sci. 8. doi:10.3389/fvets.2021.773372.

Hafeez, A., Männer, K., Schieder, C., Zentek, J., 2016. Effect of supplementation of phytogenic feed additives (powdered vs. encapsulated) on performance and nutrient digestibility in broiler chickens. Poult. Sci. 95 (3), 622–629. doi:10.3382/ps/pev368.

Hashem, N.M., El-Hady, A.A., Hassan, O., 2013. Effect of vitamin E or propolis supplementation on semen quality, oxidative status and hemato-biochemical changes of rabbit bucks during hot season. Livest. Sci. 157 (2-3), 520–526. doi:10.1016/j.livsci.2013.09.003.

Hernandez-Patlan, D., Solis-Cruz, B., Pontin, K.P., Latorre, J.D., Hernandez-Velasco, X., Merino-Guzman, R., Mendez-Albores, A., Hargis, B.M., Lopez-Arellano, R., Tellez-Isaias, G., 2019. Evaluation of ascorbic acid or curcumin formulated in a solid dispersion on Salmonella enteritidis infection and intestinal integrity in broiler chickens. Pathogens 8 (4). doi:10.3390/pathogens8040229.

Karásková, K., Suchý, P., Straková, E., 2015. Current use of phytogenic feed additives in animal nutrition: a review. Czech J. Anim. Sci. 60 (12), 521–530. doi:10.17221/8594-cjas.

Kaschubek, T., Mayer, E., Rzesnik, S., Grenier, B., Bachinger, D., Schieder, C., König, J., Teichmann, K., 2018. Effects of phytogenic feed additives on cellular oxidative stress and inflammatory reactions in intestinal porcine epithelial cells1. J. Anim. Sci. 96 (9), 3657–3669. doi:10.1093/jas/sky263.

Kaur, A.P., Bhardwaj, S., Dhanjal, D.S., Nepovimova, E., Cruz-martins, N., Kuča, K., Chopra, C., Singh, R., Kumar, H., Şen, F., Kumar, V., Verma, R., Kumar, D., 2021. Plant prebiotics and their role in the amelioration of diseases. Biomolecules 11 (3), 1–28. doi:10.3390/biom11030440.

Kholif, A.E., Olafadehan, O.A., 2021. Essential oils and phytogenic feed additives in ruminant diet: chemistry, ruminal microbiota and fermentation, feed utilization and productive performance. Phytochem. Rev. 20 (6), 1087–1108. doi:10.1007/s11101-021-09739-3.

Kil, D.Y., Stein, H.H., 2010. Board invited review: management and feeding strategies to ameliorate the impact of removing antibiotic growth promoters from diets fed to weanling pigs. Can. J. Anim. Sci. 90 (4), 447–460. doi:10.4141/cjas10028.

Latek, U., Chłopecka, M., Karlik, W., Mendel, M., 2022. Phytogenic compounds for enhancing intestinal barrier function in poultry–a review. Planta Med. 88 (03/04), 218–236. doi:10.1055/a-1524-0358.

Ma, Y., Zhang, Y., Elmhadi, M., Zhang, H., Wang, H., 2021. Thiamine alleviates high-concentrate-diet-induced oxidative stress, apoptosis, and protects the rumen epithelial barrier function in goats. Front. Vet. Sci. 8. doi:10.3389/fvets.2021.663698.

Mandey, J.S., Sompie, F.N., 2021. Phytogenic feed additives as an alternative to antibiotic growth promoters in poultry nutrition. Advanced Studies in the 21st Century Animal Nutrition. IntechOpen, London, UK doi:10.5772/intechopen.99401.

Mazanko, M.S., Gorlov, I.F., Prazdnova, E.V., Makarenko, M.S., Usatov, A.V., Bren, A.B., Chistyakov, V.A., Tutelyan, A.V., Komarova, Z.B., Mosolova, N.I., Pilipenko, D.N., Krotova, O.E., Struk, A.N., Lin, A., Chikindas, M.L., 2018. Bacillus probiotic supplementations improve laying performance, egg quality, hatching of laying hens, and sperm quality of roosters. Probiotics Antimicrob. Proteins 10 (2), 367–373. doi:10.1007/s12602-017-9369-4.

Miękus, N., Marszałek, K., Podlacha, M., Iqbal, A., Puchalski, C., Świergiel, A.H., 2020. Health benefits of plant-derived sulfur compounds, glucosinolates, and organosulfur compounds. Molecules 25 (17), 3804. doi:10.3390/molecules25173804.

Morshedy, S.A., Abdelmodather, A.M., Basyony, M.M., Zahran, S.A., Hassan, M.A., 2021. Effects of rocket seed oil, wheat germ oil, and their mixture on growth performance, feed utilization, digestibility, redox status, and meat fatty acid profile of growing rabbits. Agriculture (Switzerland) 11 (7). doi:10.3390/agriculture11070662.

Nair, A., Chattopadhyay, D., Saha, B., 2018. Plant-derived immunomodulators. In: New Look to Phytomedicine: Advancements in Herbal Products as Novel Drug Leads. Elsevier, India, pp. 435–499. doi:10.1016/B978-0-12-814619-4.00018-5.

Nicastro, H.L., Ross, S.A., Milner, J.A., 2015. Garlic and onions: their cancer prevention properties. Cancer Prev. Res. 8 (3), 181–189. doi:10.1158/1940-6207.CAPR-14-0172.

Nuutila, A.M., Puupponen-Pimiä, R., Aarni, M., Oksman-Caldentey, K.M., 2003. Comparison of antioxidant activities of onion and garlic extracts by inhibition of lipid peroxidation and radical scavenging activity. Food Chem. 81 (4), 485–493. doi:10.1016/S0308-8146(02)00476-4.

Pandey, A.K., Kumar, P., Saxena, M., 2019. Feed additives in animal health. In: Nutraceuticals in Veterinary Medicine. Springer, Singapore, pp. 345–362. doi:10.1007/978-3-030-04624-8_23.

Placha, I., Gai, F., Pogány Simonová, M., 2022. Natural feed additives in animal nutrition—their potential as functional feed. Front. Vet. Sci. 9, 1062724.

Ramdani, D., Yuniarti, E., Jayanegara, A., Chaudhry, A.S., 2023. Roles of essential oils, polyphenols, and saponins of medicinal plants as natural additives and anthelmintics in ruminant diets: a systematic review. Animals 13 (4). doi:10.3390/ani13040767.

Rao, S., Prasad, K., Rajendran, D., 2013. Recent advances in amelioration of anti-nutritional factors in livestock feedstuffs. In: Animal Nutrition Reproductive Physiology. Satish Serial Publishing House, Delhi, India, pp. 655–678.

Reyes-Camacho, D., Vinyeta, E., Pérez, J.F., Aumiller, T., Criado, L., Palade, L.M., Taranu, I., Folch, J.M., Calvo, M.A., Van Der Klis, J.D., Solà-Oriol, D., 2020. Phytogenic actives supplemented in hyperprolific sows: effects on maternal transfer of phytogenic compounds, colostrum and milk features, performance and antioxidant status of sows and their offspring, and piglet intestinal gene expression. J. Anim. Sci. 98 (1). doi:10.1093/jas/skz390.

Roy, T., George, J.G., 2020. Precision farming: a step towards sustainable, climate-smart agriculture. In: Global Climate Change: Resilient and Smart Agriculture. Springer Singapore, India, pp. 199–220. doi:10.1007/978-981-32-9856-9_10.

Saeed, M., Khan, M.S., Sultan, R., Aqib, A.I., Naseer, M.A., Muzammil, I., Farag, M.R., Abd El-Hack, M.E., Di Cerbo, A., Alagawany, M., 2022. Herb and plant-derived supplements in poultry nutrition. In: Antibiotic Alternatives in Poultry and Fish Feed. Bentham Science Publishers, Pakistan, pp. 19–32. doi:10.2174/9789815049015122010005.

Shahbakht, R.M., Nomi, Z.A., Kabeer, S.W., Anwar, M.Z., Shahid, S., Imtiaz, B., Rajar, A.A., Soomro, R.N., Rajput, S.A., 2024. The use of phytogenic feed additives in animal nutrition Pakistan scenario. Complement. Altern. Med. Feed Addit. 326–334. doi:10.47278/book.CAM/2024.401.

Shashidhara, R.G, Devegowda, G., 2003. Effect of dietary mannan oligosaccharide on broiler breeder production traits and immunity. Poult. Sci. 82 (8), 1319–1325. doi:10.1093/ps/82.8.1319.

Shehata, A.A., Attia, Y., Khafaga, A.F., Farooq, M.Z., El-Seedi, H.R., Eisenreich, W., Tellez-Isaias, G., 2022. Restoring healthy gut microbiome in poultry using alternative feed additives with particular attention to phytogenic substances: challenges and prospects. Ger. J. Vet. Res. 2 (3), 32–42. doi:10.51585/gjvr.2022.3.0047.

Singh, A.K., Choubey, M., Chaudhary, S.K., Ojha, L., Kumari, P., 2024. Phytochemicals as natural feed additives for ruminants. In: Feed Additives and Supplements for Ruminants. Springer Singapore, India, pp. 167–196. doi:10.1007/978-981-97-0794-2_8.

Soni, P.K., 2021. Rumen microbial adaptation to phyto additives and its effect on methane emission and performance of buffalo calves (PhD thesis). Indian Veterinary Research Institute, Izatnagar, India

Steiner, T., Syed, B., 2015. Phytogenic feed additives in animal nutrition. In: Medicinal and Aromatic Plants of the World: Scientific, Production, Commercial and Utilization Aspects. Springer, pp. 403–423. doi:10.1007/978-94-017-9810-5.

Swelum, A.A., Hashem, N.M., Abdelnour, S.A., Taha, A.E., Ohran, H., Khafaga, A.F., El-Tarabily, K.A., Abd El-Hack, M.E., 2021. Effects of phytogenic feed additives on the reproductive performance of animals. Saudi J. Biol. Sci. 28 (10), 5816–5822. doi:10.1016/j.sjbs.2021.06.045.

Tchoursinov, Y., Kovaliova, O., Kalyna, V., Mykolenko, S., Khomuk, N.I., 2020. Phyto-feed additives production: technological aspects and biological value. Sci. Bull. Ser. F Biotechnol. 24 (2), 43–48. https://biotechnologyjournal.usamv.ro/pdf/2020/issue_2/Art5.pdf.

Tellez-Isaias, G., Vuong, C.N., Graham, B.D., Selby, C.M., Graham, L.E., Señas-Cuesta, R., Barros, T.L., Beer, L.C., Coles, M.E., Forga, A.J., Ruff, J., Hernandez-Velasco, X., Hargis, B.M., 2021. Developing probiotics, prebiotics, and organic acids to control Salmonella spp. Ger. J. Vet. Res. 1 (3), 7–12. doi:10.51585/gjvr.2021.3.0014.

Torres-Rodriguez, A., Higgins, S.E., Vicente, J.L.S., Wolfenden, A.D., Gaona-Ramirez, G., Barton, J.T., Tellez, G., Donoghue, A.M., Hargis, B.M., 2007. Effect of lactose as a prebiotic on turkey body weight under commercial conditions. J. Appl. Poult. Res. 16 (4), 635–641. doi:10.3382/japr.2006-00127.

Torres-Sanchez, L., Lopez-Carrillo, L., Lopez-Cervantes, M., Rueda-Neria, C., Wolff, M.S., 2000. Food sources of phytoestrogens and breast cancer risk in Mexican women. Nutr. Cancer 37 (2), 134–139. doi:10.1207/S15327914NC372_3.

Tsiplakou, E., Mavrommatis, A., Gelasakis, A.I., Kalogianni, A.I., Simitzis, P.E., 2023. Effect of phytochemical feed additives on health status, milk yield, and quality characteristics in ruminants. In: Sustainable Use of Feed Additives in Livestock: Novel Ways for Animal Production. Springer International Publishing, Greece, pp. 641–663. doi:10.1007/9783031428555_21.

Wang, J., Deng, L., Chen, M., Che, Y., Li, L., Zhu, L., Chen, G., Feng, T., 2024. Phytogenic feed additives as natural antibiotic alternatives in animal health and production: a review of the literature of the last decade. Anim. Nutr. 17, 244–264. doi:10.1016/j.aninu.2024.01.012.

Yaqoob, M.U., Abd El-Hack, M.E., Hassan, F., El-Saadony, M.T., Khafaga, A.F., Batiha, G.E., Yehia, N., Elnesr, S.S., Alagawany, M., El-Tarabily, K.A., Wang, M., 2021. The potential mechanistic insights and future implications for the effect of prebiotics on poultry performance, gut microbiome, and intestinal morphology. Poult. Sci. 100 (7), 101143. doi:10.1016/j.psj.2021.101143.

Yirga, H., 2015. The use of probiotics in animal nutrition. J. Probiotics Health 03 (02). doi:10.4172/2329-8901.1000132.

Yun, S.J., Bae, G.S., Park, J.H., Song, T.H., Choi, A., Ryu, B.Y., Pang, M.G., Kim, E.J., Yoon, M., Chang, M.B., 2016. Antioxidant effects of cultured wild ginseng root extracts on the male reproductive function of boars and guinea pigs. Anim. Reprod. Sci. 170, 51–60. doi:10.1016/j.anireprosci.2016.04.002.

Zou, X., Liang, J., Sun, J., Hu, X., Lei, L., Wu, D., Liu, L., 2016. Allicin sensitizes hepatocellular cancer cells to anti-tumor activity of 5-fluorouracil through ROS-mediated mitochondrial pathway. J. Pharmacol. Sci. 131 (4), 233–240. doi:10.1016/j.jphs.2016.04.017.

Chapter 5

Animal health and welfare under functional phytofeed

Mahmoud Kamal[a], Ayman El-sayed Taha[b] and Mohamed E. Abd El-Hack[c]

[a]*Animal Production Systems Research Department, Animal Production Research Institute, Agricultural Research Center, Giza, Egypt,* [b]*Department of Public Health, College of Veterinary Medicine, King Faisal University, Al-Ahsa, Saudi Arabia,* [c]*Department of Industrial Pharmacy, College of Pharmaceutical Sciences and Drug Manufacturing, Misr University for Science and Technology (MUST), Giza, Egypt*

5.1 Introduction

The use of phytofeed additives represents a systematic development in animal nutrition, leading to improved animal health and welfare while enhancing productivity (Negm et al., 2025; Reda et al., 2025). These plant-based feed additives, derived from herbs, spices, and plant extracts, are rich in bioactive components such as essential oils, flavonoids, and phenolic acids, which possess antibacterial, antioxidant, and antiinflammatory properties (Wang et al., 2024; Abd El-Hack et al., 2025a). These compounds promote gut health, improve nutrient absorption, and strengthen immunity, thereby reducing the use of antibiotics as growth promoters (Abd El-Hack et al., 2025b; El-Abbasy et al., 2025a). This shift toward the use of natural, plant-derived compounds aligns with the global trend of reducing antibiotic utilization and promoting clean farming practices (Biswas and Kim, 2025).

Adding plant-based materials to animal feeds improved animal health by enhancing gut structure, increasing enzyme activity, and restoring microbial communities, thereby boosting productivity (Zuberi et al., 2024; Ashour et al., 2025c; Ashour et al., 2025d). Furthermore, these materials reduced stress, thereby improving immunity and overall endurance (Upadhaya and Kim, 2017; Abd El-Aziz et al., 2025). These properties of the plants contribute to improved animal welfare, reduced mortality, and enhanced metabolic health and behavioral status in cattle (Singh and Gaikwad, 2020; Kamal et al., 2024).

Integrating functional phytofeeds into animal feed aligns with clean agriculture goals, as these additives improve nutrient utilization, reduce feed waste, and mitigate the harmful effects of livestock production by reducing global

Plant-Derived Feed Additives and Animal-Based Foods: Linking Animal Health, Product Quality, and Consumer Perceptions. DOI: https://doi.org/10.1016/B978-0-443-45446-2.00011-9

warming (Nhara et al., 2025). Despite their properties, challenges remain regarding the stability, availability, and long-term effects of these compounds (Aminullah et al., 2025). Therefore continued research focusing on controlled trials and synergistic formulations is essential to enhance their applicability and fully realize their potential for improving animal health and welfare and promoting sustainable production practices.

5.2 Modulation of animal stress and cortisol

The hypothalamic–pituitary–adrenal (HPA) axis is the principal neuroendocrine system governing the body's reaction to stress in animals. In reaction to a stressor, neurons in the paraventricular nucleus of the hypothalamus secrete "corticotropin-releasing hormone (CRH)," which stimulates the pituitary gland to release "adrenocorticotropic hormone (ACTH)" (Herman et al., 2016). ACTH then causes the adrenal cortex to release glucocorticoids—mainly cortisol in most mammals and corticosterone in birds and some other species—into the blood, which are important signs of stress (Brown and Vosloo, 2017). Glucocorticoids help manage how the body reacts to stress, including metabolism, immune responses, and mental health; however, if they are constantly activated, they can weaken the immune system and harm overall health (Ataallahi et al., 2022).

Cortisol and corticosterone levels are frequently evaluated as indicators of stress in animals. Blood, saliva, hair, and feather matrices are employed to assess hormone concentrations; however, stress induced by testing may compromise the outcome, rendering noninvasive sampling approaches more preferable. Species, sex, age, and environmental context influence variations in glucocorticoid response (Karaer et al., 2023). Environmental and handling stressors, including transportation, living circumstances, and temperature fluctuations, significantly activate the HPA axis, leading to elevated cortisol levels and physiological signs of stress (Hernandez et al., 2025). Transportation-related stress raises blood cortisol levels, rectal and skin temperatures, and signs of tissue damage, such as creatine kinase, in pigs, while changes in temperature inside transport vehicles make stress worse (Lee et al., 2023; Hernandez et al., 2025). Inferior housing and extreme confinement increase stress and cortisol levels, but social housing and environmental enrichment reduce signs of stress in captive animals, especially dogs. Heat, management practices, and lack of feed and water are common cumulative stressors in livestock production that exacerbate HPA axis activation (Kruger et al., 2016).

Herbal adaptogens such as ashwagandha (*Withania somnifera*) and *Rhodiola rosea* are utilized to alleviate stress by modulating the HPA axis and reducing cortisol production (Burns, 2023). Rhodiola has demonstrated efficacy in mitigating physical and mental fatigue, burnout, and perceived stress by activating stress proteins and modulating nitric oxide signaling, resulting in reduced glucocorticoid levels (Ivanova Stojcheva and Carlos Quintela, 2022). Ashwagandha

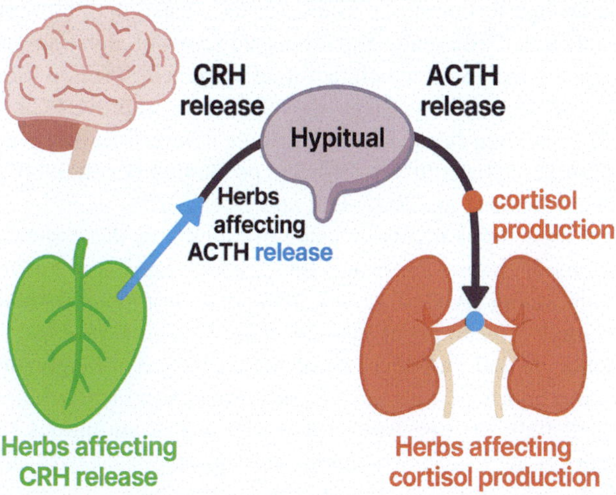

FIGURE 5.1 The hypothalamic–pituitary–adrenal axis and the herbal regulation points. *ACTH*, adrenocorticotropic hormone; *CRH*, corticotropin-releasing hormone.

is acknowledged for its anti-stress properties through HPA modulation and its antioxidant functions. Herbal medications are increasingly perceived as safe and advantageous supplements in animal diets to improve resilience against stressors (Amir et al., 2023). Research has been conducted on herbally enhanced meals to mitigate transport-induced stress. Prebiotic and natural products that include antioxidants, antiinflammatories, and adaptogenic bioactives aid in regulating cortisol and enhancing immune function under stress in species such as sheep and goats (Devaraj et al., 2024; Sejian et al., 2025). These nutritional treatments offer substantial well-being improvements with lasting benefits and minimal adverse effects.

Understanding HPA axis function and utilizing herbal adaptogens, in conjunction with enhanced ecological and handling techniques, could regulate stress in animals and alleviate physiological and behavioral stress influences, thus improving animal welfare. Fig. 5.1 illustrates the HPA axis, emphasizing the hormonal communication mechanism in which the hypothalamus secretes CRH, stimulating the pituitary to release ACTH, subsequently leading to cortisol production in the adrenal glands. Herbal regulatory sites denote areas where herbs might influence hormone release, thereby impacting stress reactivity and adrenal function.

5.3 Effects on growth, immunity, and welfare

Functional phytofeeds significantly enhance growth performance by increasing appetite, boosting gastrointestinal enzyme activities, and improving feed conversion ratios (FCR) (Mohammadi Gheisar and Kim, 2018; Abo Egila et al., 2023).

In addition, plant-based supplements enhance the digestion of nutrients, leading to improved nutritional efficiency and increased levels of alpha-amylase, lipase, maltase, and trypsin enzymes, which facilitate nutrient absorption (Seidavi et al., 2022; Kamal et al., 2025). In addition, these additives improve FCR by 3%–4% and increase BW, as plant-based feeds contribute to intestinal health and nutrient absorption by reducing harmful bacteria and promoting beneficial bacteria, thus aiding growth (Losacco et al., 2025; Oni and Oke, 2025).

Plant-based feed supplements enhance immunity, as these chemicals boost immunoglobulins, including "IgA and IgG," which are important for mucosal and systemic immunity (Abdelli et al., 2021). The immunomodulatory effects are evident through the control of cytokines and leukocyte profiles, resulting in elevated lymphocyte and monocyte numbers, indicative of heightened immune system activation and hematopoiesis. This immunostimulation helps chickens stay healthy and fight off diseases without the use of antibiotics (Huang and Lee, 2018; Putri et al., 2025). The addition of phytogenic feed leads to better health, comfort, and livability, which in turn leads to better welfare (Perera and Ravindran, 2025). Healthy, immune-boosting animals enjoy greater comfort and a higher standard of living, as proper nutrition and a strong immune system have been shown to increase their average lifespan (Oni and Oke, 2025).

Phytofeed additives in animal feed enhance digestive enzymes, boost immunity, reduce cytokines, and improve animal welfare, all of which contribute to optimal animal husbandry, care, and well-being.

5.4 Influence on animal behavior and mood-like states

Research demonstrates that functional phytofeed significantly influences animal behavior and mood, as evidenced by reduced fearfulness, feather pecking, and aggression, alongside increased curiosity and social interactions (Lundstrom et al., 2017). Behavioral modifications are estimated using tests such as body tests, movement tests, and tonic nonmovement tests, such as curiosity and fear intensity (Montalcini et al., 2025).

Functional phytofeeds comprise herbs and plant extracts that engage with essential neurobiological systems, specifically serotonergic and GABAergic signaling pathways within the central nervous system (Mendis et al., 2025). Numerous pharmaceutical plant extracts exhibit a significant affinity for receptors implicated in mood modulation, including serotonin (5-HT1A) and GABA_A receptors. This method explains the scientific basis behind its calming effect, as the results indicated that the extract of the Valeriana officinalis plant inhibits the enzymatic breakdown of gamma-aminobutyric acid (GABA), thus increasing its availability and reducing anxiety. Some herbs exhibit affinity for benzodiazepine receptors, which regulate GABAergic signaling and alleviate anxiety-related behaviors (Akotkar et al., 2024).

Behavioral assessments frequently employed clarify these benefits, with the novel object test assessing the propensity to investigate unexpected stimuli, signifying less fearfulness and heightened interest (Bari et al., 2021). The open field test assesses overall activity and exploratory behavior in a new setting (Campbell et al., 2019). The tonic immobility test assesses fearfulness by measuring the duration of immobility in animals, an antipredator response that frequently decreases with anxiolytic treatments (Tainika et al., 2023). The neurobiological basis for these behavioral effects is ascribed to herbs affecting neurotransmission (Moskaliuk et al., 2023). Phytogenic feed additives can enhance serotonergic transmission by increasing serotonin receptor activity and elevating brain serotonin concentrations, often correlated with improved mood and reduced aggression (Mendis et al., 2025). In a similar way, increased GABAergic neurotransmission through increased receptor binding or GABA availability has calming effects and reduces violent and fear-driven behavior (Lundstrom et al., 2017).

Functional phytofeed affects how animals act and feel by making them less fearful, aggressive, and stressed, and more curious and social. The effects are corroborated by the modulatory influences of herbal constituents on serotonergic and GABAergic systems, indicating potential avenues for natural behavioral modulation in animal nutrition. Table 5.1 presents the mechanisms and potential effects of phytogenic feed additives in animals' diets.

5.5 Biochemical indicators of animal well-being

The bioactive compounds in plant-based feed additives enhance the activity of antioxidant enzymes such as "SOD, CAT, and GPx," which protect cells by lowering "MDA" (Li et al., 2015; Michalak et al., 2022). These additions also improve liver function indicators, including "ALT, Aspartate aminotransferase (AST), and Alkaline phosphatase (ALP)" (Allameh et al., 2023; Kamal et al., 2023). Furthermore, adding plant extracts helps maintain normal levels of these enzymes, thus supporting their optimal function (Abd El-Hack et al., 2024; Omar et al., 2025).

Herbal supplements also affect hormones such as cortisol and adrenaline, which are important for maintaining animal mood and well-being (El-Abbasy et al., 2025b). Passionflower extract reduces stress hormones, thereby improving behavioral and biological stress regulation in mice (Adetunji et al., 2025; Yushanaji et al., 2025). Phytofeed intake additionally impacts several substantial health markers, including "blood glucose, urea, creatinine, serum proteins, and lipid." These biochemical markers indicate metabolic status, renal function, protein metabolism, and cardiovascular health (Qorbani et al., 2022). The augmentation of these biomarkers using phytogenics fosters comprehensive animal health, productivity, and well-being (El-Zaiat et al., 2023).

TABLE 5.1 Functions of phytogenic feed additives and their possible mechanisms of action.

Phytogenic feed additives	Animals	Results	Mechanism	References
Poplar and eucalyptus leaves	Buffaloes	The data include the daily milk yield, 6% FCM, and FPCM yield, digestibility coefficients for dry matter, organic matter, and neutral detergent fiber, antibody titers (log10) against the Pasteurella multocida vaccine, and skin thickness	Augment milk yield and bolster the immunity of Murrah buffaloes	Dey et al. (2021)
Astragali Radix, Rhizoma Atractylodis Macrocephalae, and Saposhnikoviae Radix	Goats	Enhance the concentrations of palmitic acid, octadecanoic acid, and arachidonic acid in the longissimus dorsi muscle, along with taste fatty acids, SFA, PUFA, the PUFA/SFA ratio, PPARγ, and CD36 expression, and the presence of Ruminococcus and Alistipes	Enhance meat quality and flavor by modulating fat metabolism and promoting beneficial microorganisms	Yang et al. (2021)
Moringa oleifera seed	Ram	Augmentation of antioxidant activity; enhancement of viability and progressive motility; diminution of sperm membrane damage	Enhance the efficacy of semen cryopreservation through antioxidant application	Carrera-Chávez et al. (2020)

(*continued on next page*)

Plantain and garlic leaf	Sheep	The results showed a reduction in mutton extract and saturated fatty acids, an enhancement in ribeye area, polyunsaturated fatty acids, linoleic acid levels, and concentrations of serum IgA, IgG1, IgG2, and IgM, along with elevated levels of TAC, SOD, GPx, and CAT	Enhance growth efficiency, health, lean mutton output, and the FAs of mutton	Redoy et al. (2020)
Albizia harveyi	Bull	Sperm motility, viability, and membrane integrity were boosted; the proportion of viable sperm cells increased; the percentages of early apoptotic and apoptotic sperm cells diminished; damage to sperm ultrastructure was reduced; total antioxidant capacity was elevated; and malondialdehyde levels dropped	Alleviate the negative impacts of the freeze-thaw procedure on cryopreserved bull semen owing to its antioxidant capacity	Sobeh et al. (2017)

(*continued on next page*)

TABLE 5.1 Functions of phytogenic feed additives and their possible mechanisms of action—cont'd

Phytogenic feed additives	Animals	Results	Mechanism	References
Moringa oleifera leaf extract	Rabbits	Rectal temperature and serum albumin levels improved, while total antioxidant capacity, testosterone levels, and seminal plasma fructose also increased. In addition, sperm parameters such as concentration, forward motility, live count, normal prevalence, progressive motility, viability, normal morphology, intact acrosome count, and integrated cell membrane status all showed enhancement	Enhancing resistance to heat, oxidation state, and semen quality throughout the summer is crucial for optimal reproductive success	El-Desoky et al. (2017)
Radix Angelicae Sinensis, Rhizome Ligustici, Semen Persicae, Rhizoma Zingiberis, and Radix Glycyrrhizae	Dairy cows	The placental retention rate, calving-to-first-service interval, and services per conception have decreased, while the calving-to-conception interval, first artificial insemination conception rate, and pregnancy occurrence within 180 days postpartum have increased	Reduce the incidence of retained placenta and improve subsequent reproductive effectiveness. It offers a proactive approach to monitoring the postpartum health of cows	Cui et al. (2014)

(continued on next page)

Yucca schidigera powder	Dairy goats	This led to improved LBW, conception rate, fertility rate, and estrus duration; it also decreased estrus resumption, TC, TG, urea, glucose, and calcium	Facilitate a beneficial modification in body weight, promote superior reproductive performance, and assist in the regulation of metabolic disturbances, not only by restoring homeostatic equilibrium but also by influencing productive traits	Hassanien (2014)
Echinacea powder extract	Broiler	The addition of 3.0 cm^3 of EPE improved growth efficiency, carcass features, and antioxidant activity to achieve beneficial health results for consumers	EPE enhanced growth efficiency, certain carcass traits, and improved antioxidant and immune activities	Ashour et al. (2025a)
Artichoke leaf extract	Broiler	Incorporating nano ALE stabilized with zinc nanoparticles at doses of 1.0 and 2.0 cm^3 into the feeds of growing chicks enhances their efficiency, carcass, and blood indicators	ALE stabilized utilizing zinc nanoparticles may serve as an unusual feed additive to enhance the health of developing chicks	Ashour et al. (2025b)
Moringa oleifera leaf powder	Broiler	ZnNPs-MLPE at 2.0 and 3.0 cm^3/L increased BW and BWG in broilers, without affecting FI or FCR. Blood tests indicated lower cholesterol, triglycerides, LDL, VLDL, and Alanine aminotransferase (ALT) levels. Moreover, IgY, IgM, and SOD rose with decreased MAD	ZnNPs-MLPE at 2.0 and 3.0 cm^3/L improved growth efficiency, antioxidant, and immunity, highlighting its potential as a natural growth enhancer for healthier poultry	El-Abbasy et al. (2025a)

(continued on next page)

TABLE 5.1 Functions of phytogenic feed additives and their possible mechanisms of action—cont'd

Phytogenic feed additives	Animals	Results	Mechanism	References
Garden cress seed powder	Broiler	GCSP supplementation did not affect growth but showed numerical improvements and altered carcasses, such as dressing percentage and abdominal fat. It enhanced immunity and antioxidative, with increased IgY, IgM, and SOD levels, and reduced MDA	These findings indicate that GCSP enhances health and leads to the production of healthier poultry products	Negm et al. (2025)
Herbal blend	Broiler	Biochemical indices and antioxidant activities improved, with effects on cecal microbial load. Also, increase total bacterial counts and *Lactobacilli*	The improved blood traits, antioxidants, and gut bacterial counts contributed to the promotion of performance	Ashour et al. (2025a)
Thyme powder and garlic powder	Broiler	The addition of phytogenic feed additives improved FI, FCR, and enhanced carcass characteristics and blood indices, which in turn can improve immunity and oxidative balance, resulting in higher immunoglobulin levels and lower oxidative stress markers	The combination of thyme powder and garlic powder at 2% each can replace antibiotics, improving broiler growth, carcass quality, and health by enhancing immune responses and antioxidant status	Ashour et al. (2025b)

(*continued on next page*)

Yizhi extract	Ducks	Reduce LDL-C levels and total amino acids in breast muscle, while enhancing umami and flavor amino acids such as arginine and glutamic acid; increase abundance of unclassified Bacteroidales and Ruminococcaceae	Increase meat nutrient profile and flavor, maintain intestinal integrity, and modify the bacterial makeup	Ji et al. (2022)
Commercial products	Broiler	Enhance slaughter weights and breast weights and reduce fat pad size, L*Y, b*, green herb flavor, and SOD2	Modulate the pathways linked to stress and antioxidants to enhance the quality of broiler meat	Orlowski et al. (2018)
Yucca schidigera extract	Laying	Increase egg quantity and mass, shell thickness, albumin, IgG, SOD, and GSH, while reducing MDA, total cholesterol, and lipid peroxidation	Enhance egg production, egg quality, immunity, and antioxidants	Alagawany et al. (2016)
Digestarom	Broiler	Enhance weight gain, FCR villus height, and total tract digestibility of dry matter, crude protein, ether extract, and Lactobacillus species; reduce crypt depth, total coliform count, coliforms, and anaerobic bacteria	Enhance the growth of broilers by optimizing digestive enzymes and altering gut flora	Murugesan et al. (2015)

ALE, *artichoke leaf extract;* BW, *body weight;* BWG, *body weight gain;* EPE, *echinacea powder extract;* MDA, *malondialdehyde;* FCM, *Fat-corrected milk;* FCR, *feed conversion ratio;* FI, *feed intake;* FPCM, *fat-protein-corrected milk;* GCSP, *garden cress seed powder;* GSH, *glutathione;* LDL, *low-density lipoprotein;* PUFA, *polyunsaturated fatty acids;* SFA, *saturated fatty acids;* SOD, *superoxide dismutase;* TC, *total cholesterol;* TG, *triglyceride;* ZnNPs-MLPE, *zinc nanoparticles-moringa oleifera leaf powder.*

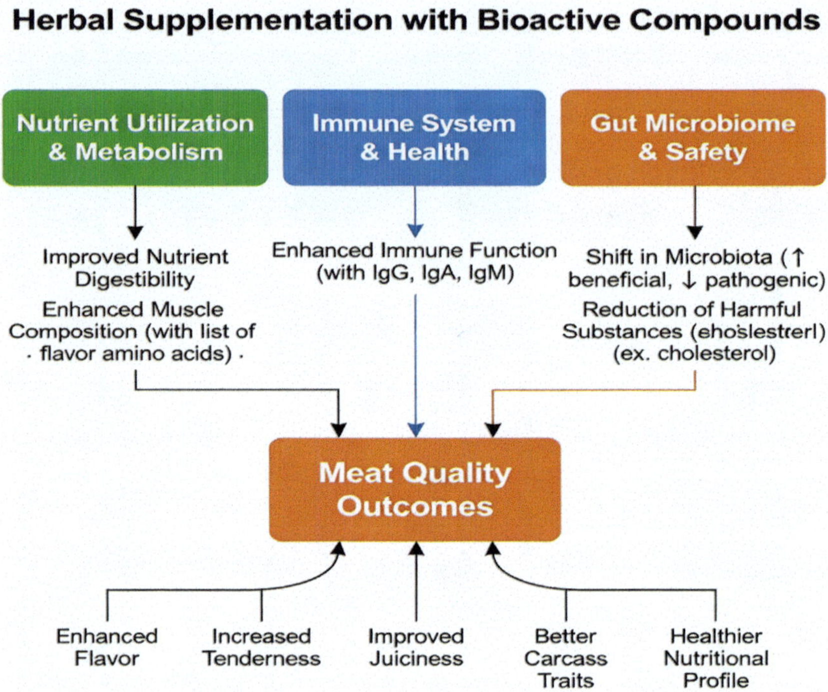

FIGURE 5.2 Pathways linking herbal supplementation with physiology and meat quality.

Functional phytofeed additives improve biochemical indicators by augmenting antioxidant defenses (SOD, CAT, GPx), diminishing oxidative stress markers (MDA), supporting hepatic function (ALT, AST, ALP), modifying stress hormones (cortisol, adrenaline), and enhancing essential metabolic biomarkers. These influences jointly enhance animal welfare, health, and productivity.

5.6 Implications for meat quality

Functional phytofeed additives can modify muscle fiber qualities and collagen features, both essential for meat softness. Dietary bioactive substances affect muscle fiber length and collagen cross-linking (Orlowski et al., 2018). Increased collagen cross-linking, commonly associated with aging or thermal stress, typically results in muscular stiffness and diminished tenderness (Holwerda and Van Loon, 2022). Phytofeeds rich in antioxidants and those containing bioactive peptides could reduce excessive cross-linking, which helps maintain muscle fiber structure and makes it softer (Pogorzelska-Nowicka et al., 2018). Polyphenols extracted from grass-fed feed additives help maintain myoglobin levels, thus inhibiting its oxidation to metmyoglobin, which leads to meat browning and an undesirable color (Zhu et al., 2024). These natural polyphenols act as antioxidants, inhibiting the oxidation of lipids and myoglobin to maintain the bright red color of fresh beef during storage and processing (Palanisamy et al., 2024).

Water-holding capacity (WHC) directly influences sensory traits such as "juiciness and texture." Phytofeed supplements boost WHC by safeguarding muscle structure, curbing oxidative damage, and stabilizing cell membranes (Prates, 2025). By easing oxidative stress, they protect protein structures, ensuring muscle cells retain fluids better—these cuts down on drip loss and cooking loss for juicier, more appealing meat (Pastsart et al., 2024). On top of that, phytofeed additives with antibacterial and antioxidant effects prolong beef shelf life by curbing microbial growth and slowing lipid oxidation (Wang et al., 2024). Essential oils and plant extracts drop the pH, halt spoilage microbes, reduce microbial counts, and minimize spoilage signs (Biswas et al., 2023), while their antioxidants fend off rancidity to maintain flavor and safety during refrigeration (Li et al., 2025) (Fig. 5.2).

The plant-based additive alters the structure of muscle fibers and collagen, regulates myoglobin, increases food's water and juice retention, and extends shelf life by inhibiting bacteria and protecting against free radicals. Polyphenols, antioxidants, and peptides facilitate these benefits by interacting with molecular pathways that reduce oxidative stress and inhibit microbial proliferation.

5.7 Conclusion

Comprehending the organization of the HPA axis and utilizing herbal supplements, alongside optimizing animal management, will facilitate the regulation of animal stress and foster their health and well-being. Phytofeed promotes poultry growth by improving feed efficiency, increasing digestive enzyme activity, and strengthening immune function through the regulation of immunoglobulins and cytokines. It positively influences behavior by diminishing fear, anger, and stress-related activities, while enhancing exploration and social interaction through its impact on the serotonergic and GABAergic systems. Phytofeed augments antioxidant defenses, diminishes oxidative stress, bolsters hepatic function, equilibrates stress hormones, and boosts metabolic biomarkers at the biochemical level. These methods contribute to improved poultry welfare and productivity. Furthermore, plant-based additives enhance meat quality by improving muscle and collagen structure, preserving color, increasing juiciness, and extending shelf life because of their antioxidant compounds. Plant-based additives provide a sustainable and organic way to improve poultry efficiency and diminish reliance on antibiotics.

References

Abd El-Hack, M.E., Aldhalmi, A.K., Kamal, M., Khafaga, A.F., Moustafa, M., Al-Shehri, M., Attia, Y.A., 2025a. Flavonoids as a phytobiotic agent: sources, classifications, biological benefits, and useful impacts on broilers and layers. Phytochem. Rev. 1–27.

Abd El-Hack, M.E., Aldhalmi, A.K., Ashour, E.A., Kamal, M., Khan, M.M.H., Swelum, A.A., 2025b. The effects of formic acid or herbal mixture on growth performance, carcass quality, blood chemistry, and gut microbial load in broiler chickens: formic acid & herbal mixture in broiler diets. Poult. Sci. 104 (6). doi:10.1016/j.psj.2025.105085.

Abd El-Hack, M.E., Majrashi, K.A., Fakiha, K.G., Roshdy, M., Kamal, M., Saleh, R.M., Khafaga, A.F., Othman, S.I., Rudayni, H.A., Allam, A.A., Moustafa, M., Tellez-Isaias, G., Alagawany, M., 2024. Effects of varying dietary microalgae levels on performance, egg quality, fertility, and blood biochemical parameters of laying Japanese quails (Coturnix coturnix Japonica). Poult. Sci. 103 (4). doi:10.1016/j.psj.2024.103454.

Abdelli, N., Solà-Oriol, D., Francisco Pérez, J., 2021. Phytogenic feed additives in poultry: achievements, prospective and challenges. Animals 11 (12), 3471. doi:10.3390/ani11123471.

Abo Egila, N.S.H., Dosoky, W.M., Khisheerah, N.S.M., Ahmed, M.H., Zahran, S.M., Almohmadi, N.H., Abusudah, W.F., Kamal, M., Moustafa, M., Tellez-Isaias, G., Al-Shehri, M., Abd El-Hack, M.E., 2023. Does dietary linseed or canola oil affect lipid metabolism, immunity, and n-3 polyunsaturated fatty acids content in quail eggs? Poult. Sci. 102 (12). doi:10.1016/j.psj.2023.103116.

Adetunji, A.O., Price, J., Owusu, H., Adewale, E.F., Adesina, P.A., Saliu, T.P., Zhu, Z., Xedzro, C., Asiamah, E., Islam, S., 2025. Mechanisms by which phytogenic extracts enhance livestock reproductive health: current insights and future directions. Front. Vet. Sci. 12. doi:10.3389/fvets.2025.1568577.

Akotkar, L., Aswar, U., Ganeshpurkar, A., Rathod, K., Bagad, P., Gurav, S., 2024. Phytoconstituents targeting the serotonin 5-HT 3 receptor: promising therapeutic strategies for neurological disorders. ACS Pharmacol. Transl. Sci. 7 (6), 1694–1710. doi:10.1021/acsptsci.4c00084.

Alagawany, M., Abd El-Hack, M.E., El-Kholy, M.S., 2016. Productive performance, egg quality, blood constituents, immune functions, and antioxidant parameters in laying hens fed diets with different levels of Yucca schidigera extract. Environ. Sci. Pollut. Res. 23 (7), 6774–6782. doi:10.1007/s11356-015-5919-z.

Allameh, A., Niayesh-Mehr, R., Aliarab, A., Sebastiani, G., Pantopoulos, K., 2023. Oxidative stress in liver pathophysiology and disease. Antioxidants 12 (9), 1653. doi:10.3390/antiox12091653.

Aminullah, N., Mostamand, A., Zahir, A., Mahaq, O., Azizi, M.N., 2025. Phytogenic feed additives as alternatives to antibiotics in poultry production: a review. Vet. World 18 (1), 141–154. doi:10.14202/vetworld.2025.141-154.

Amir, M., Vohra, M., Raj, R.G., Osoro, I., Sharma, A., 2023. Adaptogenic herbs: a natural way to improve athletic performance. Health Sci. Rev. 7, 100092. doi:10.1016/j.hsr.2023.100092.

Ashour, E.A., Al-Ardhi, S.A., Abd El-Hack, M.E., Elsherbeni, A.I., Elolimy, A.A., Madkour, M., Tufarelli, V., Swelum, A.A., 2025a. Effects of lactic acid and herbal blend as antibiotic alternatives on growth, carcass traits, blood chemistry, and microbial load in broiler chickens. Poult. Sci. 104 (5). doi:10.1016/j.psj.2025.105050.

Ashour, E.A., Aldhalmi, A.K., Elolimy, A.A., Madkour, M., Elsherbeni, A.I., Alqhtani, A.H., Khan, I.M., Swelum, A.A., 2025b. Optimizing broiler performance, carcass traits, and health: evaluating thyme and/or garlic powders as natural growth promoters in antibiotic-free diets. Poult. Sci. 104 (2). doi:10.1016/j.psj.2024.104689.

Ashour, E.A., Aldhalmi, A.K., Ismail, I.E., Kamal, M., Elolimy, A.A., Swelum, A.A., Abd El-Hack, M.E., 2025c. The effect of using Echinacea extract as an immune system stimulant and antioxidant on blood indicators, growth efficiency, and carcass characteristics in broiler chickens to produce a healthy product. Poult. Sci. 104 (1). doi:10.1016/j.psj.2024.104392.

Ashour, E.A., Aldhalmi, A.K., Kamal, M., Salem, S.S., Mahgoub, S.A., Alqhtani, A.H., Madkour, M., Elolimy, A.A., Abd El-Hack, M.E., Swelum, A.A., 2025d. The efficacy of Artichoke leaf extract conjugated with organic zinc nanoparticles on growth, carcass traits and blood biochemical parameters of broilers. Poult. Sci. 104 (1). doi:10.1016/j.psj.2024.104521.

Ataallahi, M., Nejad, J.G., Park, K.H., 2022. Selection of appropriate biomatrices for studies of chronic stress in animals: a review. J. Anim. Sci. Technol. 64 (4), 621–639. doi:10.5187/jast.2022.e38.

Bari, M.S., Allen, S.S., Mesken, J., Cohen-Barnhouse, A.M., Campbell, D.L.M., 2021. Relationship between range use and fearfulness in free-range hens from different rearing enrichments. Animals 11 (2), 1–20. doi:10.3390/ani11020300.

Biswas, O., Kandasamy, P., Nanda, P.K., Biswas, S., Lorenzo, J.M., Das, A., Alessandroni, L., Lamri, M., Das, A.K., Gagaoua, M., 2023. Phytochemicals as natural additives for quality preservation and improvement of muscle foods: a focus on fish and fish products. Food Mater. Res. 3. doi:10.48130/FMR-2023-0005.

Biswas, S., Kim, I.H., 2025. A thorough review of phytogenic feed additives in non-ruminant nutrition: production, gut health, and environmental concerns. J. Anim. Sci. Technol. 67 (3), 497–519. doi:10.5187/jast.2025.e26.

Brown, E.J., Vosloo, A., 2017. The involvement of the hypothalamopituitary-adrenocortical axis in stress physiology and its significance in the assessment of animal welfare in cattle. Onderstepoort J. Vet. Res. 84 (1). doi:10.4102/ojvr.v84i1.1398.

Burns, J., 2023. Common herbs for stress: the science and strategy of a botanical medicine approach to self-care. J. Interprof. Educ. Pract. 30, 100592. doi:10.1016/j.xjep.2022.100592.

Campbell, D.L.M., Dickson, E.J., Lee, C., 2019. Application of open field, tonic immobility, and attention bias tests to hens with different ranging patterns. PeerJ 7 (11), e8122. doi:10.7717/peerj.8122.

Carrera-Chávez, J.M., Jiménez-Aguilar, E.E., Acosta-Pérez, T.P., Núñez-Gastélum, J.A., Quezada-Casasola, A., Escárcega-Ávila, A.M., Itza-Ortiz, M.F., Orozco-Lucero, E., 2020. Effect of Moringa oleifera seed extract on antioxidant activity and sperm characteristics in cryopreserved ram semen. J. Appl. Anim. Res. 48 (1), 114–120. doi:10.1080/09712119.2020.1741374.

Cui, D., Wang, X., Wang, L., Wang, X., Zhang, J., Qin, Z., Li, J., Yang, Z., 2014. The administration of Sheng Hua Tang immediately after delivery to reduce the incidence of retained placenta in Holstein dairy cows. Theriogenology 81 (5), 645–650. doi:10.1016/j.theriogenology.2013.11.019.

Devaraj, C., Sejian, V., Shashank, C.G., Silpa, M.V., Bhatta, R., 2024. The effect of an antioxidant supplement based on selenium, electrolytes, and vitamins on relieving transportation stress in goats. Small Rumin. Res. 241, 107392. doi:10.1016/j.smallrumres.2024.107392.

Dey, A., Attri, K., Dahiya, S.S., Paul, S.S., 2021. Influence of dietary phytogenic feed additives on lactation performance, methane emissions and health status of Murrah buffaloes (Bubalus bubalis). J. Sci. Food Agric. 101 (10), 4390–4397. doi:10.1002/jsfa.11080.

El-Abbasy, M.M., Aldhalmi, A.K., Ashour, E.A., Bassiony, S.S., Kamal, M., Alqhtani, A.H., Abou-Kassem, D.E., Elolimy, A.A., Abd El-Hack, M.E., Swelum, A.A., 2025a. Enhancing broiler growth and carcass quality: impact of diets enriched with Moringa oleifera leaf powder conjugated with zinc nanoparticles. Poult. Sci. 104 (1). doi:10.1016/j.psj.2024.104519.

El-Abbasy, M.M., Aldhalmi, A.K., Ashour, E.A., Elnesr, S.S., Kamal, M., Salem, S.S., Mahgoub, S.A., Al-Shehri, M., Moustafa, M., Abd El-Hack, M.E., 2025b. Impact of dried fig fruit extract conjugated with selenium nanoparticles on productive and blood parameters of chicken. Ann. Anim. Sci. 25 (3), 1055–1063. doi:10.2478/aoas-2025-0048.

El-Aziz, A., H., A., Ghanima, M.M.A., Kamal, M., El-Hack, A., M., E., Alagawany, M., 2025. The use of bile acids supplement in poultry feed. In: Organic Feed Additives for Livestock. Academic Press, London, pp. 127–138.

El-Desoky, N.I., Hashem, N.M., Elkomy, A., Abo-elezz, Z.R., 2017. Physiological response and semen quality of rabbit bucks supplemented with Moringa leaves ethanolic extract during summer season. Animal 11 (9), 1549–1557. doi:10.1017/s1751731117000088.

El-Zaiat, H.M., Ku-Vera, J.C., Soltan, Y.A., 2023. Natural phytochemicals to enhance animal productivity and health status with low greenhouse gas emissions. Front. Vet. Sci. 10, 1280611. doi:10.3389/fvets.2023.1280611.

Hassanien, H., 2014. Effects of using Yucca schidigera powder as feed additive on productive and reproductive efficiency of Zaraibi dairy goats. Egypt. J. Sheep Goats Sci. 9 (2), 1–14.

Herman, J.P., McKlveen, J.M., Ghosal, S., Kopp, B., Wulsin, A., Makinson, R., Scheimann, J., Myers, B., 2016. Regulation of the hypothalamic-pituitary-adrenocortical stress response. Compr. Physiol. 6 (2), 603–621. doi:10.1002/cphy.c150015.

Hernandez, R.O., Rocha, A.O., Cai, C., Erasmus, M., Johnson, J.S., Brito, L.F., 2025. Effects of microclimate during transport on physiological indicators of market pig welfare: a systematic review with meta-analysis. Front. Vet. Sci. 12. doi:10.3389/fvets.2025.1657185.

Holwerda, A.M., Van Loon, L.J.C., 2022. The impact of collagen protein ingestion on musculoskeletal connective tissue remodeling: a narrative review. Nutr. Rev. 80 (6), 1497–1514. doi:10.1093/nutrit/nuab083.

Huang, C.M., Lee, T.T., 2018. Immunomodulatory effects of phytogenics in chickens and pigs—A review. Asian-Australas. J. Anim. Sci. 31 (5), 617–627. doi:10.5713/ajas.17.0657.

Ivanova Stojcheva, E., Carlos Quintela, J., 2022. The effectiveness of Rhodiola rosea L. preparations in alleviating various aspects of life-stress symptoms and stress-induced conditions—encouraging clinical evidence. Molecules 27 (12), 3902. doi:10.3390/molecules27123902.

Ji, F., Gu, L., Rong, G., Hu, C., Sun, W., Wang, D., Peng, W., Lin, D., Liu, Q., Wu, H., Dai, H., Zhou, H., Xu, T., 2022. Using extract from the stems and leaves of Yizhi (Alpiniae oxyphyllae) as feed additive increases meat quality and intestinal health in ducks. Front. Vet. Sci. 8. doi:10.3389/fvets.2021.793698.

Kamal, M., Kishk, W.H., Khalil, H.A., Abdel-Khalek, A.M., Ayoub, M.A., Swelum, A.A., Alqhtani, A.H., Ba-Awadh, H.A., Abd El-Hack, M.E., 2023. Effect of dietary chitosan supplementation on productive and physiological performance parameters of growing New Zealand white rabbits. Int. J. Biol. Macromol. 230. doi:10.1016/j.ijbiomac.2023.123166.

Kamal, M., Lele, W., Shuzhen, T., Jiandi, L., Rongyan, Q., Yanfeng, L., Wenqi, W., Xiangyu, C., Cheng, Y., 2024. Influence of dietary Salicornia europaea L. extract supplementation on feed efficiency of Altay sheep by modifying their gastrointestinal bacteria communities. Front. Microbiol. 15. doi:10.3389/fmicb.2024.1377314.

Kamal, M., Zhu, L., Abd El-Hack, M.E., Arif, M., Li, F., Cheng, Y., 2025. Functional roles of mannan and chitosan oligosaccharides on animal health and nutrition: a review. Carbohydr. Polym. Technol. Appl. 10. doi:10.1016/j.carpta.2025.100764.

Karaer, M.C., Čebulj-Kadunc, N., Snoj, T., 2023. Stress in wildlife: comparison of the stress response among domestic, captive, and free-ranging animals. Front. Vet. Sci. 10. doi:10.3389/fvets.2023.1167016.

Kruger, L.P., Nedambale, T.L., Scholtz, M.M., Webb, E.C., 2016. The effect of environmental factors and husbandry practices on stress in goats. Small Rumin. Res. 141, 1–4. doi:10.1016/j.smallrumres.2016.06.004.

Lee, G.H., Jo, W., Kang, T.K., Oh, T., Kim, K.S., 2023. Assessment of stress caused by environmental changes for improving the welfare of laboratory beagle dogs. Animals 13 (6). doi:10.3390/ani13061095.

Li, C., Zhu, Y., Wu, S., 2025. Green tea extracts as natural preservatives: enhancing shelf life and sensory quality of stewed beef chunks during refrigerated storage. LWT 233, 118493. doi:10.1016/j.lwt.2025.118493.

Li, S., Tan, H.Y., Wang, N., Zhang, Z.J., Lao, L., Wong, C.W., Feng, Y., 2015. The role of oxidative stress and antioxidants in liver diseases. Int. J. Mol. Sci. 16 (11), 26087–26124. doi:10.3390/ijms161125942.

Losacco, C., Pugliese, G., Passantino, L., Giannico, F., Ceci, E., Roselli, V., Gambacorta, L., Schiavitto, M., Laudadio, V., Piemontese, L., Tufarelli, V., 2025. Horehound (Marrubium vulgare L.) as natural dietary feed additive in rabbit: effects on productive traits, antioxidant status, caecal environment, and gut morphology. Front. Anim. Sci. 6. doi:10.3389/fanim.2025.1658188.

Lundstrom, K., Pham, H., Dinh, L., 2017. Interaction of plant extracts with central nervous system receptors. Medicines 4 (1), 12. doi:10.3390/medicines4010012.

Mendis, K., Gnanapragasam, M.A., Keerthirathne, D.H.G.A.N., Madushani, L.A.N., Piyasekara, G.M.T.M., Senanayake, S., Bandara, C., Abeytunga, T., Gunaratna, T.A.N.R., Jayakody, T., 2025. Neuropharmacology and the applications of herbal extracts found within Sri Lankan traditional anxiety remedies. Phytomed. Plus 5 (3). doi:10.1016/j.phyplu.2025.100794.

Michalak, I., Tiwari, R., Dhawan, M., Alagawany, M., Farag, M.R., Sharun, K., Emran, T.B., Dhama, K., 2022. Antioxidant effects of seaweeds and their active compounds on animal health and production–a review. Vet. Q. 42 (1), 48–67. doi:10.1080/01652176.2022.2061744.

Mohammadi Gheisar, M., Kim, I.H., 2018. Phytobiotics in poultry and swine nutrition–a review. Ital. J. Anim. Sci. 17 (1), 92–99. doi:10.1080/1828051X.2017.1350120.

Montalcini, C.M., Driver, C.C., Mendl, M.T., 2025. Relevance of state–behaviour feedbacks for animal welfare. Biol. Rev. 100 (4), 1615–1634. doi:10.1111/brv.70016.

Moskaliuk, V.S., Kozhemyakina, R.V., Khomenko, T.M., Volcho, K.P., Salakhutdinov, N.F., Kulikov, A.V., Naumenko, V.S., Kulikova, E.A., 2023. On associations between fear-induced aggression, Bdnf transcripts, and serotonin receptors in the brains of Norway rats: an influence of antiaggressive drug TC-2153. Int. J. Mol. Sci. 24 (2). doi:10.3390/ijms24020983.

Murugesan, G.R., Syed, B., Haldar, S., Pender, C., 2015. Phytogenic feed additives as an alternative to antibiotic growth promoters in broiler chickens. Front. Vet. Sci. 2. doi:10.3389/fvets.2015.00021.

Negm, M.H., Aldhalmi, A.K., Ashour, E.A., Mohamed, L.A., Kamal, M., Rashad, A., Khan, M.M.H., Abd El-Hack, M.E., Swelum, A.A., 2025. Growth, carcass traits, blood chemistry and gut microbiota in broiler chickens fed diets enriched with garden cress seed powder as a natural growth enhancer. Vet. Med. Sci. 11 (3). doi:10.1002/vms3.70397.

Nhara, R.B., Musara, C., Charambira, T., Mujati, T., Nhovoro, T., Hodzi, P., Chigede, N., Ruzengwe, F.M., 2025. Role of phytogenic feed additives on animal product quality and acceptance—a review. Trop. Anim. Health Prod. 57 (5). doi:10.1007/s11250-025-04474-9.

Omar, S., Alhotan, R., Hussein, E., Galik, B., Zahran, N., Alagawany, M., Saleh, A., 2025. The impact of microalgae and their bioactive compounds on liver well-being in rats subjected to synthetic phenolic antioxidants. Egypt. J. Vet. Sci. 56 (1), 47–60. doi:10.21608/ejvs.2024.258700.1750.

Oni, A.I., Oke, O.E., 2025. Gut health modulation in poultry through phytogenics: mechanisms, benefits, and applications. Front. Vet. Sci. 12, 1616734. doi:10.3389/fvets.2025.1616734.

Orlowski, S., Flees, J., Greene, E.S., Ashley, D., Lee, S.O., Yang, F.L., Owens, C.M., Kidd, M., Anthony, N., Dridi, S., 2018. Effects of phytogenic additives on meat quality traits in broiler chickens. J. Anim. Sci. 96 (9), 3757–3767. doi:10.1093/jas/sky238.

Palanisamy, S., Singh, A., Zhang, B., Zhao, Q., Benjakul, S., 2024. Effects of different phenolic compounds on the redox state of myoglobin and prevention of discoloration, lipid and protein oxidation of refrigerated Longtail Tuna (Thunnus tonggol) slices. Foods 13 (8), 1238. doi:10.3390/foods13081238.

Pastsart, U., Sresomjit, F., Bochuai, R., Pimpa, O., Profile, T., 2024. Water holding capacity, antioxidant activity and lipid oxidation of beef during retail display from cattle fed total mixed ration supplemented with capsicum frutescens L. and Curcuma longa L. powders. Online J. Anim. Feed Res. 14 (1), 40–46. doi:10.51227/ojafr.2024.5.

Perera, W.N.U., Ravindran, V., 2025. Role of feed additives in poultry nutrition: historical, current and future perspectives. Anim. Feed Sci. Technol. 326. doi:10.1016/j.anifeedsci.2025.116371.

Pogorzelska-Nowicka, E., Atanasov, A., Horbańczuk, J.ł, Wierzbicka, A., 2018. Bioactive compounds in functional meat products. Molecules 23 (2), 307. doi:10.3390/molecules23020307.

Prates, J.A.M., 2025. Impact of heat stress on carcass traits, meat quality, and nutritional value in monogastric animals: underlying mechanisms and nutritional mitigation strategies. Foods 14 (9), 1612. doi:10.3390/foods14091612.

Putri, M.A., Cokrowati, N., Setyono, B.D.H., 2025. Growth and hematological response of common carp (Cyprinus carpio) to dietary supplementation with water hyacinth (Eichhornia crassipes) extract. J. Fish Health 5 (3), 351–364. doi:10.29303/jfh.v5i3.7800.

Qorbani, M., Angoorani, P., Ejtahed, H.S., 2022. Dietary phytochemical index as a biomarker in nutritional studies: features and applications: the dietary phytochemical index. In: Biomarkers in Nutrition. Springer International Publishing, Cham, pp. 307–328.

Reda, F.M., Kamal, M., Mahmoud, H.K., Rudayni, H.A., Allam, A.A., Khafaga, A.F., Khan, M.M.H., Mohamed, R.S., 2025. Effect of a mixture of peppermint and clove oils on growth performance, carcass traits, blood parameters, and gut microbiota in broiler chickens. Poult. Sci. 104 (11). doi:10.1016/j.psj.2025.105718.

Redoy, M.R.A., Shuvo, A.A.S., Cheng, L., Al-Mamun, M., 2020. Effect of herbal supplementation on growth, immunity, rumen histology, serum antioxidants and meat quality of sheep. Animal 14 (11), 2433–2441. doi:10.1017/s1751731120001196.

Seidavi, A., Tavakoli, M., Asroosh, F., Scanes, C.G., Abd El-Hack, M.E., Naiel, M.A.E., Taha, A.E., Aleya, L., El-Tarabily, K.A., Swelum, A.A., 2022. Antioxidant and antimicrobial activities of phytonutrients as antibiotic substitutes in poultry feed. Environ. Sci. Pollut. Res. 29 (4), 5006–5031. doi:10.1007/s11356-021-17401-w.

Sejian, V., Devaraj, C., Shashank, C.G., Silpa, M.V., Sahoo, A., Bhatta, R., 2025. The role of prebiotic and herbal supplementation in enhancing welfare and resilience of Kenguri sheep subjected to transportation stress. Vet. Sci. 12 (5). doi:10.3390/vetsci12050442.

Singh, J., Gaikwad, D.S., 2020. Phytogenic feed additives in animal nutrition. In: Natural Bioactive Products in Sustainable Agriculture. Springer Singapore, India, pp. 273–289. doi:10.1007/978-981-15-3024-1_13.

Sobeh, M., Hassan, S.A., El Raey, M.A., Khalil, W.A., Hassan, M.A.E., Wink, M., 2017. Polyphenolics from albizia harveyi exhibit antioxidant activities and counteract oxidative damage and ultra-structural changes of cryopreserved bull semen. Molecules 22 (11). doi:10.3390/molecules22111993.

Tainika, B., Şekeroğlu, A., Akyol, A., Waithaka Ng'ang'a, Z., 2023. Welfare issues in broiler chickens: overview. World's Poult. Sci. J. 79 (2), 285–329. doi:10.1080/00439339.2023.2175343.

Upadhaya, S.D., Kim, I.H., 2017. Efficacy of phytogenic feed additive on performance, production and health status of monogastric animals-a review. Ann. Anim. Sci. 17 (4), 929–948. doi:10.1515/aoas-2016-0079.

Wang, J., Deng, L., Chen, M., Che, Y., Li, L.u, Zhu, L., Chen, G., Feng, T., 2024. Phytogenic feed additives as natural antibiotic alternatives in animal health and production: A review of the literature of the last decade. Anim. Nutr. 17, 244–264. doi:10.1016/j.aninu.2024.01.012.

Yang, H., Tan, S., Chen, S., Wu, Y., Yang, Y., Li, H., Yu, H., 2021. Effects of fermented Yupingfeng on intramuscular fatty acids and ruminal microbiota in Qingyuan black goats. Anim. Sci. J. 92 (1). doi:10.1111/asj.13554.

Yushanaji, B., Kamal, M., Baishanbieke, D., Zhu, L., Zhang, P., He, Q., Li, F., 2025. Effects of raffinose supplementation on growth performance, carcass quality, blood chemistry, gut microbiota, and intestinal morphology in Yellow-feathered broilers. Poult. Sci. 104 (11), 105888. doi:10.1016/j.psj.2025.105888.

Zhu, W., Han, M., Bu, Y., Li, X., Yi, S., Xu, Y., Li, J., 2024. Plant polyphenols regulating myoglobin oxidation and color stability in red meat and certain fish: a review. Crit. Rev. Food Sci. Nutr 64 (8), 2276–2288. doi:10.1080/10408398.2022.2122922.

Zuberi, A., Kamran, M., Younus, N., Abdel-Tawwab, M., 2024. Functional feed additives: current trends. Front. Aquac. 3, 1385508.

Chapter 6

Modulation of gut microbiota through phytofeed: implications for poultry, ruminants, and aquatic animals

Muhammad Muneeb[a], Arooj Fatima[b], Ibnath Haque Abony[c], Zeeshan Ahmad Bhutta[d], Mohamed E. Abd El-Hack[e] and Sohail Ahmad[f]

[a] *Department of Animal Nutrition, Faculty of Animal Production and Technology, University of Veterinary and Animal Sciences, Lahore, Pakistan,* [b] *Department of Veterinary Medicine, Faculty of Veterinary Sciences, University of Veterinary and Animal Sciences, Lahore, Pakistan,* [c] *Faculty of Fisheries, Bangladesh Agricultural University, Mymensingh, Bangladesh,* [d] *Laboratory of Veterinary Immunology and Biochemistry, College of Veterinary Medicine, Chungbuk National University, Cheongju, Republic of Korea,* [e] *Department of Industrial Pharmacy, College of Pharmaceutical Sciences and Drug Manufacturing, Misr University for Science and Technology (MUST), Giza, Egypt,* [f] *Department of Poultry Production, Faculty of Animal Production and Technology, University of Veterinary and Animal Sciences, Lahore, Pakistan*

6.1 Introduction

By the end of 2050, the world population is expected to reach about 10 billion, which can increase the food demands by about 70% (Ehrlich and Harte, 2015). To meet food needs, it is important to increase meat production. Surprisingly, alone meat demands will increase up to 525 million tons, and to improve this, our agriculture and food scientists are working together to make the food accessible, sustainable, and affordable. With the passage of time, there are drastic changes in the population's economic status and urbanization, which in turn increase the demand for safe and nutritious food. Pressure also increases on the animal husbandry to meet these needs. Science-based genetic selection, dietary subtherapeutics, and an increase in the growth of animals are effective methods, which, in turn, increase the efficiency of the feed (Gadde et al., 2017; Mehdi et al., 2018). Food from ruminants is also considered healthy, and therefore they are used as a source of food (Poore and Nemecek, 2018). Food from aquatic sources is also

Plant-Derived Feed Additives and Animal-Based Foods: Linking Animal Health, Product Quality, and Consumer Perceptions. DOI: https://doi.org/10.1016/B978-0-443-45446-2.00016-8

considered critical because of the presence of essential nutrients, and the sources include shrimps, finfish, mollusks, and aquatic plants (Byrd et al., 2021).

Poultry production continues to play an important role in providing safe and nutritious food. Gut health is essential in maintaining poultry health. Gut health of the poultry mainly depends on the gut microbiome and its diversity. The basic functions of the microbiome are protection against pathogens and its maturation effect on the immune system (Aruwa et al., 2021). This microbiota also helps in feed digestion and maintains the health of the intestine. Their main function is to extract nutrients from the feed and make them usable by the chicken (Rowland et al., 2018). Microbiota have a major role in the extraction, absorption, and finally in the synthesis of the biochemicals with growth promoters (Yegani and Korver, 2008). The cecal bacteria ferment soluble and resistant fibers, which reach the large intestine and produce short-chain fatty acids (SCFAs) such as acetic acid and butyric acid. The SCFAs are also essential in the physiological function of the chicken and have a role in the metabolism of different drugs and nitrogen (Lee and Hase, 2014; Cully, 2019).

The gut microbiome produces metabolites that have a major impact on the maintenance of intestinal integrity (Hashimoto et al., 2012) and helps in the control of infectious pathogens and managing stress caused by nonpathogens (Kogut, 2019), thus boosting overall immunity against pathogens. When chickens were fed a diet containing growth promoters and phytochemicals, they changed the gut microbiota of the ileum and had an important role in host immunity and gut integrity (Park et al., 2020). Microbiota in poultry also has a role in preventing the attachment of pathogenic bacteria to the intestinal enterocytes and is also helpful in the reduction of colonization of these pathogenic bacteria (Yegani and Korver, 2008).

In poultry, antibiotic growth promoters (AGPs) were previously used to promote growth and prevent infections by restricting the growth of pathogenic bacteria in the gastrointestinal tract. But their use has a major issue of antibiotic residues, which may be absorbed into the bodies of consumers and increase the probability of antibiotic resistance. For this reason, the EU and many other countries have banned the use of AGPs (Prasetyo et al., 2020). In light of these concerns, significant scientific activities have been devoted to the research, development, and application of effective alternatives to AGPs in animal production systems (Gadde et al., 2017). These alternatives include a wide variety of functional feed additives such as probiotics, prebiotics, synbiotics, phytogenic feed additives, enzymes, and essential oils (EOs), which have shown great potential to improve animal health and productivity and, at the same time, to decrease the use of antibiotics (Lillehoj et al., 2018; Nayan et al., 2026).

The main objective of this chapter is to critically review the role of the gut microbiota in the health and performance of poultry, ruminant, and aquatic species, with special focus on the modulatory role of photobiotic-based feed additives on gut microbial homeostasis. In addition, the future perspectives and emerging strategies to improve the efficacy, quality, and availability of these feed additives have also been discussed.

6.2 Gut microbiota: Composition and functions in target species

6.2.1 Poultry

The gut microbiome plays a pivotal role in maintaining health and optimizing productivity in poultry. The composition and density of microbial populations are different in various segments of the gastrointestinal tract, and the highest microbial abundance is found in the cecum (Yan et al., 2017). Poultry gut microbiota is implicated in various key physiological processes such as dietary nutrient digestion and fermentation, gut-associated immune response modulation, growth performance enhancement, and biotransformation/detoxification of mycotoxins that are often found in poultry feed (Śliżewska and Piotrowska, 2014). The first microbiome that colonizes the intestine is coliforms and fecal streptococci, and their number increases on the third day of posthatch. *Lactobacillus* species increase in number at day 40 of posthatch, and they dominate the small intestine (Rychlik, 2020). After 6–7 days, the cecum microbiota develops, and the majority of the population consists of *Clostridia*, enterobacteria, fecal *Streptococci, Pediococci*, and *Pseudomonas aeruginosa* (Donaldson et al., 2017).

The crop mainly consists of a large number of bacterial communities ranging from 1×10^8 to 1×10^9 CFU g^{-1}. The crop is mainly composed of the gram-positive bacteria, which are *Lactobacillus* species (Gong et al., 2007), and other bacteria, which are found in the crop, are *Bifidobacteria, Klebsiella pneumoniae, K. ozaenae, Escherichia coli, E. fergusonii, Enterobacter aerogenes, Eubacterium* spp., *P. aeruginosa, Micrococcus luteus, Staphylococcus lentus*, and *Sarcina* species (Rehman et al., 2007). These bacteria are mainly responsible for the fermentation of the soluble dietary fibers into SCFAs, in which acetate is the main product formed and helps in reducing pH that inhibits the growth of the pathogens (Rehman et al., 2008).

6.2.2 Ruminants

The gut microbiota of ruminants is more complex and is helpful in the digestion of the feed, mainly composed of materials of plant origin. Bacterial species such as *Firmicutes* and *Bacteroides* help make fat in the milk (O'Hara et al., 2020). In the rumen, the ecosystem is very diverse and mainly composed of bacteria, viruses, archaea, bacteriophage, fungi, protozoa, and various physiological and phylogenetic differences occur in them (Clemmons et al., 2019). The rumen microbiome also contains enzymes for the digestion of complex carbohydrates. Among these, the most abundant microbiome is bacteria and consists of 40%–90% of the microbiome (Huws et al., 2018). The microbiome helps in the conversion of carbohydrates into volatile fatty acids (VFAs) (Newbold and Ramos-Morales, 2020). *Butyrivibrio* and *Pseudobutyrivibrio* are two dominant genera within *Bacillota* that are known to produce butyrate and play a major role in plant degradation because of a cascade of carbohydrate-active enzymes (Pidcock et al., 2021). Ciliate protozoa, the largest microorganisms in the rumen, occupy up to 25% of rumen microbial biomass (Nagaraja, 2016; Firkins et al., 2020).

About 2%–4% of the rumen microbiome is archaea, 90% of the population is methanogens, and more than 10 species of archaea are methane formers (Nawaz et al., 2018). Methanogens are mainly designed to regulate rumen fermentation and help to prevent the accumulation of excess H_2, which is generated during fermentation by bacteria, fungi, and protozoa, by reducing CO_2 to form CH_4 (Krehbiel, 2014; Huws et al., 2018).

6.2.3 Aquatic animals

Gut microbiota in the finfish produce metabolites such as SCFAs, including acetate, propionate, and butyrate. These products are helpful in the physiology of the digestive tract, maintain gut health, and improve gut immunity (Nawaz et al., 2018). In finfish, complex interaction between fish and microbiome results in the maintenance of the intestinal immunity (Matamoros et al., 2013). The gut microbiome produces a variety of bioactive compounds, such as organic acids, antimicrobial salts, and secondary bile acids that together lead to the suppression and exclusion of pathogenic microorganisms from the gastrointestinal tract (Li et al., 2014; Ringø et al., 2018).

Shrimp are one of the most economically valuable species in aquaculture systems around the world (Huang et al., 2020). In shrimp, the gut microbiome is important not only for the regulation of growth, development, and general health, but also the disruption of microbial homeostasis (dysbiosis) has been closely linked to the development of white feces syndrome and other enteric disorders (Wang et al., 2020). The shrimp gut microbiota produces a wide variety of secondary metabolites, such as SCFAs, polyamines, and aryl hydrocarbon receptor-activating ligands. These microbial metabolites interact with the intestinal epithelium and promote innate immune responses and are important for reducing disease susceptibility in shrimp populations (Adak and Khan, 2019).

6.3 Phytofeed additives: Sources, classes, and bioactives

Phytofeed additives are natural supplements from plants that are added to animal feed. They contain special compounds (e.g., EOs, polyphenols, tannins, and saponins) that do more than just provide nutrition. These compounds can support digestion, improve gut microbes, and strengthen the animal's immune system (Murugesan et al., 2015). The phytogenic feed additives are mainly derived from plants, spices, and herbs. Phytogenic feeds are now used as alternatives to AGPs

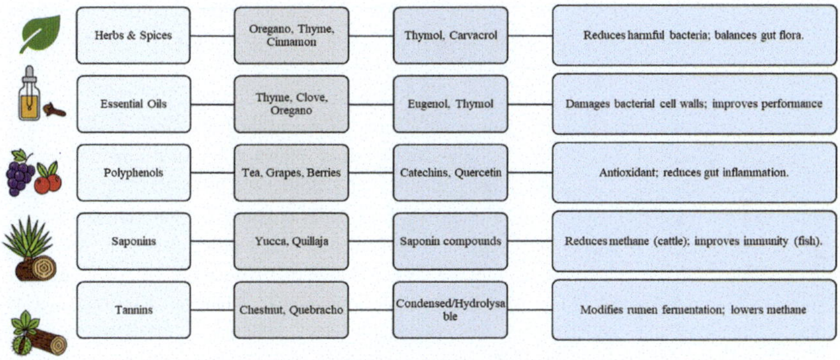

FIGURE 6.1 Classification and systemic benefits of different phytofeed additives on gut health.

and help in strengthening the gut barrier and improving the ability of the animals to defend themselves against bacterial and viral infection (Alagawany et al., 2020).

6.3.1 Main categories

Fig. 6.1 summarizes the main categories of the phytogenic feed additives.

- **Herbs and spices:** Plants such as oregano, thyme, cinnamon, and ginger provide compounds such as thymol and carvacrol, which are known to reduce harmful bacteria and help in balancing the gut microflora (Du et al., 2016; Abdelli et al., 2021).
- **Essential oils:** Oils from plants such as thyme, oregano, and clove contain active ingredients such as thymol, carvacrol, and eugenol (Mostafa et al., 2025). These compounds can damage the cell walls of harmful bacteria, support healthy gut microbes, and improve performance in poultry and fish (Dawood et al., 2022; Oni and Oke, 2025).
- **Polyphenols:** Found in tea, grapes, and berries; compounds such as catechins, resveratrol, and quercetin act as antioxidants. These compounds have the ability to influence gut health and play a significant role in the reduction of inflammation in poultry and fish (Iqbal et al., 2020).
- **Saponins:** Sourced from yucca and *Quillaja saponaria*; these compounds can reduce methane and change fermentation in the rumen of cattle, while in shrimp and fish they improve gut health and disease resistance (Dawood et al., 2022; Blue et al., 2023).
- **Tannins:** Both condensed and hydrolyzable tannins come from plants such as quebracho and chestnut. They can change rumen fermentation, lower methane, and affect gut microbes, but the results also depend on type and the plant source (Besharati et al., 2022; Fonseca et al., 2023; Klein et al., 2025) (Table 6.1).

TABLE 6.1 Examples of phytofeed additives, their sources, main compounds, and effects on animal microbiota.

Additives	Source plant	Main compounds	Effects on microbiota and animals	References
Oregano/thyme (herbs/spices)	*Origanum vulgare*, *Thymus vulgaris*	Thymol, carvacrol	• In broiler chickens, blends with thymol/carvacrol reduced *Clostridium perfringens* and improved gut structure • Microbiota shifted toward more beneficial bacteria	Du et al. (2016); Abdelli et al. (2021)
Cinnamon/ginger (herbs/spices)	*Cinnamomum* spp., *Zingiber officinale*	Cinnamaldehyde, gingerols	Used in poultry diets to support gut health and increase beneficial bacteria	Abdelli et al. (2021)
Essential oils	Thyme, oregano, clove	Thymol, carvacrol, eugenol	• These oils disrupt harmful bacteria and improve gut health in poultry • In fish, they show antibacterial and immune-boosting properties	Dawood et al. (2022); Oni and Oke (2025)

(*continued on next page*)

Polyphenols	Grapes, tea, berries	Resveratrol, catechins, quercetin	• In koi carp, polyphenols improved gut balance and health • In poultry, they act as antioxidants and support gut microbes	Iqbal et al. (2020)
Saponins (ruminants)	*Yucca schidigera, Quillaja saponaria*	Steroidal/triterpenoid saponins	In cattle, saponins reduce rumen protozoa, lower methane, and change fermentation	Yanza et al. (2024)
Saponins (aquatic)	*Q. saponaria*	Quillaja saponins	In shrimp and fish, they reduce harmful bacteria, improve growth, and support gut structure	Dawood et al. (2022); Huang et al. (2022); Blue et al. (2023)
Tannins (ruminants)	Quebracho, chestnut	Condensed and hydrolysable tannins	Can reduce methane and influence gut microbes, though effects depend on dose and type	Besharati et al. (2022); Klein et al. (2025)

6.3.2 Delivery forms

These additives can be used in different forms: As simple powders made from dried plants, as concentrated extracts, or as encapsulated products (that protect the compounds until they reach the target sites). The effects often depend on the animal species, the amount used, and how the additive is prepared. Encapsulation is one way to make these compounds more stable and effective (Nantapo and Marume, 2025).

6.4 Mechanisms of gut microbiota modulation by phytofeed

Plant-based feed additives, often called phytofeeds, can shape the gut microbes of animals in many ways. They help reduce pathogens, support the growth of beneficial bacteria, break down the communication systems of pathogens, lower gut inflammation, and change fermentation patterns. All of these actions improve animal health and performance (Oladokun et al., 2022).

6.4.1 Antimicrobial effects

Certain plant compounds such as EOs, tannins, and saponins can directly target harmful microbes while leaving most of the gut community untouched. In poultry, EO mixtures have been shown to lower *Salmonella* and *E. coli*, leading to healthier intestines (Ding et al., 2022). Feed supplements obtained from plants have been shown to reduce *Vibrio* infections and enhance survival rates in shrimp and fish (Lee et al., 2024). In ruminant production, certain phytochemicals can inhibit rumen protozoal populations, which indirectly inhibit the population of methanogenic archaea, thus leading to lower emissions of enteric methane (Ku-Vera et al., 2020).

6.4.2 Prebiotic-like stimulation

The phytogenic feed additives can also have prebiotic-like effects by selectively increasing the growth of the beneficial gut microbes. These compounds stimulate the growth of *Lactobacillus, Bifidobacterium, Ruminococcus*, and *Bacillus* species, which are key in feed digestion, nutrient uptake, and immune stimulation. The studies carried out on broilers and shrimp show that phytogenic supplementation increases the count of probiotic bacteria, improves microbial balance, and improves overall gut health (Oladokun et al., 2022; Lee et al., 2024).

6.4.3 Quorum-sensing disruption

Quorum sensing is the means of bacterial communication that allows coordination of their behaviors, such as biofilm formation, which contributes to their resistance to antimicrobial interventions and their persistence in animals. Many bioactive compounds of plant origin can disrupt the quorum-sensing signaling pathways, thus preventing the formation of biofilms. Disruption of these signaling systems makes pathogenic bacteria more vulnerable to host defense and environmental control mechanisms (Samrot et al., 2021; Gonzales et al., 2024).

6.4.4 Effects against acidity and inflammation

Several phytofeed components have great antioxidant and antiinflammatory properties that help to maintain intestinal homeostasis. By decreasing oxidative stress and inflammation, these compounds promote the maintenance of the intestinal epithelial integrity as well as provide a positive environment for beneficial microbiota. Enhanced intestinal barrier functionality increases the absorption of nutrients and reduces the predisposition to enteric infections. As an example, the use of oregano EO as supplementation has also been reported to reduce intestinal inflammation and improve the performance of epithelial barriers in sheep (Jia et al., 2022; Oladokun et al., 2022).

6.4.5 Fermentation modulation

6.4.5.1 Ruminants

In ruminants, phytofeed compounds may also be used to regulate the rumen fermentation pathway of VFA synthesis to propionate at the expense of methane formation. This metabolic shift promotes the efficiency of energy use by the host, and at the same time, it minimizes greenhouse gas emissions (Ku-Vera et al., 2020).

6.4.5.2 Poultry and aquaculture

In poultry, fish, and shrimp, phytogenic feed additives have been shown to increase the production of SCFAs, specifically acetate and butyrate. These SCFAs are direct energy sources of intestinal epithelial cells, strengthen gut barrier functions, and have antiinflammatory effects, all of which lead to a better intestinal health system and performance (Oladokun et al., 2022; Lee et al., 2024) (Fig. 6.2).

6.5 Implications for health, performance, and sustainability

The use of plant-based feed supplements has significant benefits in relation to animal health, efficiency of production, and environmental sustainability. These bioactive compounds increase the digestion and absorption of nutrients, thus lowering production cost. Improved gut health also leads to a reduced reliance on antibiotics, which is a key issue in light of the worldwide concern about antimicrobial resistance and antibiotic residue in food products of animal origin (Prasetyo et al., 2020). Animals that are fed phytogenic supplements

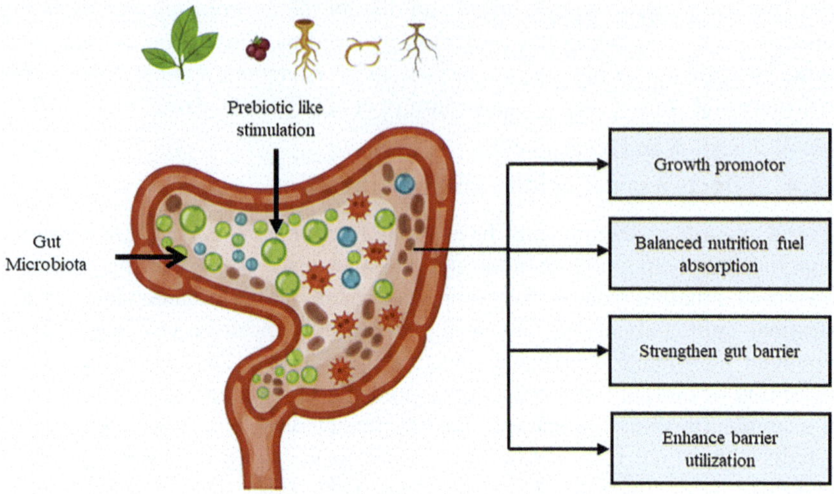

FIGURE 6.2 Mechanisms of gut microbiota modulation by phytofeed.

generally show improved immune competence, which leads to less incidence of disease, better welfare, and higher survival rate. Concurrently, product quality is positively influenced, as can be observed by increased oxidative stability of poultry meat, improved fatty acid profiles in dairy milk, and firmer, nutritionally superior flesh in fish (El-Essawy et al., 2019; Shourbela et al., 2021). Apart from the production and health benefits, phytogenic feed additives also contribute to environmental sustainability. In ruminants, certain compounds of plant origin are very effective at decreasing emissions of enteric methane, one of the main greenhouse gases linked with the production of livestock (Cremonesi et al., 2025). In aquaculture systems, these additives work by inhibiting pathogenic microbial loads in rearing water to improve biosecurity and lead to more sustainable aquaculture production practices. Collectively, these effects coincide with the increasing world demand for safe, efficient, and environmentally responsible animal production systems (Oladokun et al., 2022).

6.5.1 Poultry

Oregano EO is one of the most widely researched phytogenic additives used in poultry nutrition. Its main bioactive constituents, carvacrol and thymol, selectively enhance the growth of beneficial gut bacteria such as *Lactobacillus* spp; promoting the digestive efficiency, gut integrity, and feed utilization. These improvements have been linked to increased egg quality and general performance in laying hens (Feng et al., 2021; Gao et al., 2022).

Green tea extract is another potentially important phytogenic additive because of its high polyphenol content (especially catechins). Supplementation with green tea extract has been shown to inhibit pathogenic bacteria, including

Clostridium perfringens, the major etiological agent of necrotic enteritis. Higher inclusion levels have been linked to better gut health and less mortality in poultry (Diaz Carrasco et al., 2016).

Saponins from plants such as soybean and *Yucca schidigera* also enhance poultry health by modulating cecal fermentation and stimulating beneficial microbes' populations. In addition, saponins provide better immune responses and lipid metabolism, leading to a healthier condition and better productivity of laying hens and broilers. These effects overall diminish the reliance on AGPs and promote sustainable poultry production systems (Li et al., 2022; Dai et al., 2023).

6.5.2 Ruminants

In the case of ruminants, the rumen is home to a very complex and dynamic microbial ecosystem, which plays a central role in feed fermentation. Plant-derived tannins have been found to influence the rumen microbe population by shifting the utilization of hydrogen into propionate production instead of methane production. This metabolic shift decreases methane emissions, while at the same time increasing the amount of glucose available to the host animal, leading to better energy efficiency (Cremonesi et al., 2025). Additional studies have shown that tannins and tea polyphenols can partially inhibit the growth of methanogenic archaea without impairing the degradation of the fiber or the general efficiency of the fermentation process. Consequently, ruminant growth performance is preserved, and environmental emissions are greatly decreased (Zhao et al., 2025).

Saponins are also important in rumen modulation by lowering protozoal populations, and these are closely linked to the methane-producing archaea. Lower protozoal abundance results in lower methane production and better fermentation efficiency (Anantasook et al., 2013). EOs, such as clove oil, further increase ruminant productivity by increasing nutrient utilization and the quality of the meat. Lambs supplemented with clove oil have shown enhanced digestibility and beneficial fatty acid profiles in muscle tissue, and thus the dual positive effects of phytogenic additives on performance and nutritional value of the product (El-Essawy et al., 2019).

6.5.3 Aquatic animals

In aquaculture, gut health is of special importance because of the constant contact of fish and shrimp with microorganisms in the water. Garlic extract has shown good antimicrobial and immunomodulatory activity and protection against bacterial infections, including *Streptococcus iniae* in tilapia. Supplementation makes the gut more abundant in lactic acid bacteria that compete with pathogenic microbes and boost host immune responses (Foysal et al., 2019).

Oregano EO has also been shown to have great efficacy in aquaculture species. In *Nile tilapia*, dietary supplementation has been found to be related to improved growth performance, hematological parameters, antioxidant capacity, and immune function. These benefits result in improved resistance to stresses and diseases, which makes oregano oil especially valuable in intensive aquaculture production systems (Shourbela et al., 2021). Polyphenols also play a role in shrimp farming. They improve gut health, strengthen immunity, and reduce harmful bacteria, which lowers disease risk and improves survival rates (Iqbal et al., 2020). This reduces the need for antibiotics and makes aquaculture products safer for consumers.

6.5.4 Species-specific evidence from in vivo studies

Poultry—oregano oil improving *Lactobacillus* spp.; green tea extract reducing *C. perfringens*; saponins improving cecal fermentation profile. **Ruminants**—tannins shifting rumen microbiota to reduce methane; saponins reducing protozoal populations; EOs improving propionate-to-acetate ratio. **Aquatic animals**—garlic extract improving gut lactic acid bacteria in tilapia; oregano oil reducing *Aeromonas hydrophila* in carp; polyphenols enhancing disease resistance in shrimp (Table 6.2).

6.6 Challenges and practical considerations

6.6.1 Variability in bioactive composition

Because of the complex nature of the feed additives, it is difficult to perform comprehensive studies on animals (Yang et al., 2015). Sometimes the obtained results are not accurate because their bioactive compounds depend upon the plant, the origin where these plants grow, their storage conditions, environmental factors, and their effective dose (Windisch et al., 2008). Many feed additives are EOs, which are volatile in nature and may affect the result. In most cases, EOs may react with other compounds; therefore their effectiveness may vary. Many changes may happen to them during processing, storage, and transportation. Therefore to keep them stable is the major target (Stevanovic et al., 2018).

6.6.2 Standardization and dose optimization

Some feed additives may be absorbed before they reach the gut; therefore the desired outcome is not achieved, and their dosage needs to be adjusted. Poor solubility is another issue related to these feed additives. Fat-soluble components of these compounds are hardly absorbed in water, which reduces their availability. Because of this reason, their dosages decrease to a suboptimal level and do not produce the required results (Alagawany et al., 2020) (Fig. 6.3).

TABLE 6.2 Summary of representative studies: species, phytofeed type, and associated performance/health outcomes.

Poultry			
Phytofeed name	**Dosage**	**Effect on gut microbiome**	**References**
oregano oil	Five different treatments **1.** Simple basal diet as a control group **2.** Basal diet along with 5mg/kg flavomycin, which is used as an AGP **3.** Basal diet with 100mg/kg of oregano oil and 20mg/kg cinnamaldehyde **4.** Basal diet with 200mg/kg of oregano oil and 20mg/kg cinnamaldehyde **5.** Basal diet with 300mg/kg of oregano oil and 20mg/kg cinnamaldehyde	Increase the relative abundance of *Lactobacillus in* the group that contains a high amount of oregano oil	Gao et al. (2022)

(continued on next page)

TABLE 6.2 Summary of representative studies: species, phytofeed type, and associated performance/health outcomes—cont'd

Poultry			
Phytofeed name	**Dosage**	**Effect on gut microbiome**	**References**
Oregano oil	Four treatments **1.** Basal diet **2.** Basal diet supplemented with 100mg/kg oregano oil **3.** Basal diet supplemented with 200mg/kg oregano oil **4.** Basal diet supplemented with 400mg/kg oregano oil	At the genus level, the *Lactobacillus* accounted for the greatest proportion of the microbiota in the ileum after using oregano oil	Feng et al. (2021)
Green tea extract (GTE)	Four treatments **1.** No GTE acts as a control **2.** 250 mg GTE/kg diet **3.** Containing a 500 mg GTE/kg diet **4.** Containing a 1000 mg GTE/kg diet	Reduction in the intestinal *Clostridium Perfringens* infection, containing high amount containing a 1000 mg GTE/kg diet	Diaz Carrasco et al. (2016)

(*continued on next page*)

Soya saponin	Three treatments **1.** Basal diet **2.** Low saponin diet 50mg/kg **3.** High saponin diet 500mg/kg	• Improvement of the immune function • Lipid metabolism • Improvement of the intestinal microflora by using high saponin diet	Li et al. (2022)
Yucca saponin (YSa)	Four treatments **1.** Basal diet acts as a control **2.** YSa group (basic diet mixed with 500 mg/kg YSa) **3.** YS group (basic diet mixed with 500 mg/kg Yucca schidigera) **4.** QS group (basic diet mixed with 500 mg/kg *Q. saponaria*)	• Study found that YS is more suitable as a feed additive that ameliorates the healthy growth of broilers • Reduction in the level of IL-1β appeared in the serum of broilers using YSa treatment	Dai et al. (2023)

(*continued on next page*)

TABLE 6.2 Summary of representative studies: species, phytofeed type, and associated performance/health outcomes—cont'd

Ruminants			
Phytofeed name	**Dosage**	**Effect on gut microbiome**	**References**
Tannins	Condensed tannins at 0%, 2%, 4%, and 6% of the dry matter	• *Prevotellacae* and *Succinilactobacillus* are tolerant to condensed tannins and lower methane production • Increase in the *Firmicutes* to *Bacteroides* ratio	Cremonesi et al. (2025)

(continued on next page)

Tannins	Four treatments **1.** Basal diet acts as a control **2.** Basal diet combined with 0.4% of the tannic acids **3.** Basal diet combined with 0.4% of the tea polyphenols **4.** Basal diet + 0.2% DM tannic acid + 0.2% DM tea polyphenols	• Reduction in the methane emissions through targeted modulation of the microbiota populations in rumen	Zhao et al. (2025)
Saponins	Crude tannins and saponins at 84 and 143 g/kg	The population of protozoa and methanogens decreased	Anantasook et al. (2013)
Essential Oils (EO)	Four groups **1.** Basal diet acts as a control **2.** The basal diet plus 2 mL/day/lamb of anise EO (Anise group), **3.** 2 mL/day /lamb clove EO (Clove group) **4.** 2 mL/day/lamb thyme EO (Thyme group)	Clove EO has a considerable positive effect on nutrient digestibility and improves carcass characteristics, the meat fatty acids profile	El-Essawy et al. (2019)

(*continued on next page*)

TABLE 6.2 Summary of representative studies: species, phytofeed type, and associated performance/health outcomes—cont'd

Phytofeed name	Dosage	Effect	References
Aquatic animals			
Garlic extract	1 g garlic powder	• Recovery against *Streptococcus* infection in tilapia • *Proteobacteria* and *Tenericutes* abundance increased	Foysal et al. (2019)
Oregano essential oil	Oregano essential oil (OEO) levels at three different concentrations (0, 1, or 2 mL/kg) in their diets for 12 weeks	The introduction of dietary oregano essential oils improved growth, improved blood health, increased antioxidative capacity, and improved immune status of *Nile tilapia*	Shourbela et al. (2021)

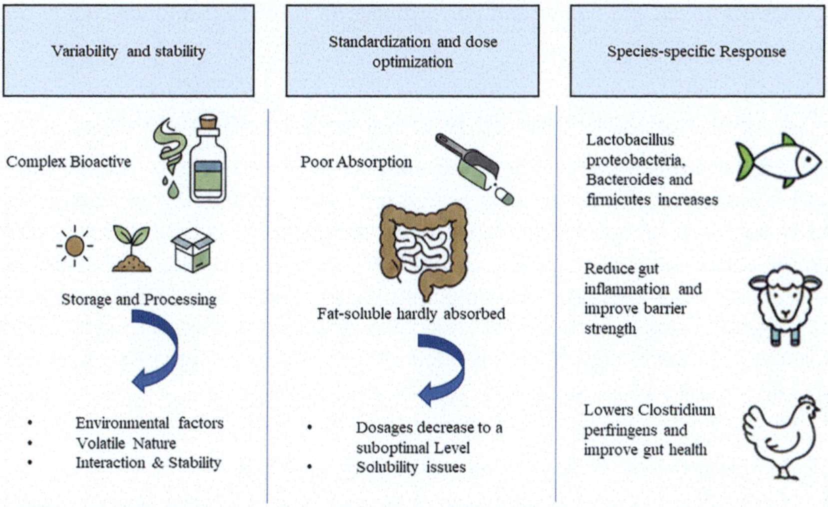

FIGURE 6.3 Phytofeed additives: Challenges and considerations.

6.6.3 Species-specific differences in microbiota response

A study conducted on fish provided methanotrophs in the diet, and observed that the relative abundance of the *Lactobacillus* species increased. Relative abundance of the *Proteobacteria, Bacteroides*, and *Firmicutes* also increased up to 70% (Yu et al., 2023). Oregano EO used as a feed additive is useful in reducing gut inflammation and improving barrier strength in the sheep population (Jia et al., 2022; Oladokun et al., 2022). Green tea extract is another effective option. Rich in polyphenols, especially catechins, lower harmful bacteria such as *C. perfringens*, the causative agent of necrotic enteritis in poultry. Birds given higher doses of green tea extract showed better gut health and lower mortality rates (Diaz Carrasco et al., 2016). So, species-specific formulations of phytofeeds should be considered, which often presents a great challenge.

6.7 Future perspectives

6.7.1 Synergistic blends of phytofeed with probiotics/prebiotics

The trend of synergistic feed additives is gaining popularity. Synergistic feed additives are mainly designed to have a combined effect rather than their individual effect, and prove to have a beneficial role (Shipradeep et al., 2012). The synergistic feed requires a combination of different feed additives such as prebiotics, probiotics, organic acids, enzymes, phyobiotics, bacteriophages, and postbiotics (De Vrese and Schrezenmeir, 2008). There are several reports that demonstrate the synergetic potential of organic acids combined with phytobiotics. Synergistic feed containing *E. faecium* strain and a commercial phytobiotic product improved gut health and performance in broilers (Smulikowska et al., 2010).

6.7.2 Nanoencapsulation for targeted delivery and stability

Many different technologies are now used for feed additives to protect them and increase their effectiveness by preventing oxidation and degradation during processing and storage. These technologies include microencapsulation, which mainly consists of polymer-based methods, such as polysaccharide–protein carriers, and other methods, including lipid-based compounds such as liposomes and vegetable oils. The basic function of nanoencapsulation is to protect the bioactive substances and effectively release these substances into their target site, especially the lower gut in animals (de Vos et al., 2010).

6.7.3 Integration with precision feeding systems

Integrated poultry operations often operate their own feed mills, supplying farms with feed tailored to various production stages. The feeds developed by these operations use computerized least-cost formulation methods that adjust ingredients based on fluctuating commodity prices. This approach helps to ensure feed efficiency and consistent nutritional quality. But it is important to maintain a balance between economic efficiency and maintaining bird health. The effective use of these feed additives in such operations can be helpful for boosting performance in animals.

6.8 Role in climate-smart livestock and aquaculture

Feed additives help maintain gut health by preventing inflammation and improving gut health. They remove harmful microbes from their gut. In ruminants, feed additives such as clove EO have a considerable positive effect on nutrient digestibility and improve carcass characteristics, meat, and fatty acids profile (El-Essawy et al., 2019). When fish are fed with oregano oil, they show stronger immunity and better stress resistance, which in turn is useful for their intensive production systems (Shourbela et al., 2021). Currently, a wide range of phytofeed additives are available, which are used not only for healthy growth but they also improve disease resistance (Calanche et al., 2020).

6.9 Conclusion

There is an increasing demand for livestock and poultry products due to an enhanced global population, urbanization, and higher economic status. To meet this growing demand, there is a major pressure on the livestock industry, which can be met either by increasing the number of animals or by boosting their productivity. As there are more animals, there is a greater demand for animal feed. Phytofeed additives improve the gut health of poultry, ruminants, and aquatic populations. These additives either modify the gut microbiome population and improve gut barrier. By using these feed additives, livestock production can become more profitable and sustainable, with reduced environmental impact and significantly lower greenhouse gas emissions that contribute to global warming and climate change.

References

Abdelli, N., Solà-Oriol, D., Francisco Pérez, J., 2021. Phytogenic feed additives in poultry: achievements, prospective and challenges. Animals 11 (12), 3471. doi:10.3390/ani11123471.

Adak, A., Khan, M.R., 2019. An insight into gut microbiota and its functionalities. Cell. Mol. Life Sci. 76 (3), 473–493. doi:10.1007/s00018-018-2943-4.

Alagawany, M., Elnesr, S.S., Farag, M.R., Tiwari, R., Yatoo, M.I., Karthik, K., Michalak, I., Dhama, K., 2020. Nutritional significance of amino acids, vitamins and minerals as nutraceuticals in poultry production and health–a comprehensive review. Vet. Q. 41 (1), 1–29. doi:10.1080/01652176.2020.1857887.

Anantasook, N., Wanapat, M., Cherdthong, A., Gunun, P., 2013. Changes of microbial population in the rumen of dairy steers as influenced by plant containing tannins and saponins and roughage to concentrate ratio. Asian-Australas. J. Anim. Sci. 26 (11), 1583–1591. doi:10.5713/ajas.2013.13182.

Aruwa, C.E., Pillay, C., Nyaga, M.M., Sabiu, S., 2021. Poultry gut health – microbiome functions, environmental impacts, microbiome engineering and advancements in characterization technologies. J. Anim. Sci. Biotechnol. 12 (1). doi:10.1186/s40104-021-00640-9.

Besharati, M., Maggiolino, A., Palangi, V., Kaya, A., Jabbar, M., Eseceli, H., De Palo, P., Lorenzo, J.M., 2022. Tannin in ruminant nutrition: review. Molecules 27 (23). doi:10.3390/molecules27238273.

Blue, C.E.C., Emami, N.K., White, M.B., Cantley, S., Dalloul, R.A., 2023. Inclusion of Quillaja saponin clarity Q manages growth performance, immune response, and nutrient transport of broilers during subclinical necrotic enteritis. Microorganisms 11 (8). doi:10.3390/microorganisms11081894.

Byrd, K.A., Thilsted, S.H., Fiorella, K.J., 2021. Fish nutrient composition: a review of global data from poorly assessed inland and marine species. Pub. Health Nutr. 24 (3), 476–486. doi:10.1017/S1368980020003857.

Calanche, J.B., Beltrán, J.A., Hernández Arias, A.J., 2020. Aquaculture and sensometrics: the need to evaluate sensory attributes and the consumers' preferences. Rev. Aquac. 12 (2), 805–821. doi:10.1111/raq.12351.

Clemmons, B.A., Voy, B.H., Myer, P.R., 2019. Altering the gut microbiome of cattle: considerations of host-microbiome interactions for persistent microbiome manipulation. Microb. Ecol. 77 (2), 523–536. doi:10.1007/s00248-018-1234-9.

Cremonesi, P., Severgnini, M., Battelli, M., Monistero, V., Penati, M., Gazzonis, A.L., Castiglioni, B., Rapetti, L., Manfredi, M.T., Addis, M.F., 2025. Reducing enteric methane emission in dairy goats: impact of dietary inclusions of quebracho tannin extract on ruminal microbiota. Front. Microbiol. 16. doi:10.3389/fmicb.2025.1595924.

Cully, M., 2019. Microbiome therapeutics go small molecule. Nat. Rev. Drug Discov. 18 (8), 569–572. doi:10.1038/d41573-019-00122-8.

Dai, Z., Wang, H., Liu, J., Zhang, H., Li, Q., Yu, X., Zhang, R., Yang, C., 2023. Comparison of the effects of Yucca saponin, Yucca schidigera, and Quillaja saponaria on growth performance, immunity, antioxidant capability, and intestinal flora in broilers. Animals 13 (9), 1447. doi:10.3390/ani13091447.

Dawood, M.A.O., El Basuini, M.F., Yilmaz, S., Abdel-Latif, H.M.R., Alagawany, M., Kari, Z.A., Razab, M.K.A.A., Hamid, N.K.A., Moonmanee, T., Doan, H.V., 2022. Exploring the roles of dietary herbal essential oils in aquaculture: a review. Animals 12 (7). doi:10.3390/ani12070823.

De Vrese, M., Schrezenmeir, J., 2008. Probiotics, prebiotics, and synbiotics. Adv. Biochem. Eng. Biotechnol. 111, 1–66. doi:10.1007/10_2008_097.

de Vos, P., Faas, M.M., Spasojevic, M., Sikkema, J., 2010. Encapsulation for preservation of functionality and targeted delivery of bioactive food components. Int. Dairy J. 20 (4), 292–302. doi:10.1016/j.idairyj.2009.11.008.

Diaz Carrasco, J.M., Redondo, L.M., Redondo, E.A., Dominguez, J.E., Chacana, A.P., Fernandez Miyakawa, M.E., 2016. Use of plant extracts as an effective manner to control Clostridium perfringens induced necrotic enteritis in poultry. Biomed. Res. Int. 2016, 1–15. doi:10.1155/2016/3278359.

Ding, Y., Hu, Y.i, Yao, X., He, Y., Chen, J., Wu, J., Wu, S., Zhang, H., He, X.i, Song, Z., 2022. Dietary essential oils improves the growth performance, antioxidant properties and intestinal permeability by inhibiting bacterial proliferation, and altering the gut microbiota of yellow-feather broilers. Poult. Sci. 101 (11), 102087. doi:10.1016/j.psj.2022.102087.

Donaldson, E.E., Stanley, D., Hughes, R.J., Moore, R.J., 2017. The time-course of broiler intestinal microbiota development after administration of cecal contents to incubating eggs. PeerJ 2017 (7), e3587. doi:10.7717/peerj.3587.

Du, E., Wang, W., Gan, L., Li, Z., Guo, S., Guo, Y., 2016. Effects of thymol and carvacrol supplementation on intestinal integrity and immune responses of broiler chickens challenged with Clostridium perfringens. J. Anim. Sci. Biotechnol. 7 (1), 19. doi:10.1186/s40104-016-0079-7.

Ehrlich, P.R., Harte, J., 2015. Opinion: to feed the world in 2050 will require a global revolution. Proc. Nat. Acad. Sci. U.S.A. 112 (48), 14743–14744. doi:10.1073/pnas.1519841112.

El-Essawy, A., Abdou, A., Khattab, I., Abdel-Wahed, A., 2019. Effect of addition of anise, clove and thyme essential oils on Barki lambs performance, digestibility, rumen fermentation, carcass characteristics and intramuscular fatty acids. Egypt. J. Nutr. Feeds 22 (3), 465–477. doi:10.21608/ejnf.2019.79415.

Feng, J., Lu, M., Wang, J., Zhang, H., Qiu, K., Qi, G., Wu, S., 2021. Dietary oregano essential oil supplementation improves intestinal functions and alters gut microbiota in late-phase laying hens. J. Anim. Sci. Biotechnol. 12 (1). doi:10.1186/s40104-021-00600-3.

Firkins, J.L., Yu, Z., Park, T., Plank, J.E., 2020. Extending Burk Dehority's perspectives on the role of ciliate protozoa in the rumen. Front. Microbiol. 11. doi:10.3389/fmicb.2020.00123.

Fonseca, N.V.B., Cardoso, A.S., Bahia, A.S.R.d.S., Messana, J.D., Vicente, E.F., Reis, R.A., 2023. Additive tannins in ruminant nutrition: an alternative to achieve sustainability in animal production. Sustainability (Switzerland) 15 (5). doi:10.3390/su15054162.

Foysal, M.J., Alam, M., Momtaz, F., Chaklader, M.R., Siddik, M.A.B., Cole, A., Fotedar, R., Rahman, M.M., 2019. Dietary supplementation of garlic (Allium sativum) modulates gut microbiota and health status of tilapia (Oreochromis niloticus) against Streptococcus iniae infection. Aquac. Res. 50 (8), 2107–2116. doi:10.1111/are.14088.

Gadde, U., Kim, W.H., Oh, S.T., Lillehoj, H.S., 2017. Alternatives to antibiotics for maximizing growth performance and feed efficiency in poultry: a review. Anim. Health Res. Rev. 18 (1), 26–45. doi:10.1017/S1466252316000207.

Gao, F., Zhang, L., Li, H., Xia, F., Bai, H., Piao, X., Sun, Z., Cui, H., Shi, L., 2022. Dietary oregano essential oil supplementation influences production performance and gut microbiota in late-phase laying hens fed wheat-based diets. Animals 12 (21), 3007. doi:10.3390/ani12213007.

Gong, J., Si, W., Forster, R.J., Huang, R., Yu, H., Yin, Y., Yang, C., Han, Y., 2007. 16S rRNA gene-based analysis of mucosa-associated bacterial community and phylogeny in the chicken gastrointestinal tracts: from crops to ceca. FEMS Microbiol. Ecol. 59 (1), 147–157. doi:10.1111/j.1574-6941.2006.00193.x.

Gonzales, M., Kergaravat, B., Jacquet, P., Billot, R., Grizard, D., Chabrière, É., Plener, L., Daude, D., 2024. Disrupting quorum sensing as a strategy to inhibit bacterial virulence in human, animal, and plant pathogens. Pathog. Dis. 82. doi:10.1093/femspd/ftae009.

Hashimoto, T., Perlot, T., Rehman, A., Trichereau, J., Ishiguro, H., Paolino, M., Sigl, V., Hanada, T., Hanada, R., Lipinski, S., Wild, B., Camargo, S.M.R., Singer, D., Richter, A., Kuba, K., Fukamizu, A., Schreiber, S., Clevers, H., Verrey, F., Rosenstiel, P., Penninger, J.M., 2012. ACE2 links amino acid malnutrition to microbial ecology and intestinal inflammation. Nature 487 (7408), 477–481. doi:10.1038/nature11228.

Huang, K.C., Lee, J.W., Shiu, Y.L., Ballantyne, R., Liu, C.H., 2022. Micro-aid liquid 10 promotes growth performance and health status of white shrimp, Litopenaeus vannamei. J. Mar. Sci. Eng. 10 (1). doi:10.3390/jmse10010049.

Huang, Z., Zeng, S., Xiong, J., Hou, D., Zhou, R., Xing, C., Wei, D., Deng, X., Yu, L., Wang, H., Deng, Z., Weng, S., Kriengkrai, S., Ning, D., Zhou, J., He, J., 2020. Microecological Koch's postulates reveal that intestinal microbiota dysbiosis contributes to shrimp white feces syndrome. Microbiome 8 (1). doi:10.1186/s40168-020-00802-3.

Huws, S.A., Creevey, C.J., Oyama, L.B., Mizrahi, I., Denman, S.E., Popova, M., Muñoz-Tamayo, R., Forano, E., Waters, S.M., Hess, M., Tapio, I., Smidt, H., Krizsan, S.J., Yáñez-Ruiz, D.R., Belanche, A., Guan, L., Gruninger, R.J., McAllister, T.A., Newbold, C.J., Roehe, R., Dewhurst, R.J., Snelling, T.J., Watson, M., Suen, G., Hart, E.H., Kingston-Smith, A.H., Scollan, N.D., Do Prado, R.M., Pilau, E.J., Mantovani, H.C., Attwood, G.T., Edwards, J.E., McEwan, N.R., Morrisson, S., Mayorga, O.L., Elliott, C., Morgavi, D.P., 2018. Addressing global ruminant agricultural challenges through understanding the rumen microbiome: Past, present, and future. Front. Microbiol. 9. doi:10.3389/fmicb.2018.02161.

Iqbal, Y., Cottrell, J.J., Suleria, H.A.R., Dunshea, F.R., 2020. Gut microbiota-polyphenol interactions in chicken: a review. Animals 10 (8), 1–18. doi:10.3390/ani10081391.

Jia, L.i, Wu, J., Lei, Y.u, Kong, F., Zhang, R., Sun, J., Wang, L., Li, Z., Shi, J., Wang, Y., Wei, Y., Zhang, K.e, Lei, Z., 2022. Oregano essential oils mediated intestinal microbiota and metabolites and improved growth performance and intestinal barrier function in sheep. Front. Immunol. 13. doi:10.3389/fimmu.2022.908015.

Klein, M.L., Erikson, C.B., McCabe, C.J., Huang, L., Rodrigues, J.L.M., Mitloehner, F.M., 2025. Limited effects of tannin supplementation on the dairy cattle fecal microbiome with modulation of metabolites. Front. Microbiol. 16. doi:10.3389/fmicb.2025.1570127.

Kogut, M.H., 2019. The effect of microbiome modulation on the intestinal health of poultry. Anim. Feed Sci. Technol. 250, 32–40. doi:10.1016/j.anifeedsci.2018.10.008.

Krehbiel, C.R., 2014. Invited review: applied nutrition of ruminants: fermentation and digestive physiology. Prof. Anim. Sci. 30 (2), 129–139. doi:10.15232/s1080-7446(15)30100-5.

Ku-Vera, J.C., Jiménez-Ocampo, R., Valencia-Salazar, S.S., Montoya-Flores, M.D., Molina-Botero, I.C., Arango, J., Gómez-Bravo, C.A., Aguilar-Pérez, C.F., Solorio-Sánchez, F.J., 2020. Role of secondary plant metabolites on enteric methane mitigation in ruminants. Front. Vet. Sci. 7. doi:10.3389/fvets.2020.00584.

Lee, W.J., Hase, K., 2014. Gut microbiota-generated metabolites in animal health and disease. Nat. Chem. Biol. 10 (6), 416–424. doi:10.1038/nchembio.1535.

Lee, Y.S., Umam, K., Kuo, T.F., Yang, Y.L., Feng, C.S., Yang, W.C., 2024. Functional and mechanistic studies of a phytogenic formulation, Shrimp Best, in growth performance and vibriosis in whiteleg shrimp. Sci. Rep. 14 (1). doi:10.1038/s41598-024-62436-x.

Li, P., Gao, M., Fu, J., Yan, S., Liu, Y., Mahmood, T., Lv, Z., Guo, Y., 2022. Dietary soya saponin improves the lipid metabolism and intestinal health of laying hens. Poult. Sci. 101 (4), 101663. doi:10.1016/j.psj.2021.101663.

Li, T.i, Shen, P., Liu, W., Liu, C., Liang, R., Yan, N.a, Chen, J., 2014. Major polyphenolics in pineapple peels and their antioxidant interactions. Int. J. Food Prop. 17 (8), 1805–1817. doi:10.1080/10942912.2012.732168.

Lillehoj, H., Liu, Y., Calsamiglia, S., Fernandez-Miyakawa, M.E., Chi, F., Cravens, R.L., Oh, S., Gay, C.G., 2018. Phytochemicals as antibiotic alternatives to promote growth and enhance host health. Vet. Res. 49 (1). doi:10.1186/s13567-018-0562-6.

Matamoros, S., Gras-Leguen, C., Le Vacon, F., Potel, G., hetiere, M., 2013. Development of intestinal microbiota in infants and its impact on health. Trends Microbiol. 21 (4), 167–173. doi:10.1016/j.tim.2012.12.001.

Mehdi, Y., Létourneau-Montminy, M.P., Gaucher, M.L., Chorfi, Y., Suresh, G., Rouissi, T., Brar, S.K., Côté, C., Ramirez, A.A., Godbout, S., 2018. Use of antibiotics in broiler production: global impacts and alternatives. Anim. Nutr. 4 (2), 170–178. doi:10.1016/j.aninu.2018.03.002.

Mostafa, M.S., Tushar, Z.H., Hashem, M.A., Rahman, M.M., 2025. Role of natural antioxidants on active packaging of meat and meat products: a review. Appl. Food Res. 5 (2). doi:10.1016/j.afres.2025.101437.

Murugesan, G.R., Syed, B., Haldar, S., Pender, C., 2015. Phytogenic feed additives as an alternative to antibiotic growth promoters in broiler chickens. Front. Vet. Sci. 2. doi:10.3389/fvets.2015.00021.

Nagaraja, T.G., 2016. Microbiology of the rumen. Rumenology. 39–61. doi:10.1007/978-3-319-30533-2_2.

Nantapo, C.W.T., Marume, U., 2025. Strategic technologies to improve phytogenic feed additive efficacy in pigs and poultry. Anim. Nutr. 23, 286–303. doi:10.1016/j.aninu.2024.06.010.

Nawaz, A., Bakhsh javaid, A., Irshad, S., Hoseinifar, S.H., Xiong, H., 2018. The functionality of prebiotics as immunostimulant: Evidences from trials on terrestrial and aquatic animals. Fish Shellfish Immunol. 76, 272–278. doi:10.1016/j.fsi.2018.03.004.

Nayan, M.N.I., Faraji, M.S., Hasan, M.Z., Abony, I.H., Saiyara, U., Islam, M.S., 2026. Enhanced efficacy of synbiotics compared to antibiotics in promoting growth, intestinal health, and immune response in stinging catfish. Aquacult. Nutr. 2026 (1). doi:10.1155/anu/2158993.

Newbold, C.J., Ramos-Morales, E., 2020. Review: ruminal microbiome and microbial metabolome: effects of diet and ruminant host. Animal 14 (S1), s78. doi:10.1017/s1751731119003252.

O'Hara, E., Neves, A.L.A., Song, Y., Guan, L.L., 2020. The Role of the gut microbiome in cattle production and health: driver or passenger? Annu. Rev. Anim. Biosci. 8, 199–220. doi:10.1146/annurev-animal-021419-083952.

Oladokun, S., Clark, K.F., Adewole, D.I., 2022. Microbiota and transcriptomic effects of an essential oil blend and its delivery route compared to an antibiotic growth promoter in broiler chickens. Microorganisms 10 (5). doi:10.3390/microorganisms10050861.

Oni, A.I., Oke, O.E., 2025. Gut health modulation through phytogenics in poultry: mechanisms, benefits, and applications. Front. Vet. Sci. 12. doi:10.3389/fvets.2025.1616734.

Park, I., Oh, S., Lillehoj, E.P., Lillehoj, H.S., 2020. Dietary supplementation with magnolia bark extract alters chicken intestinal metabolite levels. Front. Vet. Sci. 7. doi:10.3389/fvets.2020.00157.

Pidcock, S.E., Skvortsov, T., Santos, F.G., Courtney, S.J., Sui-Ting, K., Creevey, C.J., Huws, S.A., 2021. Phylogenetic systematics of butyrivibrio and pseudobutyrivibrio genomes illustrate vast taxonomic diversity, open genomes and an abundance of carbohydrate-active enzyme family isoforms. Microb. Genom. 7 (10). doi:10.1099/mgen.0.000638.

Poore, J., Nemecek, T., 2018. Reducing food's environmental impacts through producers and consumers. Science 360 (6392), 987–992. doi:10.1126/science.aaq0216.

Prasetyo, A.F, Ulum, M.Y.M., Prasetyo, B., Sanyoto, J.I., 2020. Performance of broiler chickens after termination of feed with additional antibiotic grow promoters (AGP) in Jember district. J. Peternak. 17 (1), 25. doi:10.24014/jupet.v17i1.7536.

Rehman, H., Böhm, J., Zentek, J., 2008. Effects of differentially fermentable carbohydrates on the microbial fermentation profile of the gastrointestinal tract of broilers. J. Anim. Physiol. Anim. Nutr. (Berl.) 92 (4), 471–480. doi:10.1111/j.1439-0396.2007.00736.x.

Rehman, H.U., Vahjen, W., Awad, W.A., Zentek, J., 2007. Indigenous bacteria and bacterial metabolic products in the gastrointestinal tract of broiler chickens. Arch. Anim. Nutr. 61 (5), 319–335. doi:10.1080/17450390701556817.

Ringø, E., Hoseinifar, S.H., Ghosh, K., Doan, H.V., Beck, B.R., Song, S.K., 2018. Lactic acid bacteria in finfish-an update. Front. Microbiol. 9. doi:10.3389/fmicb.2018.01818.

Rowland, I., Gibson, G., Heinken, A., Scott, K., Swann, J., Thiele, I., Tuohy, K., 2018. Gut microbiota functions: metabolism of nutrients and other food components. Eur. J. Nutr. 57 (1), 1–24. doi:10.1007/s00394-017-1445-8.

Rychlik, I., 2020. Composition and function of chicken Gut microbiota. Animals 10 (1), 103. doi:10.3390/ani10010103.

Samrot, A.V., Mohamed, A.A., Faradjeva, E., Si Jie, L., Hooi Sze, C., Arif, A., Chuan Sean, T., Michael, E.N., Yeok Mun, C., Ng, X., Qi, P., Mok, L., Kumar, S.S., 2021. Mechanisms and impact of biofilms and targeting of biofilms using bioactive compounds—a review. Medicina (Mex) 57 (8), 839. doi:10.3390/medicina57080839.

Shipradeep, S.K., Khare, R.S., Ojha, S., Kundu, K., Kundu, S., 2012. Development of probiotic candidate in combination with essential oils from medicinal plant and their effect on enteric pathogens: a review. Gastroenterol. Res. Pract. 2012, 1–6. doi:10.1155/2012/457150.

Shourbela, R.M., El-Hawarry, W.N., Elfadadny, M.R., Dawood, M.A.O., 2021. Oregano essential oil enhanced the growth performance, immunity, and antioxidative status of Nile tilapia (Oreochromis niloticus) reared under intensive systems. Aquaculture 542. doi:10.1016/j.aquaculture.2021.736868.

Śliżewska, K., Piotrowska, M.ał, 2014. Reduction of Ochratoxin A in chicken feed using probiotic. Ann. Agric. Environ. Med. 21 (4), 676–680. doi:10.5604/12321966.1129913.

Smulikowska, S., CzerwiÅ„ski, J., Mieczkowska, A., 2010. Effect of an organic acid blend and phytase added to a rapeseed cake-containing diet on performance, intestinal morphology, caecal microflora activity and thyroid status of broiler chickens. J. Anim. Physiol. Anim. Nutr. (Berl.) 94 (1), 15–23. doi:10.1111/j.1439-0396.2008.00876.x.

Stevanovic, Z.D., Bosnjak-Neumuller, J., Pajic-Lijakovic, I., Raj, J., Vasiljevic, M., 2018. Essential oils as feed additives-future perspectives. Molecules 23, 1717. doi:10.3390/molecules23071717.

Wang, Y., Wang, K., Huang, L., Dong, P., Wang, S., Chen, H., Lu, Z., Hou, D., Zhang, D., 2020. Fine-scale succession patterns and assembly mechanisms of bacterial community of Litopenaeus vannamei larvae across the developmental cycle. Microbiome 8 (1). doi:10.1186/s40168-020-00879-w.

Windisch, W., Schedle, K., Plitzner, C., Kroismayr, A., 2008. Use of phytogenic products as feed additives for swine and poultry1. J. Anim. Sci. 86 (suppl_14), E140. doi:10.2527/jas.2007-0459.

Yan, W., Sun, C., Yuan, J., Yang, N., 2017. Gut metagenomic analysis reveals prominent roles of Lactobacillus and cecal microbiota in chicken feed efficiency. Sci. Rep. 7 (1). doi:10.1038/srep45308.

Yang, C., Chowdhury, M.A., Huo, Y., Gong, J., 2015. Phytogenic compounds as alternatives to in-feed antibiotics: potentials and challenges in application. Pathogens 4 (1), 137–156. doi:10.3390/pathogens4010137.

Yanza, Y.R., Irawan, A., Jayanegara, A., Ramadhani, F., Respati, A.N., Fitri, A., Hidayat, C., Niderkorn, V., Cieslak, A., Szumacher-Strabel, M., Hidayat, R., Tanuwiria, U.H., 2024. Saponin extracts utilization as dietary additive in ruminant nutrition: a meta-analysis of in vivo studies. Animals 14 (8). doi:10.3390/ani14081231.

Yegani, M., Korver, D.R., 2008. Factors affecting intestinal health in poultry. Poult. Sci. 87 (10), 2052–2063. doi:10.3382/ps.2008-00091.

Yu, M., Li, X., Wang, J., Longshaw, M., Song, K., Wang, L., Zhang, C., Lu, K., 2023. Substituting fish meal with a bacteria protein (Methylococcus capsulatus, Bath) grown on natural gas: Effects on growth non-specific immunity and gut health of spotted seabass (Lateolabrax maculatus). Anim. Feed Sci. Technol. 296. doi:10.1016/j.anifeedsci.2022.115556.

Zhao, R., Sun, J., Lin, Y., Yan, H., Zhang, S., Huo, W., Chen, L., Liu, Q., Wang, C., Guo, G., 2025. Effects of dietary tannic acid and tea polyphenol supplementation on rumen fermentation, methane emissions, milk protein synthesis and microbiota in cows. Microorganisms 13 (8), 1848. doi:10.3390/microorganisms13081848.

Chapter 7

Product quality and functional composition of animal foods

Mahmoud Kamal[a], Mahmoud Moustafa[b] and Mohamed E. Abd El-Hack[c]
[a]*Animal Production Systems Research Department, Animal Production Research Institute, Agricultural Research Center, Giza, Egypt,* [b]*Department of Biology, College of Science, King Khalid University, Abha, Kingdom of Saudi Arabia,* [c]*Department of Industrial Pharmacy, College of Pharmaceutical Sciences and Drug Manufacturing, Misr University for Science and Technology (MUST), Giza, Egypt*

7.1 Introduction

The use of plant-based feed additives in animal nutrition is a strong trend toward improving animal health and increasing livestock production (Negm et al., 2025; Reda et al., 2025). Feed additives derived from herbs, spices, and plant extracts contain many beneficial compounds, such as essential oils, flavonoids, and phenolic acids, which exhibit antibacterial, antioxidant, and anti-inflammatory properties (Wang et al., 2024; Abd El-Hack et al., 2026). These natural compounds improve gut health, facilitating digestion and nutrient absorption, and boosting the immune system. This, in turn, reduces the utilization of conventional antibiotics as growth (Abd El-Hack et al., 2025b; El-Abbasy et al., 2025a, 2025b). This shift toward organic compounds aligns with global efforts to combat antimicrobial resistance and promote sustainable livestock productivity (Biswas and Kim, 2025).

Animal-based foods are important in meals worldwide because they provide basic proteins and minerals such as vitamin B12, iron, and zinc, which are crucial for health and growth (Wood et al., 2024). The concept of functional foods has received considerable publicity, with animal products recognized as a vital source of these nutrients. Many compounds found in animal products, such as bioactive peptides, polyunsaturated fatty acids (PUFA), carotenoids, and antioxidants, can help boost the immune system and reduce inflammation, thus benefiting overall health (Cuchillo-Hilario et al., 2024; Ashour et al., 2025a,2025b). Animal compounds also improve the impacts of beneficial gut bacteria, leading to a positive influence on human health (Oancea et al., 2023; Kamal et al., 2024).

Plant-Derived Feed Additives and Animal-Based Foods: Linking Animal Health, Product Quality, and Consumer Perceptions. DOI: https://doi.org/10.1016/B978-0-443-45446-2.00004-1

Despite the increasing demand for plant-based foods, livestock products contain a unique combination of nutrients and bioactive compounds that are difficult to obtain from plant-based foods alone (Hajzer et al., 2025; Kamal et al., 2025a). Animal-derived products, including milk, meat, and eggs fortified with omega-3 fatty acids, show health benefits; however, further clinical and mechanistic studies are needed to confirm these (Mataraci and Kumcuoglu, 2025; Skenderidou et al., 2025). This chapter aims to assess the quality and functional animal foods, focusing on their biochemical traits, health ramifications, and potential for improving human nutrition and health outcomes during advanced animal nutrition and food science.

7.2 Influence of phytofeed on nutritional profiles

Plant additives significantly influence the biological value of animal products by altering their protein, fat, and micronutrient concentrations (Nastoh et al., 2024; Abd El-Hack et al., 2025a). Plant extracts improve intestinal structure (by increasing villus length and crypt depth) and activate digestive enzymes (Abd El-Aziz et al., 2025; Yushanaji et al., 2025). Furthermore, the flavonoids provided in these additives boost gut welfare and antioxidant enzymes, leading to enhanced growth efficiency and increased meat and dairy yields (Abd El-Hack et al., 2023). In addition, these additives may improve bioavailability through their effects on rumen fermentation or gut microbiota, thereby enhancing feed efficiency (El-Aziz et al., 2025; Kamal et al., 2025b).

Specific herbs may change the Fatty Acids (FAs) and micronutrient compositions of agricultural products, especially in poultry meat. Integrating herbs into meat preserves its dietary value and makes it more visually charming by increasing the good PUFA and Monounsaturated Fatty Acids (MUFA) and lowering the Saturated Fatty Acids (SFA) (Jachimowicz et al., 2022). Additive components provide antioxidant properties that safeguard unsaturated lipids from oxidation; hence, they preserve meat quality. These extracts are associated with elevated concentrations of vitamins and minerals owing to their alteration of mineral absorption and enhancement of antioxidants (Nastoh et al., 2024; Mohai Ud Din et al., 2025). These organic improvements improve the nourishing quality of livestock feed, enhance animal health, and reduce reliance on antibiotics in response to boosting consumer interest.

7.3 Enhancement of bioactive and antioxidant content

Adding phytochemicals or their metabolites to animal products to enhance their active and antioxidant traits is a key goal in animal nutrition and food science. "Phytochemicals," molecules synthesized by plants, are known for their potent antioxidant properties that promote animal health and product quality (Lee et al., 2017; Basiouni et al., 2023) When animals consume phytogenic feed additives

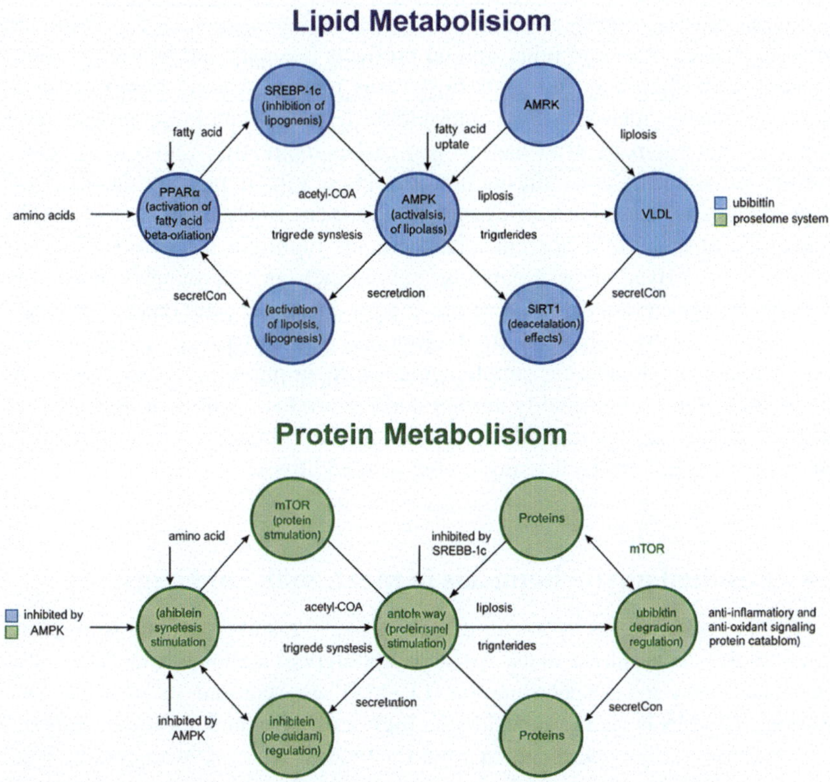

FIGURE 7.1 Diagrams of lipid and protein pathways influenced by herbal additives.

that are high in polyphenols, flavonoids, and other antioxidants, these bioactive molecules or their metabolites can accumulate in tissues that humans consume, such as meat, eggs, and milk. This effect leads to an increase in the antioxidant potential of animal foods, which improves their nutritional value and functional properties (El-Zaiat et al., 2023; Cuchillo-Hilario et al., 2024). The addition of herbal extracts, such as *Echinacea purpurea* or tomato powder, has been shown to increase the activity of antioxidant enzymes in broilers and lower the levels of lipid peroxidation, which improves the quality of the meat and its oxidative stability (El-Zaiat et al., 2023; Ashour et al., 2025a). Fig. 7.1 illustrates the effects of herbal supplements on lipid and protein pathways.

This diagram integrates the key regulatory molecules and pathways modulated by herbal additives on lipid and protein metabolism, showing interplay between lipid metabolism regulators (SREBP-1c, PPARα, AMPK, and SIRT1) and protein regulation (mTOR, ubiquitin-proteasome) through antiinflammatory and antioxidant actions.

Phytochemicals improve the oxidative stability and shelf life of animal products by eliminating reactive oxygen species and stopping lipid oxidation, which is a major cause of food spoilage and degradation (Biswas et al., 2023). Oxidative rancidity makes meat and other animal products less safe and less tasty, which shortens their shelf life and costs businesses money. Adding phytochemicals to animal tissues increases their antioxidant levels, which helps protect them from oxidative damage after harvest and makes them stiffer, hold more water, and remain stable in color during storage. These effects prolong the shelf life of products such as meat and eggs, keeping them fresh (Jairath et al., 2024; Mathobela et al., 2025). Recent advancements highlight the incorporation of natural plant-derived antioxidants into animal diets as a sustainable alternative to synthetic preservatives, in accordance with consumer preferences for clean-label and naturally enhanced food products. These findings underscore the dual benefits of phytogenic supplementation in improving animal health and producing safer, higher-quality animal-derived products with a prolonged shelf life. Table 7.1 summarizes the research on the enhancement of antioxidant capacity in animal products using herbal feed additives.

7.4 Residual phytochemicals in meat, milk, and eggs

Phytochemical residues in "meat, milk, and eggs" result from the use of plant-based additives in animal diets. Advanced analytical techniques such as liquid chromatography-mass spectrometry (LC-MS) are required to identify these residues. LC-MS is characterized by its high sensitivity and ability to analyze a wide variety of phytochemicals in complex food matrices (Chuang et al., 2021; Jia et al., 2024). Detection methods need to be improved for sample preparation and analysis to accurately determine and measure phytochemical residues, which is important for food safety and compliance with regulations. Global monitoring systems use these methods to check for the presence and levels of residues, ensuring that they remain below safety standards to protect consumers.

The safety evaluation of phytochemical residues focuses on determining the maximum residue limits (MRLs) and assessing their potential toxicological impacts on consumers. The European Medicines Agency (EMA) and other regulatory bodies set MRLs after conducting extensive research on how people are exposed to pharmacologically active compounds, such as important phytochemicals (Lin et al., 2019). While allowing the use of phytogenic chemicals in animal husbandry, these restrictions aim to reduce harmful exposure to humans. Safety assessments include checking the levels of residues in food and setting withdrawal periods to ensure that the levels of residues are below certain limits when the food is consumed (Negatu et al., 2021). Biologically derived substances, including certain immunological veterinary treatments, may be exempt from MRL evaluations if they are deemed devoid of pharmacological effects.

Regulatory concerns regarding phytochemical residues include national and international control strategies established by competent authorities to oversee

TABLE 7.1 Illustration of the antioxidant capacity of animal products when using herbal feed additives.

Herbal additive	Animal	Effects	References
Echinacea powder extract	Broiler	The addition of 3.0 cm^3 of EPE enhanced the antioxidant, leading to positive health outcomes for consumers	Ashour et al. (2025a)
Artichoke leaf extract	Broiler	Incorporating 2.0 cm^3 of nano ALE stabilized with ZnNPs increased the IgY, IgM, and SOD while decreasing MDA	Ashour et al. (2025a)
Moringa oleifera leaf powder	Broiler	Adding with MLP at 2.0 and 3.0 cm^3/L increased IgY, IgM, and SOD while reducing MDA	El-Abbasy et al. (2025a)
Garden cress seed powder	Broiler	It enhanced broiler immunity and antioxidative capability, evidenced by elevated IgY, IgM, and SOD, alongside decreased MDA	Negm et al. (2025)
Artemisia annua	Broiler	Increased antioxidant indices in liver and serum; reduced oxidative stress via cholesterol biosynthesis inhibition	Pan et al. (2025)
Huanglian (Berberine)	Broiler	Enhanced antioxidant capacity by neutralizing free radicals | Broiler chicken liver and serum	Pan et al. (2025)
Proanthocyanidin (grape oligomers)	Broiler	Highest efficacy in maintaining antioxidant defenses | Broiler chicken liver and serum	Pan et al. (2025)

(continued on next page)

TABLE 7.1 Illustration of the antioxidant capacity of animal products when using herbal feed additives—cont'd

Herbal additive	Animal	Effects	References
Traditional Vietnamese herb mix	Laying hens	Increased serum and liver activities of GSH-Px, SOD, T-AOC; decreased MDA content	Phan and Nguyen (2024)
Mulberry leaf extract	Laying hens	Increased SOD activity, decreased MDA levels, improved oxidative markers, including gene expression	Lee et al. (2017)
Rosemary extract	Layers	Boosted antioxidant capacity in layer diets	Alem (2024)
Albizia harveyi	Bull	Sperm motility, viability, and membrane integrity improved, with increased viable sperm cells and decreased early apoptotic and apoptotic sperm cells. Sperm ultrastructure damage was reduced, total antioxidant capacity rose, and malondialdehyde levels decreased	Sobeh et al. (2017)
Plantain and garlic leaf	Sheep	The results showed an increasing level of TAC, SOD, GPx, and CAT	Redoy et al. (2020)

ALE, *artichoke leaf extract;* CAT, *catalase;* EPE, *echinacea powder extract;* GPx, *glutathione peroxidase;* GSH, *glutathione;* MDA, *malondialdehyde;* MLP, *Moringa oleifera leaf powder;* SOD, *superoxide dismutase;* TAC, *total antioxidant capacity;* ZnNPs, *zinc nanoparticles.*

residue testing in meat, milk, and eggs (Šerniene et al., 2013). In the European Union, national control plans follow laws such as EU Regulation 2022/1644, which lists the compounds and product group combinations that need to be monitored to ensure that biocidal and veterinary medicine residues remain below acceptable levels (Brocca and Salvatore, 2023). These frameworks require strict rules for sampling, monitoring, and enforcement to eliminate food products with harmful residues. These findings protect public health and maintain confidence in the safety of foods of animal origin (Alenezi et al., 2024). Regulatory events are strengthening standards and measures for waste control. Table 7.2 lists the common phytochemicals and chemical residues present in meat, milk, and eggs. These substances are derived from veterinary drugs, environmental pollutants, pesticides, and dietary phytochemicals.

7.5 Functional properties of animal-based foods

The functional traits of animal-based foods, "digestibility and bioavailability," play vital roles in assessing their nutritional value and health benefits (Modupalli and Rahman, 2025). Animal proteins typically reveal improved digestibility and complete amino acids, therefore facilitating human nutrition (Diao et al., 2025). Furthermore, Yao et al. (2025) suggested that enzymatic and processing modifications of animal proteins can improve their digestibility and functional properties, thereby increasing the availability of amino acids and overall nutritional value.

Herbal additives utilized in animal feed impact the sensory qualities of animal products, such as "taste, aroma, and texture" (Biswas and Kim, 2025). These additives may boost the "taste of feed," potentially leading to raised FI and improved growth efficiency (Mullenix et al., 2024). They also alter the taste and aroma of meat by modifying its oxidative stability and microbial composition during processing and storage (Naveen et al., 2025). Furthermore, mixtures coated with plant and algal materials lead to improvements in the sensory properties of poultry meat, as well as enhanced oxidative stability (Mullenix et al., 2024; Wong et al., 2025). Likewise, animal proteins showed higher uptake and enhanced bioavailability relative to plant proteins (Lotfi et al., 2022). Animal proteins usually supply all essential amino acids in adequate amounts to satisfy human requirements. Conversely, certain plant proteins lack specific critical amino acids such as lysine, leucine, and methionine (MacArthur et al., 2021). The efficiency ratios for processing animal proteins typically range from 90% to 95%, whereas those for digesting plant proteins are between 75% and 80% (Ajomiwe et al., 2024; Zhou et al., 2025). This variance is partly attributable to the existence of antinutritional factors in plants and the more intricate protein structures that complicate digestion and conversion into amino acids.

Animal products offer notable advantages to consumers because of their ease of digestion and bioavailability, as well as their positive effects on sensory

TABLE 7.2 The common residual phytochemicals detected in meat, milk, and eggs.

Phytochemical residue type	Description	Meat	Milk	Eggs	References
Organochlorine pesticides (OCPs)	Pesticides such as DDT, endosulfan, and dieldrin	Yes (accumulative in fat)	Yes (α-endosulfan, β-endosulfan, DDT)	Yes (DDT, dieldrin, toxaphene)	Tilahun et al. (2016), Sarkar et al. (2017), Arif et al. (2021)
Polychlorinated biphenyls (PCBs)	Environmental contaminants, accumulative in fat	Yes	Yes	Yes	Tilahun et al. (2016), Sarkar et al. (2017), Arif et al. (2021)
Antibiotic residues	Veterinary drugs such as β-lactams, fluoroquinolones	Yes	Yes (oxytetracycline)	Yes (detected in some eggs)	Zakaria et al. (2023), Rinky et al. (2025)
Plant-derived antioxidants	Anthocyanins, rosmarinic acid, carvacrol, cinnamaldehyde	Yes (reduce oxidation)	Possibly in trace amounts	Possibly in trace amounts	Estévez (2021), Prommachart et al. (2021), van Vliet et al. (2021)
Natural egg white proteins	Ovalbumin, ovotransferrin, lysozyme	N/A	N/A	Yes (natural antioxidant proteins)	Nimalaratne and Wu (2015)
Fat-soluble vitamins and carotenoids	Nutritional phytochemicals that accumulate in fat	Yes	Yes	Yes (carotenoids in yolk)	van Vliet et al. (2021)
Myristicin and nutmeg derivatives	Nutmeg-derived phytochemicals with potential residues	Possible in meat	Possible in milk	Possible in eggs	Prommachart et al. (2021), Oleszek et al. (2023)

perception (Abo Egila et al., 2023; Wang et al., 2024). These products meet nutritional needs while also possessing desirable flavor, aroma, and texture, making them more appealing and satisfying to consumers (Nhara et al., 2025). The use of medicinal additives in animal feed is an effective method for improving quality and may contribute to improved health because of their role as natural antioxidants (Kamal et al., 2023; Farag et al., 2025).

7.6 From farm to fork, linking phytofeed to consumer value

The significance of phytogenics in animal diets is increasingly noted for distinguishing products from farm to customer and enhancing value for the client (Janjić et al., 2022; Onbaşilar et al., 2025). The livestock sector utilizes these chemicals to improve well-being, growth, and food safety (Martinez et al., 2022). Nonetheless, they continue to face possible hurdles concerning dose and extraction methods (Obianwuna et al., 2024). Herbal additives enhance consumer confidence and differentiate poultry products in the market, as essential oils are used to improve immunity and digestion, leading to improved nutritional and sensory traits (Kırkpınar et al., 2025). In Europe, studies on black pepper flour in laying hen diets show improvements in egg yolk quality, increasing the value of nutritionally enhanced egg products (Harrington et al., 2023; Aminullah et al., 2025; Nhara et al., 2025). Because of these strict standards, medicinal additives are used to maintain animal welfare and meet those requirements (Aminullah et al., 2025).

The natural additives market is expected to experience significant growth in the coming decade, driven by nutritional suggestions aimed at differentiating natural animal products (Wang et al., 2024; Taleb et al., 2025). These additives provided a trend focused on the consumer, animal welfare, decreasing antibiotic utilization, and improving meat and egg quality (Behboodi et al., 2025; Ren et al., 2025). This trend fosters consumer confidence and a willingness to pay greater prices, also supporting the global change toward natural livestock yield. Continuous innovation in feed phytogenics will further impact food quality and security, showcasing how strategic feeding can differentiate products in a competitive market. Fig. 7.2 illustrates a flowchart that explains how absorption, metabolism, tissue accumulation, transfer into food products, and consumer exposure to beneficial plant chemicals from herbs work together.

This flowchart delineates the steps of absorption, metabolism, tissue accumulation, transfer into consumable goods, and consumer exposure to bioactive phytochemicals derived from herbs.

7.7 Conclusion

The quality and nutritional value of animal products are essential to meeting dietary needs and consumer preferences for healthy products. Advances in animal nutrition, particularly through the incorporation of bioactive compounds and

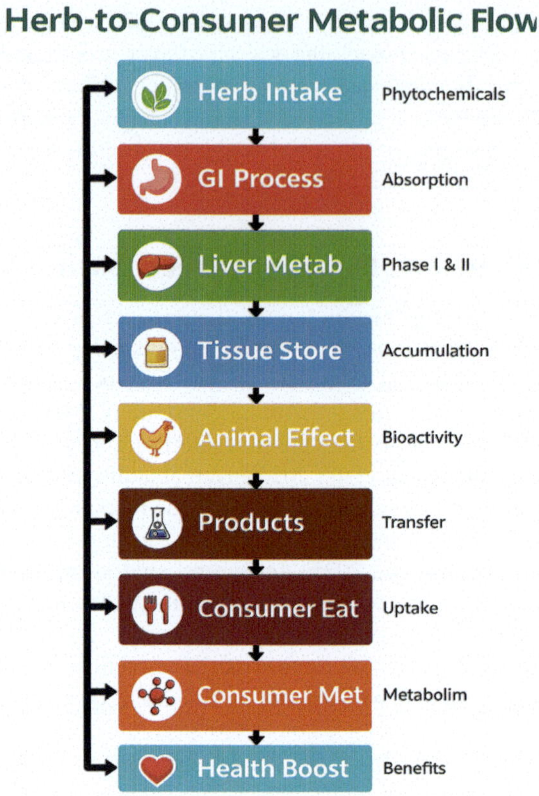

FIGURE 7.2 Flowchart of metabolism: From herb ingestion to tissue deposition to consumer effects.

plant supplements, have shown great potential in improving the nutritional value and functional properties of animal products, including "meat, milk, and eggs." By improving feeding methods and animal welfare, producers can enhance the sensory attributes and health benefits of these products, thus fostering environmentally friendly agriculture and public health. The integration of scientific progress in animal nutrition and management ensures that animal products serve as nutritious foods, enhancing consumer health while addressing the need for healthier, ethically produced options.

References

Abd El-Aziz, A.H., Ghanima, M.M.A., Kamal, M., Abd El-Hack, M.E., Alagawany, M., 2025. The use of bile acids supplement in poultry feed. In: Organic Feed Additives for Livestock. Academic Press, London, pp. 127–138. doi:10.1016/B978-0-443-13510-1.00008-6.

Abd El-Hack, M.E., Aldhalmi, A.K., Kamal, M., Khafaga, A.F., Moustafa, M., Al-Shehri, M., Świątkiewicz, S., Khalifa, N.E., Attia, Y.A., 2026. Flavonoids as a phytobiotic agent: sources, classifications, biological benefits, and useful impacts on broilers and layers. Phytochem. Rev. 25 (1), 331–357. doi:10.1007/s11101-025-10129-2.

Abd El-Hack, M.E., Allam, A.A., Aldhalmi, A.K., Kamal, M., Arif, M., Alawam, A.S., Rudayni, H.A., Taha, A.E., Ashour, E.A., Khan, M.M.H., 2025a. Nutritional value and therapeutic applications of Chenopodium album for human and livestock health: a comprehensive review. Food Biosci. 68. doi:10.1016/j.fbio.2025.106624.

Abd El-Hack, M.E., Aldhalmi, A.K., Ashour, E.A., Kamal, M., Khan, M.M.H., Swelum, A.A., 2025b. The effects of formic acid or herbal mixture on growth performance, carcass quality, blood chemistry, and gut microbial load in broiler chickens: formic acid & herbal mixture in broiler diets. Poult. Sci. 104 (6). doi:10.1016/j.psj.2025.105085.

Abd El-Hack, M.E., de Oliveira, M.C., Attia, Y.A., Kamal, M., Almohmadi, N.H., Youssef, I.M., Khalifa, N.E., Moustafa, M., Al-Shehri, M., Taha, A.E., 2023. The efficacy of polyphenols as an antioxidant agent: an updated review. Int. J. Biol. Macromol. 250. doi:10.1016/j.ijbiomac.2023.126525.

Abo Egila, N.S.H., Dosoky, W.M., Khisheerah, N.S.M., Ahmed, M.H., Zahran, S.M., Almohmadi, N.H., Abusudah, W.F., Kamal, M., Moustafa, M., Tellez-Isaias, G., Al-Shehri, M., Abd El-Hack, M.E., 2023. Does dietary linseed or canola oil affect lipid metabolism, immunity, and n-3 polyunsaturated fatty acids content in quail eggs? Poult. Sci. 102 (12). doi:10.1016/j.psj.2023.103116.

Ajomiwe, N., Boland, M., Phongthai, S., Bagiyal, M., Singh, J., Kaur, L., 2024. Protein nutrition: understanding structure, digestibility, and bioavailability for optimal health. Foods 13 (11), 1771. doi:10.3390/foods13111771.

Alem, W.T, 2024. Effect of herbal extracts in animal nutrition as feed additives. Heliyon 10 (3), e24973. doi:10.1016/j.heliyon.2024.e24973.

Alenezi, M.S., Tartor, Y.H., El-Sherbini, M., Pet, E., Ahmadi, M., Abdelkhalek, A., 2024. Antibiotic residues in milk and milk-based products served in Kuwait hospitals: multi-Hazard Risk Assessment. Antibiotics 13 (11). doi:10.3390/antibiotics13111073.

Aminullah, N., Mostamand, A., Zahir, A., Mahaq, O., Azizi, M.N., 2025. Phytogenic feed additives as alternatives to antibiotics in poultry production: a review. Vet. World 18 (1), 141–154. doi:10.14202/vetworld.2025.141-154.

Ashour, E.A., Aldhalmi, A.K., Ismail, I.E., Kamal, M., Elolimy, A.A., Swelum, A.A., Abd El-Hack, M.E., 2025a. The effect of using Echinacea extract as an immune system stimulant and antioxidant on blood indicators, growth efficiency, and carcass characteristics in broiler chickens to produce a healthy product. Poult. Sci. 104 (1). doi:10.1016/j.psj.2024.104392.

Ashour, E.A., Aldhalmi, A.K., Kamal, M., Salem, S.S., Mahgoub, S.A., Alqhtani, A.H., Madkour, M., Elolimy, A.A., Abd El-Hack, M.E., Swelum, A.A., 2025b. The efficacy of Artichoke leaf extract conjugated with organic zinc nanoparticles on growth, carcass traits and blood biochemical parameters of broilers. Poult. Sci. 104 (1). doi:10.1016/j.psj.2024.104521.

Basiouni, S., Tellez-Isaias, G., Latorre, J.D., Graham, B.D., Petrone-Garcia, V.M., El-Seedi, H.R., Yalçın, S., El-Wahab, A.A., Visscher, C., May-Simera, H.L., Huber, C., Eisenreich, W., Shehata, A.A., 2023. Anti-inflammatory and antioxidative phytogenic substances against secret killers in poultry: current status and prospects. Vet. Sci. 10 (1). doi:10.3390/vetsci10010055.

Behboodi, H.-R., Asghari-Moghadam, M., Mehri, M., 2025. Segmented regression modeling of a phytogenic blend on growth and immune function in broilers: a dose–response study. Poul. Sci. 104 (12), 105913. doi:10.1016/j.psj.2025.105913.

Biswas, O., Kandasamy, P., Nanda, P.K., Biswas, S., Lorenzo, J.M., Das, A., Alessandroni, L., Lamri, M., Das, A.K., Gagaoua, M., 2023. Phytochemicals as natural additives for quality preservation and improvement of muscle foods: a focus on fish and fish products. Food Mater. Res. 3. doi:10.48130/FMR-2023-0005.

Biswas, S., Kim, I.H., 2025. A thorough review of phytogenic feed additives in non-ruminant nutrition: production, gut health, and environmental concerns. J. Anim. Sci. Technol. 67 (3), 497–519. doi:10.5187/jast.2025.e26.

Brocca, D., Salvatore, S., 2023. Report for 2021 on the results from the monitoring of veterinary medicinal product residues and other substances in live animals and animal products. EFSA Support. Publ. 20 (2). doi:10.2903/sp.efsa.2023.en-7886.

Chuang, W.Y., Lin, L.J., Shih, H.D., Shy, Y.M., Chang, S.C., Lee, T.T., 2021. The potential utilization of high-fiber agricultural by-products as monogastric animal feed and feed additives: a review. Animals 11 (7). doi:10.3390/ani11072098.

Cuchillo-Hilario, M., Fournier-Ramírez, M.I., Díaz Martínez, M., Benavides, S.M.ño, Calvo-Carrillo, M.C., Carrillo Domínguez, S., Carranco-Jáuregui, M.E., Hernández-Rodríguez, E., Mora-Pérez, P., Cruz-Martínez, Y.R., Delgadillo-Puga, C., 2024. Animal food products to support human nutrition and to boost human health: the potential of feedstuffs resources and their metabolites as health-promoters. Metabolites. 14 (9). doi:10.3390/metabo14090496.

Diao, X., Xue, D., Mao, X., Li, S., Yang, K., Ke, W., Li, C., 2025. Animal-derived and plant-derived proteins in diets have different in vitro digestive characteristics and bioaccessibility. Food Res. Int. 211, 116377. doi:10.1016/j.foodres.2025.116377.

El-Abbasy, M.M., Aldhalmi, A.K., Ashour, E.A., Bassiony, S.S., Kamal, M., Alqhtani, A.H., Abou-Kassem, D.E., Elolimy, A.A., Abd El-Hack, M.E., Swelum, A.A., 2025a. Enhancing broiler growth and carcass quality: impact of diets enriched with Moringa oleifera leaf powder conjugated with zinc nanoparticles. Poult. Sci. 104 (1). doi:10.1016/j.psj.2024.104519.

El-Abbasy, M.M., Aldhalmi, A.K., Ashour, E.A., Elnesr, S.S., Kamal, M., Salem, S.S., Mahgoub, S.A., Al-Shehri, M., Moustafa, M., Abd El-Hack, M.E., 2025b. Impact of dried fig fruit extract conjugated with selenium nanoparticles on productive and blood parameters of chicken. Ann. Anim. Sci. 25 (3), 1055–1063. doi:10.2478/aoas-2025-0048.

El-Aziz, A.A., Ghanima, M.A., Kamal, M., Aldhalmi, A.K., Khan, M.M.H., Abd El-Hack, M.E., 2025. Quercetin as a bioactive feed additive: nutritional, antioxidant, and gut health implications in poultry and rabbits. Vet. Anim. Sci. 29. doi:10.1016/j.vas.2025.100487.

El-Zaiat, H.M., Ku-Vera, J.C., Soltan, Y.A., 2023. Natural phytochemicals to enhance animal productivity and health status with low greenhouse gas emissions. Front. Vet. Sci. 10, 1280611.

Estévez, M., 2021. Critical overview of the use of plant antioxidants in the meat industry: opportunities, innovative applications and future perspectives. Meat Sci. 181, 108610. doi:10.1016/j.meatsci.2021.108610.

Farag, S.A., Dosoky, W.M., Moussa, A.A., Kamal, M., Elolimy, A.A., El-Mekkawy, M.M.F., Abd El-Hack, M.E., Swelum, A.A., 2025. Impact of date palm (Phoenix dactylifera) pollen supplementation on growth performance, carcass traits, cecal microbial composition, and blood parameters in Japanese quail (Coturnix coturnix Japonica). Poult. Sci. 104 (7). doi:10.1016/j.psj.2025.105164.

Hajzer, Z.E., Alibrahem, W., Helu, N.K., Oláh, C., Prokisch, J., 2025. Functional foods in clinical trials and future research directions. Foods. 14 (15), 2675. doi:10.3390/foods14152675.

Harrington, D., Leach, M., Van Der Klis, J.D., 2023. Improving the sustainability of laying hen production using plant-based feed additives. In: Sustainable Use of Feed Additives in Livestock: Novel Ways for Animal Production. Springer International Publishing, Denmark, pp. 561–608. doi:10.1007/9783031428555_19.

Jachimowicz, K., Winiarska-Mieczan, A., Tomaszewska, E., 2022. The impact of herbal additives for poultry feed on the fatty acid profile of meat. Animals 12 (9), 1054. doi:10.3390/ani12091054.

Jairath, G., Biswas, A.K., Mal, G., Suman, S.P., 2024. Bioactive compounds in meat: their roles in modulating palatability and nutritional value. Meat Muscle Biol. 8 (1). doi:10.22175/mmb.16992.

Janjić, J., Savić, K.Š., Marković, R., Šefer, D., Nedić, D., Đurić, S., Vejnović, B., Mirilović, M., 2022. Influence of phytobiotics in feed on the cost-effectiveness of broiler production during fattening. Meat Technol. 63 (1), 51–58. doi:10.18485/meattech.2022.63.1.5.

Jia, Q., Liao, G., Chen, L., Qian, Y., Yan, X., Qiu, J., 2024. Pesticide residues in animal-derived food: current state and perspectives. Food Chem. 438. doi:10.1016/j.foodchem.2023.137974.

Kamal, M., Zhu, L., Abd El-Hack, M.E., Arif, M., Li, F., Cheng, Y., 2025a. Functional roles of mannan and chitosan oligosaccharides on animal health and nutrition: a review. Carbohydr. Polym. Technol. Appli. 10. doi:10.1016/j.carpta.2025.100764.

Kamal, M., Sindi, R.A., El-Azzazi, F.E., Kishk, W.H., Khalil, H.A., Abdel-Khalek, A.M., Ayoub, M.A., Tufarelli, V., Abd El-Hack, M.E., 2023. Sexual behaviour response, testicular development and semen quality of New Zealand white rabbit bucks as influenced by dietary chitosan. Reprod. Domest. Anim. 58 (3), 405–413. doi:10.1111/rda.14300.

Kamal, M., Linlin, K., Gao, J., Xinrui, Z., Xinming, C., Haibo, W., Lulu, D., El-Hack, M.E.A., Mahrose, K., Cheng, Y., 2025b. Effects of Saccharomyces cerevisiae and Bacillus subtilis on in vitro fermentation in the rumen of Hu sheep. J. Sci. Food Agric. 105 (1), 498–506. doi:10.1002/jsfa.13848.

Kamal, M., Lele, W., Shuzhen, T., Jiandi, L., Rongyan, Q., Yanfeng, L., Wenqi, W., Xiangyu, C., Cheng, Y., 2024. Influence of dietary Salicornia europaea L. extract supplementation on feed efficiency of Altay sheep by modifying their gastrointestinal bacteria communities. Front. Microbiol. 15. doi:10.3389/fmicb.2024.1377314.

Kırkpınar, F., Açıkgöz, Z., Niyonshuti, E., 2025. Current Situation, Future Goals, and Strategies of the Global Feed and Feed Additives Industries. IntechOpen, London doi:10.5772/intechopen.1011333.

Lee, M.T., Lin, W.C., Yu, B., Lee, T.T., 2017. Antioxidant capacity of phytochemicals and their potential effects on oxidative status in animals—a review. Asian-Australas. J. Anim. Sci. 30 (3), 299–308. doi:10.5713/ajas.16.0438.

Lin, H., Zhao, S., Fan, X., Ma, Y., Wu, X., Su, Y., Hu, J., 2019. Residue behavior and dietary risk assessment of chlorothalonil and its metabolite SDS-3701 in water spinach to propose maximum residue limit (MRL). Regul. Toxicol. Pharmacol. 107, 104416. doi:10.1016/j.yrtph.2019.104416.

Lotfi, K., Mohammadi, S., Mirzaei, S., Asadi, A., Akhlaghi, M., Saneei, P., 2022. Dietary total, plant and animal protein intake in relation to metabolic health status in overweight and obese adolescents. Sci. Rep. 12 (1). doi:10.1038/s41598-022-14433-1.

MacArthur, M.R., Mitchell, S.J., Treviño-Villarreal, J.H., Grondin, Y., Reynolds, J.S., Kip, P., Jung, J., Trocha, K.M., Ozaki, C.K., Mitchell, J.R., 2021. Total protein, not amino acid composition, differs in plant-based versus omnivorous dietary patterns and determines metabolic health effects in mice. Cell Metab. 33 (9), 1808. doi:10.1016/j.cmet.2021.06.011.

Martinez, D.A., Ponce-De-Leon, C.L., Vilchez, C., 2022. Meta-analysis of commercial-scale trials as a means to improve decision-making processes in the poultry industry: a phytogenic feed additive case study. arXiv doi:10.48550/arxiv.2201.01818.

Mataraci, C.E., Kumcuoglu, S., 2025. Extraction of valuable bioactive compounds from Aronia (Aronia melanocarpa) fruit and their use in the production of fast-dissolving oral strips. J. Food Sci. 90 (10). doi:10.1111/1750-3841.70561.

Mathobela, R.M., Chikwanha, O.C., Katiyatiya, C.L.F., Semwogerere, F., Molotsi, A.H., Marufu, M.C., Strydom, P.E., Mapiye, C., 2025. Preslaughter practices, pork physicochemical attributes and fatty acid profiles of pigs raised and slaughtered on smallholder urban farms in the Cape Metropole, South Africa. Vet. Anim. Sci. 28. doi:10.1016/j.vas.2025.100454.

Modupalli, N., Mahfuzur Rahman, M.d, 2025. Enhancing the digestibility and functional properties of food proteins using non-thermal plasma. Sustain. Food Proteins 3 (2). doi:10.1002/sfp2.70008.

Mohai Ud Din, R., Eman, S., Zafar, M.H., Chong, Z., Saleh, A.A., Husien, H.M., Wang, M., 2025. Moringa oleifera as a multifunctional feed additive: synergistic nutritional and immunomodulatory mechanisms in livestock production. Front. Nutr. 12. doi:10.3389/fnut.2025.1615349.

Arif, A.M, Javed, I., Ayaz, M., Abdullah, M., Imran, M., Shahbaz, M., Gondal, T.A, Ali, M., Iqbal, Z., Iqbal, Z., Salehi, B., Sharifi-Rad, J., Martins, N., 2021. Organochlorine pesticide residues in raw milk samples collected from dairy farms and urban areas of Lahore district, Pakistan. J. Food Sci. Technol. 58 (1), 129–137. doi:10.1007/s13197-020-04522-2.

Mullenix, G.J., Greene, E.S., Ramser, A., Maynard, C., Dridi, S., 2024. Effect of a microencapsulated phyto/phycogenic blend supplementation on growth performance, processing parameters, meat quality, and sensory profile in male broilers. Front. Vet. Sci. 11. doi:10.3389/fvets.2024.1382535.

Nastoh, N.A., Waqas, M., Çınar, A.A., Salman, M., 2024. The impact of phytogenic feed additives on ruminant production: a review. J. Anim. Feed Sci. 33 (4), 431–453. doi:10.22358/jafs/191479/2024.

Naveen, P., Kour, H., Malik, R., Naliyapara, H., Gabbur, A., Rana, P., Goyal, A., 2025. Influence of different taste feed additives on feed intake and growth performance in growing goats. Flavour Fragr. J. 40 (3), 440–444. doi:10.1002/ffj.3842.

Negatu, B., Dugassa, S., Mekonnen, Y., 2021. Environmental and health risks of pesticide use in Ethiopia. J. Health Pollut. 11 (30). doi:10.5696/2156-9614-11.30.210601.

Negm, M.H., Aldhalmi, A.K., Ashour, E.A., Mohamed, L.A., Kamal, M., Rashad, A., Khan, M.M.H., Abd El-Hack, M.E., Swelum, A.A., 2025. Growth, carcass traits, blood chemistry and gut microbiota in broiler chickens fed diets enriched with garden cress seed powder as a natural growth enhancer. Vet. Med. Sci. 11 (3). doi:10.1002/vms3.70397.

Nhara, R.B., Musara, C., Charambira, T., Mujati, T., Nhovoro, T., Hodzi, P., Chigede, N., Ruzengwe, F.M., 2025. Role of phytogenic feed additives on animal product quality and acceptance—a review. Trop. Anim. Health Prod. 57 (5). doi:10.1007/s11250-025-04474-9.

Nimalaratne, C., Wu, J., 2015. Hen egg as an antioxidant food commodity: a review. Nutrients 7 (10), 8274–8293. doi:10.3390/nu7105394.

Oancea, A.-G., Dragomir, C., Untea, A., Saracila, M., Turcu, R., Cismileanu, A., Boldea, I., Radu, G.L, 2023. The effects of Brewer's Spent Yeast (BSY) inclusion in dairy sheep's diets on ruminal fermentation and milk quality parameters. Agriculture 13 (8), 1605. doi:10.3390/agriculture13081605.

Obianwuna, U.E., Chang, X., Oleforuh-Okoleh, V.U., Onu, P.N., Zhang, H., Qiu, K., Wu, S., 2024. Phytobiotics in poultry: revolutionizing broiler chicken nutrition with plant-derived gut health enhancers. J. Anim. Sci. Biotechnol. 15 (1). doi:10.1186/s40104-024-01101-9.

Oleszek, M., Kowalska, I., Bertuzzi, T., Oleszek, W., 2023. Phytochemicals derived from agricultural residues and their valuable properties and applications. Molecules 28 (1), 342. doi:10.3390/molecules28010342.

Onbaşilar, İ., Yalçin, S., Eser, H., Ramay, M.S., Yalçin, S., Özsoy, B., Erbay Elibol, F.K., Taban, S., Koçoğlu, S.T., Torlak, E., 2025. Combined use of essential oils with organic acids in modifying performance, intestinal health, caecal microflora, and selected blood and bone parameters in broilers. Ankara Univ. Vet. Fak. Derg. 72 (3), 377–386. doi:10.33988/auvfd.1613810.

Pan, H., Bi, J., Hu, H., Huang, Y., Li, A., Zhang, H., Wan, Y., Zhan, K., Wang, K., Zhao, Z., Bai, X., 2025. Chinese herbal medicine improves antioxidant capacity of chicken liver at high stocking density involved gut-liver microbiota axis based on multi-omics technologies. Poult. Sci. 104 (5), 105015. doi:10.1016/j.psj.2025.105015.

Phan, T. H. P., Nguyen, D. H., 2024. Production performance, antioxidant capacity, immunity and odorous gas emission of excreta of laying hens supplemented with traditional Vietnamese herbal. Livestock Research for Rural Development 36. http://www.lrrd.org/lrrd36/3/3625ndho.html.

Prommachart, R., Cherdthong, A., Navanukraw, C., Pongdontri, P., Taron, W., Uriyapongson, J., Uriyapongson, S., 2021. Effect of dietary anthocyanin-extracted residue on meat oxidation and fatty acid profile of male dairy cattle. Animals 11 (2), 322. doi:10.3390/ani11020322.

Reda, F.M., Kamal, M., Mahmoud, H.K., Rudayni, H.A., Allam, A.A., Khafaga, A.F., Khan, M.M.H., Mohamed, R.S., 2025. Effect of a mixture of peppermint and clove oils on growth performance, carcass traits, blood parameters, and gut microbiota in broiler chickens. Poult. Sci. 104 (11). doi:10.1016/j.psj.2025.105718.

Redoy, M.R.A., Shuvo, A.A.S., Cheng, L., Al-Mamun, M., 2020. Effect of herbal supplementation on growth, immunity, rumen histology, serum antioxidants and meat quality of sheep. Animal 14 (11), 2433–2441. doi:10.1017/s1751731120001196.

Ren, J., Ren, S., Yang, H., Ji, P., 2025. Effects of phytogenic feed additive on production performance, slaughtering performance, meat quality, and intestinal flora of white-feathered broilers. Vet. Sci. 12 (5), 396. doi:10.3390/vetsci12050396.

Rinky, F., Rahman, A., Reza, S., Nowar, A., Alim, S.R., 2025. Residual antibiotics in milk samples: assessing the risk and prevalence in Bangladesh. Heliyon 11 (1). doi:10.1016/j.heliyon.2024.e41422.

Sarkar, A., Hossain, M.M., Sarkar, A.C., 2017. Heavy metal residue in eggs of chicken (*Gallus gallus domesticus*) available in Sylhet. In: Proceedings of the Fifth International Conference on Chemical Engineering (ICChE 2017): Bioengineering. Bangladesh University of Engineering and Technology (BUET, Dhaka, Bangladesh, pp. 146–150.

Šerniene, L., Stimbirys, A., Daunoras, G., 2013. Trends in monitoring residues of pharmacologically active group B substances in products of animal origin in Lithuania from 1999 to 2008. Food Addit. Contam.: Part B 6 (3), 187–195. doi:10.1080/19393210.2013.782353.

Skenderidou, I., Leontopoulos, S., Skenderidis, P., 2025. Functional food ingredients enhancing immune health. Int. J. Mol. Sci. 26 (17). doi:10.3390/ijms26178408.

Sobeh, M., Hassan, S.A., El Raey, M.A., Khalil, W.A., Hassan, M.A.E., Wink, M., 2017. Polyphenolics from albizia harveyi exhibit antioxidant activities and counteract oxidative damage and ultra-structural changes of cryopreserved bull semen. Molecules 22 (11). doi:10.3390/molecules22111993.

Taleb, H.M., Mahrose, K., Abdel-Halim, A.A., Kasem, H., Ramadan, G.S., Fouad, A.M., Khafaga, A.F., Khalifa, N.E., Kamal, M., Salem, H.M., Alqhtani, A.H., Swelum, A.A., Arczewska-Włosek, A., Świątkiewicz, S., Abd El-Hack, M.E., 2025. Using artificial intelligence to improve poultry productivity-a review. Ann. Anim. Sci. 25 (1), 23–33. doi:10.2478/aoas-2024-0039.

Tilahun, A., Jambare, L., Teshale, A., Getachew, A., 2016. Review on chemical and drug residue in meat. World J. Agric. Sci. 12 (3), 196–204.

van Vliet, S., Provenza, F.D., Kronberg, S.L., 2021. Health-promoting phytonutrients are higher in grass-fed meat and milk. Front. Sustain. Food Syst. 4. doi:10.3389/fsufs.2020.555426.

Wang, J., Deng, L., Chen, M., Che, Y., Li, L., Zhu, L., Chen, G., Feng, T., 2024. Phytogenic feed additives as natural antibiotic alternatives in animal health and production: a review of the literature of the last decade. Anim. Nutr. 17, 244–264. doi:10.1016/j.aninu.2024.01.012.

Wong, Y.H, Khin, A.A, Lim, C.S, 2025. Effect of different additives incorporation in chicken diets on consumers' sensory preference. Adv. Agri. Food Res. J. 6 (1). doi:10.36877/aafrj.a0000562.

Wood, J.D., Giromini, C., Givens, D.I., 2024. Animal-derived foods: consumption, composition and effects on health and the environment: an overview. Front. Anim. Sci. 5. doi:10.3389/fanim.2024.1332694.

Yao, L., Yu, P., Qiuyu, H., Yuanxue, J., Jing, Z., 2025. Advances in understanding the nutritional value and digestive characteristics of plant-based meat. Shipin Kexue 46 (15) 357-338.

Yushanaji, B., Kamal, M., Baishanbieke, D., Zhu, L., Zhang, P., He, Q., Li, F., 2025. Effects of raffinose supplementation on growth performance, carcass quality, blood chemistry, gut microbiota, and intestinal morphology in yellow-feathered broilers. Poult. Sci. 104 (11), 105888. doi:10.1016/j.psj.2025.105888.

Zakaria, A.M., Elmeligy, E., Abdulkarim, A., Mohammed, H.H., Khalphallah, A., Ali, M.A., Khalifa, M.I., 2023. Monitoring antimicrobial residues in table eggs in Aswan governorate markets and their impact on egg quality and public health. Open Vet. J. 13 (5), 523–531. doi:10.5455/OVJ.2023.v13.i5.4.

Zhou, H., Suryamiharja, A., 2025. Protein showdown: comparison of plant-based and animal-based foods. Bulletin 1575.

Chapter 8

Consumer health and well-being: Biological mechanisms, translational evidence, and educational communication implications

Islam Mohamed Youssef[a], Abrar H. Omar[b] and Mohamed E. Abd El-Hack[c]

[a]*Animal Production Systems Research Department, Animal Production Research Institute, Agricultural Research Center, Giza, Egypt,* [b]*Department of English, College of Languages and Translation, Misr University for Science and Technology (MUST), Giza, Egypt,* [c]*Department of Industrial Pharmacy, College of Pharmaceutical Sciences and Drug Manufacturing, Misr University for Science and Technology (MUST), Giza, Egypt*

8.1 Introduction

The diet and mental health relationship has shifted from the importance of nutrient sufficiency to a more comprehensive systems-level approach that encompasses the gut–brain–immune axis. Recent findings suggest that food constituents affect neurochemistry, inflammation, and stress responses in a complex manner that involves microbial metabolites, immune factors, and neuromodulatory mechanisms (Silva et al., 2020; Clerici et al., 2025). At the same time, studies on dietary patterns have repeatedly shown that high-quality, antiinflammatory diets such as the Mediterranean diet are associated with a lower risk of developing depressive symptoms and enhanced psychological resilience (Yin et al., 2021; Tavakoly et al., 2025). These developments have triggered interest in functional foods that can specifically affect mood, cognition, and stress responsiveness via biologically plausible mechanisms.

Food items obtained from animals play an important and unique part within this framework. In addition to providing high-protein, high-quality food products that contain many essential micronutrients (i.e., vitamins B12, iron, choline, and zinc), animal-derived food items also provide an important vehicle for the delivery of bioactive peptides (e.g., dairy peptides such as alpha-casozepine) and long-chain omega-3 fatty acids, carotenoid compounds, and/or microbial-active

Plant-Derived Feed Additives and Animal-Based Foods: Linking Animal Health, Product Quality, and Consumer Perceptions. DOI: https://doi.org/10.1016/B978-0-443-45446-2.00017-X

compounds (Derbyshire and Obeid, 2020; Yosaee et al., 2022; Zinkow et al., 2025). Some examples would include: Evidence for the potential anxiolytic effects of alpha-casozepine as mediated via gamma-aminobutyric acid (GABA), GABA activity in animal models, the potential influence of several species of fermentative bacteria present in some fermented dairy products on depressive and anxiety symptoms via either microbial-to-microbial communication, or through mechanisms that modulate the immune system, and the potential visual and cognitive performance support from eggs enriched with lutein and zeaxanthin (Nolan et al., 2016).

Novel concepts in farming take this idea even further. Pasture-based diverse and herbal-fed agriculture has been found to influence the composition of phytochemicals, carotenoids, and fatty acids in milk, meat products, and eggs. This effectively transfers plant-based bioactive compounds into the human food web (Steinshamn et al., 2008; van Vliet et al., 2021). This transfer of bioactive compounds from food products suggests that animal matrices have the potential to function as a delivery system for agents that are neuroactive or antiinflammatory in nature. These mechanisms are also in line with current human psychoneuroimmunology models that highlight the importance of immune–microbiome interplay in regulating human moods and cognitive processes (Teixeira et al., 2024). However, to make these mechanisms credible claims for human health requires rigorous testing. The psychobiotic research paradigm indicates that clinical effects are limited in scope and vary significantly in magnitude, especially in response to strain specificity, baseline inflammatory levels, and expectancy factors (Sharma et al., 2021; Asad et al., 2025). Moreover, bioaccessibility, digestive stability, and matrix interactions are essential factors that dictate the biological activity of feed-derived phytochemicals or peptides following ingestion (Dima et al., 2020). The need for transparency and rigorous controls in human trials must occur before any claims about health can be promoted by following ethical/regulatory influences based upon WHO and WMA guidelines.

This chapter examines the indirect consumer health benefits of animal-derived foods, especially those raised using herbal-fed or pasture-diverse systems, from the perspective of the gut–immune–brain axis. Through the integration of nutritional neuroscience, microbiome science, functional food technology, and behavioral research, this chapter seeks to shed light on the biological plausibility, evidence base, and challenges of translation related to the role of animal products in mental health.

8.2 Mood and cognition modulation by food components

8.2.1 Nutrients linked to brain performance

The two amino acids, tryptophan and tyrosine, serve as precursors (starting points) for the creation of some important hormones in the body: Serotonin and catecholamines (dopamine, norepinephrine), respectively. Because these hormones are necessary for mood and executive functioning (the ability to analyze, process, and execute decisions), there is a specific connection between the amounts of tryptophan/tyrosine we consume in our diets and the availability of these hormones in our bodies (Jenkins et al., 2016). When someone does not have enough tryptophan in their system, they will have lower levels of serotonin (a hormone) in their brain, and they may have more difficulty executing tasks than when there is an adequate amount of tryptophan in their system. Also, when there are low baseline levels of serotonin in a person's body and they consume large amounts of tryptophan (tryptophan loading), they may perform better on memory-related tasks but take longer (and have slower reaction times) to perform fine motor tasks because of mild sedation (Silber and Schmitt, 2010). This relationship between baseline serotonin levels and the effects of tryptophan loading indicates that these effects will change depending upon the level of baseline serotonin and the type of task performed. Likewise, tyrosine availability is regulated to produce different effects on reward-related behaviors and motivation under periods of stress and/or high cognitive demand, coinciding with the dopamine function of effort-based decision-making and cognitive flexibility (Salamone and Correa, 2024). Overall, these findings support the premise that the availability of amino acids can have a transient impact on the production of neurotransmitters and subsequently on behavior; nonetheless, enhancement of cognition from dietary levels in healthy, nonstressed adults is variable and context-dependent (Ghallab et al., 2024).

B vitamins, specifically B6, B9, and B12, play a role in regulating one-carbon metabolism and homocysteine levels and have an effect on methylation pathways, which is a critical aspect of neuroplasticity and vascular function (Djuric et al., 2024). Meta-analytic findings of randomized controlled trials suggest that supplementing B-complex vitamins in the older population can result in some modest improvement in global cognitive function, although removing outliers and low-quality studies (Berg et al., 2025) and domain-specific effects (episodic memory and executive function) are generally of small magnitude with different levels of certainty across reviews (Fekete et al., 2023). In addition, a meta-analysis focused on B12 supplementation finds that it would not have a positive effect on cognitive and depressive symptoms without overt deficiency, which illustrates the need for targeted addressing of deficiency rather than universal supplementation (Han et al., 2025).

Omega-3 long-chain polyunsaturated fatty acids (LC-PUFAs), particularly EPA/DHA, incorporate into neuronal membranes and regulate neuroinflammation, synaptic function, and hypothalamic–pituitary–adrenal axis reactivity (Zinkow et al., 2025). Recent studies have shown that depressive anxiety symptoms may be relieved with higher levels of EPA in fish oil and should be included as a supplemental form of treatment in combination with other medications; however there is not an even variation of clinical trial results based on the dosages used, the ratio of EPA to DHA in the oil and the baseline characteristics of the subject group (Barros et al., 2025). In cognitive measure meta-analysis studies, there were minor improvements in cognition, both in total and in some specific cognitive functions, indicating that the effects of omega-3 may not be so clear-cut as "the more omega-3 taken, the greater the improvement" (Zhou et al., 2022). There is a possible strong mechanistic basis for the use of omega-3 fatty acids in clinical practice, due to having either an antineuroinflammatory or a proneuroplastic mechanism; however, their actual improvements for patients are small, and depend on the patient group. Therefore, it is logical for omega-3 oils to be considered part of an overall multifaceted lifestyle change, as opposed to being used in isolation.

Magnesium is a cofactor for many enzymatic processes, including those regulating NMDA receptors. Magnesium has been shown to have significant effects in reducing depressive symptoms in RCTs and meta-analyses with rapid onset and good tolerability profiles, although sample sizes are small and populations diverse. More research is necessary. Zinc is integral for synaptic transmission and neurogenesis. It has been found to have a negative correlation with depression risk. A comparative meta-analysis indicates that zinc supplementation reduces depressive symptom scores, especially when used as a monotherapy rather than as an adjunct to antidepressants, implying unique pharmacodynamic effects that may be diminished in combination therapies (Yosaee et al., 2022). In conclusion, the amino acid precursors can rapidly affect neurotransmission; B vitamins can potentially aid older or lower-status people in their cognitive capacity (Porter et al., 2016); omega-3 fatty acids can potentially aid in depression and cognitive capacity (Kalkman et al., 2021); and magnesium and zinc may demonstrate antidepressant properties that need to be investigated (Nechifor, 2025). Precision nutrition approaches that consider baseline nutrient levels, inflammation levels, and clinical phenotypes will most likely produce optimal results (Khiroya et al., 2023).

8.2.2 Food-derived neuroactive compounds

A novel category of peptides derived from food has been identified as potentially able to cross the blood–brain barrier (BBB) and reach the brain, where they may influence systems related to neuroinflammation, oxidative stress, Aβ aggregation, and neurotransmitters (Cheng et al., 2024). In vitro transport studies utilizing models of the BBB, perfusion studies in situ, and the use of in vivo tracers have demonstrated that specific di- and tripeptides can cross the BBB and possibly provide neuroprotective effects and improve memory in animal models. Preliminary clinical trials have also demonstrated cognitive improvements using whey-derived tryptophan–tyrosine-containing peptides when used in tired adults, as well as further confirmatory studies (Kita et al., 2018). Narrative and systematic reviews of a range of food-derived peptides with purported action within the central nervous system (CNS) have been published, while there have also been a number of ongoing issues identified in terms of the bioavailability, stability, and structure-activity relationships associated with these types of peptides to support their ultimate translational efficacy (Katayama and Nakamura, 2019).

The various fruits and vegetables mentioned above contain flavonoids; flavonoids are polyphenolic compounds present in these and other plant-based foods. The gut microbiome breaks down some of the flavonoids to yield phenolic metabolites, which then circulate throughout the body, including the brain, to increase BDNF signaling, improve cerebrovascular health, and reduce neuroinflammation (Cheng et al., 2022). According to a meta-analysis of 80 randomized trials examining the effects of different flavonoid dietary sources, cocoa, berries, and ginkgo have the greatest impact on cognitive functioning, particularly long-term memory and processing speed (Neshatdoust et al., 2016). In human studies, higher serum levels of BDNF correlate with superior cognitive abilities; these correlations correspond to the consumption of foods rich in flavonoids or cocoa flavanols. These data support earlier studies demonstrating that CREB-BDNF signaling protects the brain and enhances memory (Amidfar et al., 2024). The gut microbiome is crucial: More than 90% of dietary flavonoids are metabolized by microbes prior to absorption, and converging evidence indicates that exercise combined with flavonoids may work together through microbiome-mediated antiinflammatory and immunomodulatory metabolites (Cheatham et al., 2022).

SCFAs such as acetate, propionate, and butyrate, derived from the fermentation of fiber components by gut microbiota, play the role of mediators in the gut–brain axis in association with the integrity of the BBB, including TJ barrier maintenance through G-protein-coupled receptors (GPCR) signaling and HDAC inhibition, glial cell development, and microglial homeostasis (Fock and Parnova, 2023). In recent reviews, SCFAs play the role of protectors in maintaining the integrity of the BBB through replenishment of junctional proteins and activation of NF-κB and Nrf-2 signaling pathways; in another review, SCFAs are introduced as the key component in the "SCFAs-microglia pathway" in the context of neuroimmune regulation in CNS disorders (Cao et al., 2025). In the context of Alzheimer's disease, SCFAs play the role of modulators in Aβ oligomerization and glial hyperactivation, offering the key to the neuroprotective role of dietary fiber (Skawratananond et al., 2025).

Molecules from food interact with standard neurotransmitter systems. GABAergic transmission has been modified or altered by many natural compounds, including phytochemicals and flavonoids, which interact with receptors or change the synthesis or metabolism of GABA. These substances often provide anxiolytic and/or antidepressant-like effects in preclinical models, which demonstrates potential pathways through which food could modulate an inhibitory tone (Ríos et al., 2022). The glutamate- and GABA-based balance, which is critical for excitation- and inhibition-based homeostasis and cognitive function, is also associated with neurodegenerative diseases and mood disorders; reviews of the glutamate and GABA balance demonstrate how microglial receptors and neuronal–glial interactions in Alzheimer's disease alter the balance and provide targets for dietary metabolites' potential indirect effects (Czapski and Strosznajder, 2021). In addition to modifying neuro-transmission, many different types of flavonoids and peptides have been shown to coalesce within the CREB/BDNF signaling path to promote synaptic plasticity, increase the density of spines and encourage the genesis of new neurons, and create a mechanistic link to the effects associated with eating diets rich in plants/peptides and cognitive function in subjects participating in clinical studies. Lastly, the brain–gut–microbiome axis adds a receptor-level dimension: Microbes that make GABA and peripheral GABA receptors found in the gut and brain suggest that microbiota shaped by diet may signal through GABAergic pathways that affect mood and cognition. This is a new area of research that has been mapped by transcriptomic and physiological evidence (Amidfar et al., 2014; Neshatdoust et al., 2016). In conclusion, food-derived peptides, flavonoids, and short-chain fatty acids (SCFAs) affect neural function by interacting with receptors (GABA, glutamate, GPCRs), having epigenetic/HDAC effects, and signaling neurotrophins. The microbiome often acts as a necessary link for bioactivation (Fock and Parnova, 2023). Nutrients and bioactive compounds in animal foods with potential mood impact are exhibited in Table 8.1.

8.2.3 Epidemiological evidence

Increasingly, prospective cohorts and randomized trials are examining the relationship between comprehensive dietary patterns and mental health. Observational syntheses consistently link higher adherence to the Mediterranean diet (MedDiet) with a reduced incidence of depression (Lassale et al., 2019). A study of longitudinal cohorts by means of a meta-analysis revealed a 33% risk reduction for the highest adherence group in comparison to the lowest adherence group, with similar effects for low dietary inflammatory index scores, thus implying the resilience effects of antiinflammatory dietary patterns (Tavakoly et al., 2025). A large cohort study conducted in Sweden involving approximately 49,000 women and a follow-up of 20 years revealed dose-response effects for the reduction of clinically diagnosed depression in association with adherence to the Mediterranean diet, with the strongest effects in subjects with severe depression and of older age (Yin et al., 2021). A recent meta-analysis of randomized controlled trials (RCTs) shows that MedDiet interventions lower the severity of depressive symptoms in young and middle-aged adults with depression (SMD ≈ -0.53). However, there is a lot of variation in how these interventions are carried out, which makes it hard to predict how effective they will be (Encardone, 2025). A 2-year randomized trial involving previously depressed patients did not decrease the risk of recurrence but did mitigate subsyndromal symptoms at various time points, which may be significant for quality of life, even in the absence of preventing full relapse (Bockting et al., 2015). On the other hand, in a recent SR that only included RCTs focusing on depressive disorders, no definite impact of short-term MedDiet in comparison to controls was found in terms of depression severity; nevertheless, the certainty of this evidence was rated very low, indicating that there are clear methodological limitations that need to be addressed in terms of longer-term trials with mechanistic endpoints (Bizzozero-Peroni et al., 2025).

According to data from systematic reviews, there is evidence that adherence to the dietary approaches to stop hypertension (DASH) diet can improve

TABLE 8.1 Nutrients and bioactive compounds in animal foods with potential mood impact.

Animal food	Key nutrient/bioactive	Proposed mechanism related to mood	References
Milk (pasture/herbal-fed)	Omega-3 fatty acids (EPA/DHA)	Antiinflammatory effects; modulation of neuronal membrane fluidity; serotonin signaling	Kidd (2007), Zinkow et al. (2025)
Milk peptides (α-casozepine)	Bioactive peptides	GABA-A receptor modulation; anxiolytic-like effects	Benoit et al. (2020), Lalonde (2024)
Fermented dairy	Probiotic strains (*Lactobacillus, Bifidobacterium*)	Microbiome modulation; SCFA production; cytokine regulation	Gao et al. (2025)
Eggs (enriched)	Choline	Acetylcholine synthesis; neurodevelopment; methylation pathways	Mujica-Coopman et al. (2024)
Eggs	Lutein and zeaxanthin	Antioxidant protection; cognitive support	Wallace (2018)
Red meat	Vitamin B12, iron, zinc	Neurotransmitter synthesis; prevention of anemia-related fatigue	Kolarš et al. (2025)

psychological outcomes. Multiple observational studies have linked better adherence to DASH to fewer depressive symptoms, and other RCTs have found that participants in specific populations (e.g., those with type 2 diabetes) saw improvements in quality of life or emotional symptoms (Daneshzad et al., 2022; Tan et al., 2024). In addition, higher levels of adherence to the DASH diet were correlated with lower levels of depression and stress in case–control type analyses (e.g., those following up with individuals post-COVID) and this may also be attributed to the foods/diet overall being made up primarily of fruits/vegetables, whole grains, nuts and low levels of sodium and sugar—which may potentially reduce chronic inflammation and improve vascular health, both of which have positive effects on brain health (Khorasanchi et al., 2022).

According to a significant meta-analysis of randomized controlled studies on dietary improvement in both clinical and non-clinical samples, findings indicate very small but statistically significant reductions in depression (Hedges' g = 0.28–0.32) and do not show a consistent effect on anxiety, which suggests that dietary factors may impact more significantly depression than other mood disorders; hence, individuals who receive dietary interventions should have individualized programs that address sex, initial dietary behavior, and inflammatory status (Firth et al., 2019). Although the results have been mixed (because of differences in measurement of adherence, traditional patterns of food consumption, and the complex nature of mood disorders), overall there is considerable support for the incorporation of antiinflammatory, plant-based diets (MedDiet/DASH) into the overall treatment of mood disorders in conjunction with physical activity and sleep (Esgunoglu et al., 2024). While there continue to be mixed results due to variability in adherence measurement and diet, as well as different cultural foods and the complicated nature of mood disorders, the majority of the available research supports using antiinflammatory, plant-based diets (such as the Mediterranean or DASH diets) as part of a comprehensive treatment approach to mood disorders, along with exercise and proper sleep habits (Esgunoglu et al., 2024). Future studies will utilize methodologic designs focused on mechanisms—such as the acute crossover study comparing the effects of a Mediterranean diet to a Western diet on mood and cognitive function following meals that are matched on macronutrient composition—and will employ multiomic analyses to facilitate the resolution of variability in the results of prior studies and the identification of patient subgroups that respond favorably.

8.3 Psychobiotic potential of animal-origin foods

8.3.1 Hypothetical pathways

Theoretically, "psychobiotic" animal foods may emerge when herbivores graze on botanically diverse pastures or receive specialized phytofeeds, resulting in the concentration of plant-derived metabolites (terpenoids, phenolics, carotenoids) and microbially derived products in milk, meat, or eggs that, upon human consumption, activate gut–brain signaling (Caradus et al., 2024). Studies on grazing systems indicate that a greater diversity of plants in pastures results in higher concentrations of phytochemicals (such as terpenoids, phenols, and carotenoids) in meat and milk when contrasted with feedlot diets. This could enhance the antiinflammatory and antioxidant benefits for consumers, extending beyond merely fatty acids (ω-3, CLA) (van Vliet et al., 2021). In poultry, reformulating feed and giving them access to pasture changes the yolk's carotenoids and antioxidants. This results in eggs with higher lutein/zeaxanthin levels and better ω-6: ω-3 ratios—molecules that are known to play a role in visual and cognitive physiology and may protect the brain from oxidative stress (Sergin et al., 2022). Controlled supplementation has shed light on transfer efficiency: Hens receiving xanthophylls have been found to have a marked improvement in the amount of lutein/zeaxanthin in eggs (more effectively if in an external form), which could provide a mechanism to enrich animal products, which could have an influence on macular pigment carotenoid levels in the human brain (Nolan et al., 2016; Zaheer, 2017).

In ruminants, clover-rich forages, particularly red clover, increase milk isoflavones and the mammalian metabolite equol, indicating that phytoestrogenic phenolics can move from feed to rumen to milk and into human diets; these changes depend on the type of forage and how much concentrate is used (Steinshamn et al., 2008). These phytochemicals may influence the gut–brain axis via anti-inflammatory signaling, endocrine mimicry, microbial modulation, or direct receptor interactions following host or microbial biotransformation (Krajčová et al., 2010). The progress in the field of psychobiotics suggests that dietary influences can shape microbial metabolites (e.g., SCFAs), cytokine profiles, and neuroactive transmitter reservoirs that can be targeted by phytochemical-rich animal products (Sarkar et al., 2016). Although the supplementation of resveratrol in dairy cows has been mainly explored in terms of health implications in animals and heat stress resilience, this compound can be used to explain the concept that phytochemicals can be used in animal diets to modify the microbial populations in the rumen and consequently impact the product composition and human response, an area that is currently progressing in human-centric studies (Chen et al., 2023; Rezaei Ahvanooei et al., 2023).

The theory suggests that "herbal-fed" animal matrices function as carrier systems for neuroactive or neuroprotective molecules, including carotenoids, isoflavones, and terpenoids, along with their microbially transformed derivatives. Upon ingestion, these substances modulate host immunity, barrier function, and neuromodulatory pathways (including BDNF and the GABAergic/glutamatergic balance) through both gut microbiome-dependent and gut microbiome-independent mechanisms (Nair et al., 2025). There is evidence of food-chain transfer (milk phytoestrogens; fortified eggs), but the dosage, bioavailability, and clinical effects on mood and cognition have yet to be determined in rigorously controlled human trials that connect farm biodiversity, product metabolomes, microbiome alterations, and brain outcomes.

8.3.2 Functional animal products

Milk proteins yield bioactive peptides that demonstrate anxiolytic-like effects in preclinical models; the most extensively characterized is α-casozepine, a decapeptide derived from casein that interacts with GABA. A benzodiazepine receptor modulates anxiety-like behavior in rodents, with emerging human translational indicators from hydrolysate substances (e.g., alleviation of sleep or anxiety symptoms) (Benoit et al., 2020). This study reviews food-protein-derived anxiolytic peptides and explores potential mechanisms, including the GABAergic system and c-Fos patterns that differ from diazepam. This supports the concept of "calming dairy" products; however, there is a scarcity of research employing standardized endpoints and dose-response studies in diagnosed mood conditions (Hafeez et al., 2021).

Fermented dairy products may provide an additional dimension—that is, they provide a psychobiotic effect. Meta-analysis shows that probiotics may have a positive effect on reducing symptoms of depression among clinical populations, while recent studies have shown a relationship between consumption of fermented dairy products and decreased risk of developing depression in certain groups of people (Asad et al., 2025); however, the evidence for this association is mixed and dependent on the specific type of product used; there are prospective studies that have shown no relationship between stress and depression that also demonstrate a positive relationship between the two variables (Luo et al., 2023). In a recent triple-blind RCT using healthy female participants, a probiotic fermented yogurt product was found to alter metabolites and connectivity within the hippocampus as well as beta-diversity within the gut microbiome; these findings suggest a potential central mechanism, although there were no statistically significant changes in the subjects' level of depression (Karbownik et al., 2022; Marx et al., 2026).

Choline (a precursor to acetylcholine) and lutein/zeaxanthin are both found in eggs. They are believed to help with cognitive function as we age; however, strong evidence has shown the importance of choline for the development of a baby's brain. Therefore functional concepts based on eggs, including docosahexaenoic acid (DHA) and lutein-enriched eggs, can be viewed as biologically plausible approaches to promoting cognitive health, although research is limited at this time regarding the effects on mood (Nolan et al., 2016). Tryptophan and tyrosine are found in meats; however, one must consider the role of iron and B12 when discussing "mood-modulating meat" and the relationship to lipids and their potential to alter an individual's neuromuscular system. Grass-fed meat can provide omega-3s and phytochemicals that may also help to modify the tone of an individual's neuroimmunological system (Derbyshire and Obeid, 2020).

Some of the largest challenges are: (1) Standardizing bioactives by season, breed, and feed; (2i) ensuring they are accessible and bioavailable (peptide survival, carotenoid uptake); (3) ensuring they are effective in fermented foods (strain and substrate specificity); (4) conducting studies with preregistered outcomes (mood, anxiety, cognition) and mechanistic endpoints (SCFAs, cytokines, BDNF, and neuroimaging); and (5) overcoming regulatory and claim barriers that demand more than data demonstrating association. Not all milk peptides are advantageous; β-casomorphin-7 (derived from A1 β-casein digestion) exhibits opioid activity and is controversially associated with gut–brain effects (González-Rodríguez et al., 2025; Robinson et al., 2025).

8.3.3 Placebo and expectation effects

Any psychobiotic component in animal-derived foods will be confounded with the effects of expectancy (placebo/nocebo responses), which are potent variables influencing mood and subjective well-being and may render nutritional trials less valid (Sharma et al., 2021). The general literature on psychobiotics has already highlighted that the clinical impact is relatively small on average and highly dependent on the specific strain and context, rendering participant beliefs and branding cues such as "grass-fed," "probiotic," and "omega-3-enriched" highly relevant (Kezer et al., 2025). In the realm of food psychology, health-halo and labeling studies have demonstrated that when a food is regarded as "healthy," "organic," or "functional," it influences taste expectations, reward processing, and self-reported satiety or energy, mechanisms that may transiently enhance mood regardless of biochemical factors; thus double-blind, sensory-controlled studies and evaluations of expectancy are essential (Pham et al., 2019).

In RCTs involving psychobiotics, there is a lot of variability and publication bias. Expectancy influences subjective mood through unblinded fermented food interventions, which can significantly exaggerate the amount of benefit received. However, expectancy can also—even in open-label contexts—provide a pragmatic improvement in mood through reducing stress and modifying behavior (Asad et al., 2025). When functional animal products are marketed using narratives such as "calming milk peptides" and "brain-support eggs," those narratives may improve adherence with and perceived benefits of those products while also producing clinically important outcomes as a result of a minor biological effect in some users; yet, those same narratives could produce a nocebo effect for those same products if presented in a manner suggesting a potential for harm (e.g., A1 β-casein/BCM-7 discussions), creating an additional layer with which to interpret (Robinson et al., 2025).

The broad-based evaluation will require the use of multimodal endpoints pertaining to health, such as biomarker data (e.g., levels of SCFA, bioactivity of cytokine, and carotenoid status), microbiome profiles, neuroimaging or neurocognitive tests, and blind taste-matched controls for the purpose of assessing expectancy effects via psychological measurement instruments (e.g., questionnaires) (Vartanian et al., 2025). The recent study with a focus on the hippocampus and yogurt product illustrates a shift toward the use of brain-mechanistic outcome measures, as well as rigorous control over other potential sources of bias in trials, including changes in subtle symptoms of mental disorder, and it becomes a basis for the evaluation of animal-based psychobiotic products, such as lutein eggs, equol dairy products, and grass-fed milk (Hajzer et al., 2025). In the end, adding behavioral science to nutrition psychiatry will help tell the difference between real effects of nutrients on the microbiome and brain and improvements that are based on expectations. This will allow for credible claims and smart product design in this new field. Summary of psychobiotic mechanisms and hypothesized pathways is presented in Table 8.2.

8.4 Bioactive compound availability postdigestion

8.4.1 Stability in the gut environment

The extent to which a bioactive compound in the diet eventually reaches absorption sites depends on its bioaccessibility, which is the extent to which it is released from the food matrix and its resistance to degradation during the oral, gastric, and small intestinal phases (Dima et al., 2020). The INFOGEST static and semidynamic models of in vitro digestion have shown that pH, enzyme activity (salivary amylase, pepsin, and pancreatin), bile, and dilution rate are critical to the stability and micellization of lipophilic substances and the solubilization of hydrophiles, and these models were designed to provide a standardized comparison of such factors across foods (Mat et al., 2016; Xavier and Mariutti, 2021). Chemical structure determines the pH and enzyme sensitivity of many polyphenols. Multistage digestion studies show that some phenolics, such as quercetin glycosides and chlorogenic acid, are relatively stable in the acidic stomach but break down and change a lot in the small intestine, which is close to neutral to alkaline; this also changes the measured antioxidant capacity. Contrary to this, some subclasses have been observed to show an increase in their content following gastric digestion because of the release of the matrix (Li et al., 2023; Zhao et al., 2025). These reactions are not pH-dependent alone but are affected by targeted delivery systems and encapsulating agents such as chitosan/alginate systems and protein–polysaccharide systems, which protect phenolic compounds from degradation and affect their release in the intestine (Tang et al., 2025).

Food matrix constituents can also influence the amount of water-soluble material that can be absorbed by the gastrointestinal tract, including the effects of the fibers, proteins, and lipids contained in food matrices. Hydrogen bonding, hydrophobic interactions, and electrostatic interactions contribute to the formation of polysaccharide and polyphenol complexes. These interactions will affect the chemical stability and diffusivity of the complexes and can enhance the bioactivity of the complexes while decreasing the bioaccessibility of the complexes (i.e., delayed-release reservoirs remain in the colon) (Shahidi and Athiyappan, 2025). Recently, there has been growing data showing that nonextractable polyphenol-fiber conjugates (cell wall-bound phenolics, high molecular weight tannins/proanthocyanidins) are not able to be absorbed in the small

TABLE 8.2 Summary of psychobiotic mechanisms and hypothesized pathways.

Mechanism	Biological pathway	Brain-relevant outcome	References
SCFA production	Fermentation → acetate, propionate, butyrate	Blood–brain barrier integrity; antiinflammatory signaling	Silva et al. (2020)
Cytokine modulation	Reduced IL-6, TNF-α	Reduced neuroinflammation	Ishijima and Nakajima (2021)
Tryptophan metabolism	Kynurenine pathway regulation	Serotonin availability	Li et al. (2017)
Vagus nerve signaling	Gut microbial signaling → afferent vagal stimulation	Stress resilience	Kurhaluk et al. (2025)
HPA axis modulation	Cortisol regulation	Reduced anxiety/stress	Elnazer and Baldwin (2014)

intestine and are transported to the colon as substrates for microbial catabolism, thus changing the site and/or timing of their bioactivity (Fernandes et al., 2023).

The dairy matrix provides additional examples: Structure (e.g., gel versus fluid) and composition (casein:whey ratio, presence of the milk–fat–globule membrane) influence disintegration, lipid emulsification, and the release/micellization of lipophilic bioactives (carotenoids, tocopherols), modifying postprandial kinetics in vivo (Turgeon and Brisson, 2020). A human crossover study demonstrated significantly reduced β-carotene bioaccessibility and bioavailability from a high-fat dairy breakfast in contrast to an olive-oil matrix, indicating matrix-dependent sequestration and/or enhanced metabolic conversion as critical limiting factors (Kruger et al., 2024). Reviews on carotenoids in milk also emphasize the role of forage, processing, and product structure in carotenoid stability and bioaccessibility in digestion, again emphasizing the active role of the matrix, rather than being a passive carrier (Conboy Stephenson et al., 2021). Gastric resistance to acidity and pepsin is compound-specific; intestinal stability is the limiting factor for phenolics, and interactions between the matrix (fiber, protein, and lipid) can either protect or entrap bioactives, depending on where (small intestine vs. colon) and how (parent vs. transformed molecules) they will exert their action (Cirkovic Velickovic and Stanic-Vucinic, 2018).

8.4.2 Metabolite formation and absorption

For numerous plant bioactives, particularly polyphenols, the absorbed chemical entities are not the original compounds but rather microbially and host-metabolized derivatives exhibiting distinct pharmacokinetics and bioactivities (Gade and Kumar, 2023). A small number of glycosylated phenolics are broken down and absorbed in the small intestine. A lot of it goes to the colon, where resident microbiota break it down into simpler phenolic acids, urolithins, and other small molecules that are often easier to absorb and can enter circulation to act systemically (gut–liver–brain axis) (Alqudah and Claesen, 2024). The exact mechanisms (e.g., ellagitannins to urolithin A producers, flavan-3-ol to phenyl-γ-valerolactones to phenylpropionic acids) are determined by the microbial enzyme gene set of the individual, which leads to high variability of the response (Sallam et al., 2021). Fermentation and food processing can prebiotransform polyphenols to increase the solubility and bioavailability of these compounds, which are more easily absorbed, while simultaneously impacting the microbiota, events that have been summarized in recent reviews of fermented foods (Alharbi, 2026).

Following transepithelial passage via passive diffusion, carriers, or paracellular mechanisms, many phenolics and many other bioactives are subjected to first-pass metabolism in enterocytes and hepatocytes. Phase I reactions, primarily CYP-mediated oxidation, and Phase II conjugations—glucuronidation (UGTs), sulfation (SULTs), and O-methylation (COMT) produce polar conjugates that circulate to target tissues or are excreted via biliary or urinary routes

(Hoda et al., 2019) Extensive Phase II conjugation is a major cause of low plasma levels of the parent compound, and conjugates can keep or gain bioactivity at different targets (Hoda et al., 2019) Sulfation is important because human data on green tea EGCG shows that EGCG-4"-sulfate is a major circulating metabolite with plasma exposure similar to free EGCG, thanks to SULT1A1/1A3 activity in the liver and intestine (Hayashi et al., 2022). Moreover, mechanisms associated with Sulfotransferase (SULT) biology emphasize the importance of tissue localization and regulation through nuclear receptors and have demonstrated that Phase II activity (and genes) play a major role in the bioavailability of both natural products and xenobiotics such as drugs and chemicals (Xie and Xie, 2020). SCFAs, which are produced by the fermentation of dietary fiber by intestinal bacteria, are absorbed into colon cells via monocarboxylate-transporter (MCT) systems, serve as an energy source for epithelial tissue metabolism, and activate GPCRs to modulate the immune system and neuroendocrine activity, thereby demonstrating that microbial-derived compounds are bioavailable (Ríos-Covián et al., 2016; Silva et al., 2020). To sum up, microbial metabolism modulates the profile of materials that traverse the portal vein—host Phase I and Phase II metabolism establish the concentration and half-life of materials in circulation; therefore understanding the bioavailability of bioactive metabolites is essential for linking dietary intake with their physiological effects.

8.4.3 Challenges in measuring bioavailability

However, it is still difficult to accurately quantify the actual bioavailability of food bioactives because the processes of matrix degradation, luminal conversion, epithelial uptake, microbiota metabolism, and first-pass conjugation all occur between the time of ingestion and the time of systemic bioavailability (Song et al., 2025). To standardize the initial steps, the INFOGEST static protocol and its semidynamic extension establish community-consensus parameters (pH, electrolyte/enzyme activities, bile, volumes, and times) for evaluating bioaccessibility and digestion endpoints; these have significantly enhanced interstudy comparability, while recognizing that static models fail to represent gastric kinetics (Sousa et al., 2023). Dynamic upper-GI simulators (e.g., TIM-1) and microbiome reactors (e.g., SHIME, with proximal/distal colon and mucosal modules) more accurately replicate secretion, peristalsis, transit, and microbial fermentation, facilitating kinetic sampling and long-term dosing; however, they are expensive and necessitate specialized expertise (Zhu et al., 2024). Studies comparing in vitro models emphasize that predictions are always dependent on the level of complexity used and that it is not possible to find a system that accurately reflects in vivo physiology. Hence, tiered testing strategies from static to semidynamic to dynamic are used for cell-based uptake tests (Kiran et al., 2025).

For cellular uptake, Caco-2 monolayers are still considered the gold standard for measuring intestinal permeability; however, there are still some major issues: different labs use different levels of expression for transporters/enzymes, they are best for passive diffusion, and interfaces such as mucus, the immune system, and microbiota are not well represented (Kus et al., 2023). The current drive for improving in vitro testing encourages standardization of cultures and validation criteria, as well as conditioning of INFOGEST digests such as ultrafiltration. Even with improved vitro pipelines, in vivo variability (genetics, microbiota enterotypes, diet context) and first-pass hepatic/conjugative metabolism make it hard to go directly from "bioaccessibility/permeability" to systemic exposure (Camporesi et al., 2026). Physiologically based pharmacokinetic (PBPK) modeling and integrated in vitro–in silico workflows utilizing outputs from TIM/simulator of the human intestinal microbial ecosystem (SHIME) and cell systems have demonstrated enhanced prediction of human plasma profiles; however, they necessitate high-quality kinetic and distribution data to extend applicability beyond specific case studies (Havenaar and Bellmann, 2022).

There are still some gaps in the standardization of the processes. For instance, the digestion conditions are not reported in the same way; there are no matrices or controls; the endpoints are not the same (bioaccessibility, transepithelial flux, AUC of parent, and conjugates in the bloodstream), and there is a lack of integration of different types of omics in human studies (Likpalimor et al., 2025). Statements from the community on the dairy matrix and the effects of the matrix emphasize the importance of considering the structure of the food product, as opposed to the dose, when assessing bioavailability (Turgeon and Brisson, 2020; Weaver, 2021). Collectively, progress depends on synchronized in vitro digestion, suitable dynamic simulators for mechanisms, standardized epithelial models, and model-informed analysis in well-structured human studies that measure conjugated metabolites alongside parent compounds, ensuring that "availability postdigestion" accurately represents the body's perception. Fig. 8.1 shows the multilevel pathway from the intake of animal-based foods. These foods cause the release of bioactive molecules in the gut, which in turn stimulate mediators of the microbiome and immune systems, such as SCFAs and cytokines. These mediators then stimulate the CNS through the vagus nerve and blood signaling, finally affecting the brain to regulate neurotransmitters, neuroinflammation, and the HPA axis, leading to enhanced mood, cognitive functions, and stress resilience.

8.5 Gut–immune–brain axis interactions

8.5.1 Immune signaling pathways

Through a bidirectional communication system, there is a connection between the brain, the well-being of the immune system, and the gastrointestinal system, aided by circulating cytokines; this is done via the various effects of cytokines on microglia, which are the immune cells of the CNS (Fung, 2020). By entering the peripheral tissues of the immune system (gut mucosa, etc.), proinflammatory

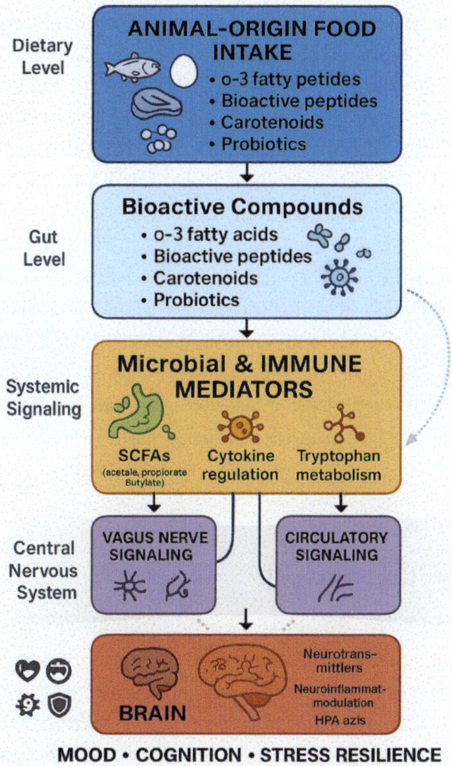

FIGURE 8.1 Flowchart of digestion → absorption → brain effect of bioactives.

cellular mediators (cytokines) such as IL-6 and TNF-α gain access to the CNS via multiple routes: they can enter directly by traveling across the fenestrated capillary walls of the circumventricular organs, they can also be transported across the BBB via active transport, or they may enter indirectly by means of vagal or humoral signaling therefore activating microglia via proinflammatory actions (Garzoni et al., 2013; Ronaldson and Davis, 2020). In addition, regulatory cytokines such as IL-10 can create a state that maintains homeostasis (i.e., M2-like state) and decrease the levels of inflammation in the CNS (Lobo-Silva et al., 2016). These three pieces of evidence suggest that metaflammation (systemic, low-grade inflammation) has an important role in activating microglia and modifying the process of synaptic pruning, as well as influencing neurotrophic signaling; thus linking peripheral immune tone to neuroinflammation and behavior (Wang et al., 2026).

Mechanically, SCFAs produced by colonic fermentation of dietary fiber affect immune set-points by promoting Treg cell development, improving epithelial barrier function, and binding to G-protein-coupled receptors, thus lowering systemic cytokine flux (e.g., IL-6 and TNF-α) and subsequently affecting microglia and the BBB (Kim, 2021; Tan et al., 2023). SCFAs can also provide energy to colonocytes and have epigenetic effects (e.g., butyrate acting as an HDAC inhibitor), which have been linked to the maturation of microglia and the reprogramming of the immune system to fight inflammation. This suggests a possible biochemical link between diet and microbiota composition and the immune tone in the CNS (Stein and Riber, 2023).

Clinically, elevated systemic IL-6 and TNF-α correlate with cognitive and affective symptoms in inflammatory disorders, and human cohort/intervention studies demonstrate that diet- or microbiota-modulated alterations in inflammatory markers are associated with brain-relevant outcomes (Vanlaer et al., 2025). Thus to sum up, it can be said that the proinflammatory cytokines such as IL-6 and TNF-α, and antiinflammatory cytokines such as IL-10 are balanced in a manner which is influenced by microbiota, barrier function, and diet to determine whether neuroinflammatory signals are enhanced or reduced, and microglia are the key players here.

8.5.2 Neurotransmitter crosstalk

The gut microbiota adds to the host neurochemical environment by either creating or modulating neurotransmitters and their precursors, which have effects on brain function via neural, endocrine, immune, and metabolic systems (Oleskin et al., 2017). Certain *Lactobacillus* and *Bifidobacterium* species are able to produce GABA from glutamate via the activity of the enzyme glutamate decarboxylase, whereas *Enterococcus, Escherichia*, and *Streptococcus* species can influence serotonin (5-HT) and dopamine levels via tryptophan metabolism and the action of aromatic amino acid decarboxylases and through the regulation of 5-HT release from enterochromaffin cells (Dicks, 2022). Although the neurotransmitters created by microbes might not be able to easily cross the BBB, the effect of neurotransmitters that are created by microbes at the periphery on vagal afferents, immune cells, and intestinal endocrine cells, together with the conversion from neurotransmitter to bioactive metabolite, can have an effect on central neurotransmission (Oleskin et al., 2017).

For example, microbial metabolism of the amino acid tryptophan can regulate the relative activity of the serotonergic system versus the kynurenine system, impacting microglial activation and glutamatergic activity (Roth et al., 2021); SCFAs and secondary bile acids can regulate the activity of the dopaminergic system and GABAergic system through receptor activation and epigenetic modification (Dong et al., 2025). The concept of "psychobiotics" brings these findings together, arguing that probiotics, prebiotics, and synbiotics can impact mood and cognitive function through the regulation of neurotransmitter activity in concert with the immune system and endocrine system (Al-Adham et al., 2025). The concept of psychobiotics has been broadened in recent RCTs to include fermented dairy products: An 8-week triple-blind RCT in healthy women found that a probiotic yogurt product altered hippocampal glutathione levels and functional connectivity, as well as altering the microbiome beta-diversity in the gut microbiome. These neurochemical changes occurred in the absence of behavioral changes, indicating that neurochemical "footprints" occur before behavioral changes: "Behavioral changes are likely to be secondary to neurochemical changes" (Marx et al., 2026). The above-mentioned data altogether support a model where GABA, serotonin, and dopamine pathways in microbes, together with tryptophan/kynurenine equilibrium and SCFAs, fine-tune the excitatory–inhibitory balance in the CNS, stress responses, and neuroplasticity by crosstalk at multiple nodes.

8.5.3 Consumer health implications

There are three main domains in which gut–immune–brain mechanisms translate into something that can be relevant to consumers as follows: Stress resilience, cognitive performance, and anxiety/depression symptoms (Ribeiro et al., 2025). SCFAs and psychobiotic interventions can mitigate hyper-reactivity of the HPA axis by reducing interleukin (IL)-6/tumor necrosis factor (TNF)-alpha production through immune quiescence, vagal signaling and enhancing gut barrier function during times of stress, which will ultimately promote psychological and inflammatory resilience against both psychological and inflammatory stressors (Kurhaluk et al., 2025). There is some evidence from observational and interventional studies indicating that (1) there may be an association between the consumption of fermented foods/dairy products and probiotics and reduced depressive symptoms, and (2) that depressed individuals show measurable differences in at least some aspects of their brain function (e.g., hippocampal metabolites/connectivity) compared with nondepressed individuals, including some potentially small cognitive enhancements. However, these data are variable and typically show small effect sizes (Crocetta et al., 2024).

Fermented dairy products, i.e., yogurt/cheese, have been associated with reduced risk of depression according to a meta-analysis of cohort studies. The previously discussed RCT of probiotics + yogurt also showed neurometabolic modulation in the CNS. The above-mentioned findings provide further evidence for the brain-relevant effects of fermented dairy. A systematic review and meta-analysis of RCTs in patients with depression showed significant improvement of depressive symptoms by probiotics with a large SMD for depression and a moderate SMD for anxiety. However, heterogeneity and strain-specificity caution against universal recommendations (Luo et al., 2023; Asad et al., 2025; Marx et al., 2026). Expectancy ("health-halo") and product matrix can affect how people think about the benefits; so, we need blinded, strain-defined psychobiotic trials with inflammation (IL-6/TNF-α/CRP), tryptophan–kynurenine, SCFAs, and BDNF/neuroimaging endpoints to separate biological effects from expectancy effects and find responders (for example, people with inflammatory phenotypes or certain microbiome enterotypes) (Silva et al., 2020). The most

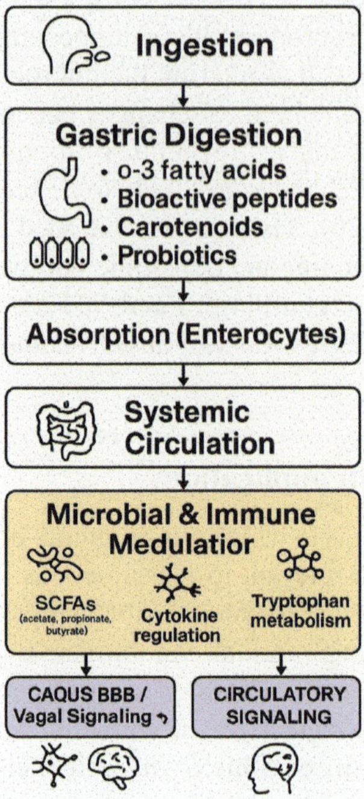

FIGURE 8.2 Gut–immune–brain axis in the context of animal-origin food intake.

evidence-based advice for consumers and professionals today focuses on: (1) Dietary fiber patterns that promote SCFA production; (2) fermented foods with known strains; and (3) overall antiinflammatory dietary quality—all of which target the immune "bridge" that connects the gut to the brain. As this field develops, precision nutrition that uses microbiome and inflammatory profiling should make psychobiotic strategies for mood, stress, and cognition more personalized. Fig. 8.2 shows the entire process from ingestion to gastric digestion of bioactive compounds (omega-3 fatty acids, bioactive peptides, carotenoids, and probiotics), absorption of these compounds by enterocytes into the systemic circulation, and then function as microbial and immune modulators (production of SCFAs, regulation of cytokines, and metabolism of tryptophan) that finally reach the brain via two separate routes: CAQUS BBB/vagal signaling and circulatory signaling.

8.6 Hypotheses for future human studies

8.6.1 Trial designs

We hypothesize that randomized, double-blind, placebo-controlled trials evaluating animal-derived foods from herbal/pasture-diverse feeding systems will yield significant effects on gut–immune–brain endpoints compared to taste- and macromatched conventional controls. Reviews indicate that species-diverse pasture feeding enriches meat and milk with terpenoids, phenolics, and carotenoids beyond ω-3 differences, yet direct human data are scarce—explicitly calling for systematic human trials to link feed-driven phytochemical profiles to consumer health outcomes (Ryan et al., 2016; van Vliet et al., 2021). To ensure internal validity, interventions must: (1) Standardize matrix and sensory attributes (fat, protein, salt, flavor) and obscure product identity; (2) preregister protocols and randomization; and (3) include expectancy evaluations to measure potential health-halo effects commonly associated with “grass-fed,” “natural,” or “probiotic” labels. (Comparable nutrition-brain trials demonstrate that double/triple-blind food designs with brain readouts are both feasible and sensitive.) The main idea is that feeding dairy and meat with herbs will lower systemic IL-6/TNF-α/CRP and make mood and cognitive composites better. This will happen through changes in the microbiome and SCFA dynamics (Giollabhui et al., 2021).

Meta-analytic, longitudinal evidence associates IL-6/CRP with subsequent depressive symptoms, endorsing inflammation as a mechanistic target for mood-related nutrition trials. Trials should include stratification for sex and baseline inflammatory status, randomizing participants to enriched or conventional products for at least 8–12 weeks in duration, including mechanistic substudy components such as metabolomics and microbiomics. Cortisol outcome measures are common in stress-related nutrition RCTs but are limited due to heterogeneous sampling methods. We suggest using standardized methods for diurnal cortisol or cortisol-awakening-response assessments: Wake, +30 min wake-up, +60 min wake-up, midday, and evening time points (Tribble et al., 2015; Ryan et al., 2016).

8.6.2 Endpoints and biomarkers

We hypothesize that dietary interventions resulting in diminished systemic inflammation and enhanced microbiota profiles will produce beneficial alterations in established psychological metrics and biomarkers. The main psychological outcomes should be Profile of Mood States (POMS) (total mood disturbance; tension, vigor, fatigue, etc.), Hospital Anxiety and Depression Scale (HADS) (anxiety/depression subscales), and Positive and Negative Affect Schedule (PANAS) (positive/negative affect). These are all well-established psychometrics for nutrition studies that calculate the same thing frequently (Watson et al., 1988; Jovanović, 2015). Derived cognitive measures may include attention and working memory tests, with the addition of optional neuroimaging techniques such as resting-state functional magnetic resonance imaging (rs-fMRI), depending on the availability of resources. High-sensitivity CRP, IL-6, TNF-α, diurnal salivary cortisol, CAR, and diurnal slope will be assessed, as there is evidence that inflammatory markers may predict risk for mood and that salivary cortisol sampling is required for the proper interpretation of results (Giollabhui et al., 2021).

Investigative mediators encompass tryptophan–kynurenine ratios and fecal SCFAs, aiming to elucidate the mechanistic link between microbial metabolism and the central excitatory–inhibitory balance as well as microglial tone within psychoneuroimmunology frameworks (Morris et al., 2017). For the gut microbiota, we recommend shotgun metagenomics or 16S rRNA sequencing with alpha-/beta-diversity, taxa, and pathway inference, along with standardized sample collection and processing. Dynamic in vitro models such as SHIME can be used to investigate fermentation mechanisms in controlled settings prior to or concurrently with human trials (Teixeira et al., 2024; Zhu et al., 2024). Predefined mediation models (product → microbiome/SCFAs/inflammation/cortisol → POMS/HADS/PANAS) will evaluate causal pathways (Ryan et al., 2016). Lastly, it is essential to quantify and modify expectancy in analyses to distinguish between biological and belief-driven effects, which is a critical concern in functional food trials.

8.6.3 Ethical and regulatory considerations

Every study will abide by the World Medical Association's Declaration of Helsinki with respect to each situation, including obtaining informed consent from the subject, weighing the risk against the benefit, protecting vulnerable persons involved in the study, getting approval on the research protocol from a separate committee, and disclosing information on the conduct and results of the study (World Medical Association, 2024). In addition, the entire food chain from production to consumption must meet stringent food safety requirements, because the research subjects are receiving the experimental product as food rather than as a medicine. In addition, the production and management of the study products must comply with the General Principles of Food Hygiene as established by the Codex Alimentarius (CXC 1-1969-as amended in 2020/2022) and documented by having a Hazard Analysis and Critical Control Point (HACCP) system for records to validate that the test and control products will be produced in a safe manner (World Health Organization, 2023). If study dissemination or post-trial commercialization includes nutrition or health claims, investigators and partners must follow Regulation (EC) No 1924/2006 in the EU/UK and FDA rules in the U.S. For example, 21 CFR 101.93 for structure/function statements (30-day notice; substantiation; required disclaimer) (Verhagen et al., 2010).

Unless and until authorized health claims are made, labels in publications and materials should not make disease claims and should use neutral terms (e.g., "pasture-diverse dairy"). In addition to consent and labeling, ethics necessitate data privacy, adverse event surveillance, and regulations commensurate with low-risk food trials; furthermore, there must be community transparency regarding sourcing and husbandry. This level of protection allows researchers to carry out high-quality, blind, placebo-controlled food-based RCTs in a way that is compliant with ethical standards, food-safe, and considers any governing regulations. Consequently, it has become more feasible for animal products fed herbs to be taken from the laboratory/paddock and into the clinical setting (Smith et al., 2015; Murdoch and Caulfield, 2018).

8.6.4 Educational and health consequences of English language teaching

The research on the gut–immune–brain axis, psychobiotics, and functional animal products also points out the importance of communicative responsibility in addition to the complexities of the subject. As the subject matter gets more technical in terms of the claims made on the health benefits of "microbiome modulation," "antiinflammatory pathways," "bioactive peptides," and "structure-function claims," the scope for misunderstanding and overgeneralization also increases. Health literacy, which is the ability to access, understand, appraise, and apply health information, has been cited as an important factor in health outcomes (Nutbeam, 2000;Sørensen et al., 2012).

Decreased health literacy is consistently linked to suboptimal health decisions, diminished compliance with medical recommendations, and increased vulnerability to misleading or exaggerated health assertions (Berkman et al., 2011). In the realm of functional foods and nascent psychobiotic products, consumers are required to analyze probabilistic evidence, differentiate correlation from causation, and comprehend regulatory distinctions between sanctioned health claims and structure-function statements (Verhagen et al., 2010). These interpretive demands highlight the significance of organized educational interventions.

The ESP and CLIL frameworks offer models to help educators create language education programs that include current nutritional science (Dudley-Evans, 1998; Coyle et al., 2010). Task-based learning activities using authentic research abstracts, randomized controlled trial reports, and regulatory documents can help students develop their disciplinary literacy and critical appraisal skills in medical, pharmaceutical, agricultural, and life sciences programs. These models support the recommendations within health education that call for an emphasis on critical evaluation of scientific information versus just acquiring this information passively (Kickbusch et al., 2013).

Learners can be engaged more deeply when they engage in real-world learning applications that take complex biomedical ideas and translate them for the public using project-based instructional strategies. Research also shows that when people receive structured training on how to simplify complex information, they not only improve their comprehension of the information but also improve their resistance to misinformation (National Academies of Sciences, Engineering, and Medicine, 2017). In addition, students' ability to interpret functional food marketing claims may be improved through instruction on the recognition of placebo and expectancy design effects, and this will help reinforce their ability to reason from the evidence (Sharma et al., 2021).

From a curriculum-development standpoint, collaboration between nutrition scientists and experts in English language pedagogy may enhance the creation of interdisciplinary modules that incorporate:

1. Critical reading of epidemiological and randomized controlled trial evidence
2. Discourse analysis of health-claim framing
3. Ethical communication standards in line with WHO and WMA guidance
4. Reflective evaluation of consumer perception and health-halo effects

Such integration not only enhances academic English proficiency but also contributes to broader public health objectives by improving consumer decision-making capacity. As such, with the progression from functional animal products through the research pipeline and into the marketplace, the educational component becomes an important supplement to biological proof. In that regard, health literacy and science communication can be understood as a bridge between agriculture, research, and consumer engagement.

8.7 Conclusion

The intersection of agricultural innovation, nutrition research, and consumer health is exemplified by animal-derived foods that are produced in an environment of herbal feeds and pasture diversity. The systems underpinning the production of milk, meat, eggs, etc. can be a means to modulate fatty acids, carotenoids, bioactive peptides, and other derivatives to influence the dynamics of the gut–immune–brain axis and downstream pathways involved with inflammation, neuroplasticity, and stress. Evidence from converging fields such as microbiome science, psychoneuroimmunology, and nutritional biochemistry supports mechanistic plausibility. However, while biological plausibility exists, it does not provide sufficient evidence to support meaningful health claims. Therefore well-designed, controlled human studies are critical for establishing dose-response connections, bioavailability of metabolites related to their composition, interindividual differences, and clinically important sizes of response to treatment effect. Precision nutrition approaches, including inflammatory profiling, microbiome characterization, and metabolomics, may help to classify subgroups of responders and refine translational efforts.

Correspondingly important is the communicative aspect related to the development of functional foods. Indeed, as products with high levels of omega-3 fatty acids, carotenoids, bioactive peptides, and probiotics are introduced into consumer markets, clear and evidence-based communication represents a scientific imperative rather than a marketing choice. The misinterpretation of probabilistic evidence, expectancy effects, and health-halo effects can lead to misperceptions regarding the value of these products if regulatory and ethical standards are not stringently applied.

In this context, combining health literacy and educational approaches with the translational pathway for health information is part of the foundation of public health rather than an ancillary consideration. By enhancing the consumer's ability to critically evaluate scientific claims, understand the results of controlled trials (especially RCTs), and to understand how to differentiate between correlation and causation, we are providing an indirect but significant contribution to public health. Cooperation between agricultural scientists, nutrition scientists, health practitioners, and individuals who design learning materials will help promote the development of both innovative food products (functional animal-derived foods) as well as providing an increased understanding of evidence-based nutrition. The future success of functional animal-derived foods is contingent on a balanced approach to validation, regulation, and education. When agricultural practices are aligned with rigorous biomedical evidence and supported by transparent, literacy-informed communication strategies, the potential exists to narrow the gap between production systems and verifiable improvements in consumer well-being.

References

Al-Adham, I.S.I., Ali Agha, A.S.A., Al-Remawi, M., AlAkayleh, F., Al-Sheikh, A., Aburub, F., Collier, P.J., 2025. Probiotics, psychobiotics, paraprobiotics, and postbiotics in gut-brain axis modulation: multi-omics and AI-driven precision nutrition for cognitive health. Probiotics Antimicrob. Proteins doi:10.1007/s12602-025-10886-8.

Alharbi, N.A., 2026. Polyphenol metabolites in fermented foods: biotransformation, bioavailability, and functional roles. Front. Nutr. 13. doi:10.3389/fnut.2026.1767453.

Alqudah, S., Claesen, J., 2024. Mechanisms of gut bacterial metabolism of dietary polyphenols into bioactive compounds. Gut Microbes 16 (1). doi:10.1080/19490976.2024.2426614.

Amidfar, M., Garcez, M.L., Askari, G., Bagherniya, M., Khorvash, F., Golpour-Hamedani, S., de Oliveira, J., 2024. Role of BDNF Signaling in the neuroprotective and memory-enhancing effects of flavonoids in Alzheimer's disease. CNS Neurol. Disord. Drug Targets 23 (8), 984–995. doi:10.2174/1871527323666230912090856.

Asad, A., Kirk, M., Zhu, S., Dong, X., Gao, M., 2025. Effects of prebiotics and probiotics on symptoms of depression and anxiety in clinically diagnosed samples: systematic review and meta-analysis of randomized controlled trials. Nutr. Rev. 83 (7), e1504. doi:10.1093/nutrit/nuae177.

Barros, M.I., Brandão, T., Irving, S.C., Alves, P., Gomes, F., Correia, M., 2025. Omega-3 polyunsaturated fatty acids and cognitive decline in adults with non-dementia or mild cognitive impairment: an overview of systematic reviews. Nutrients 17 (18). doi:10.3390/nu17183002.

Benoit, S., Chaumontet, C., Schwarz, J., Cakir-Kiefer, C., Boulier, A., Tomé, D., Miclo, L., 2020. Anxiolytic activity and brain modulation pattern of the α-casozepine-derived pentapeptide YLGYL in mice. Nutrients 12 (5), 1497. doi:10.3390/nu12051497.

Berg, J.N., Grant, R.E.C., Siervo, M.H.N., Stephan, B.C.M., Tully, P.J., 2025. Efficacy of B vitamin supplementation on global cognitive function in older adults: a systematic review and meta-analysis. Nutr. Rev. 83 (12), 2256–2267. doi:10.1093/nutrit/nuaf155.

Berkman, N.D., Sheridan, S.L., Donahue, K.E., Halpern, D.J., Crotty, K., 2011. Low health literacy and health outcomes: an updated systematic review. Ann. Intern. Med. 155 (2), 97–107. doi:10.7326/0003-4819-155-2-201107190-00005.

Bizzozero-Peroni, B., Martínez-Vizcaíno, V., Fernández-Rodríguez, R., Jiménez-López, E., Núñez de Arenas-Arroyo, S., Saz-Lara, A., Díaz-Goñi, V., Mesas, A.E., 2025. The impact of the Mediterranean diet on alleviating depressive symptoms in adults: a systematic review and meta-analysis of randomized controlled trials. Nutr. Rev. 83 (1), 29–39. doi:10.1093/nutrit/nuad176.

Bockting, C.L., Hollon, S.D., Jarrett, R.B., Kuyken, W., Dobson, K., 2015. A lifetime approach to major depressive disorder: the contributions of psychological interventions in preventing relapse and recurrence. Clin. Psychol. Rev. 41, 16–26. doi:10.1016/j.cpr.2015.02.003.

Camporesi, G., Canzian, C., Bordoni, A., 2026. Optimizing INFOGEST digest conditioning for reliable in vitro assessment of nutrient bioavailability using Caco-2 cell models. Nutrients 18 (2), 339. doi:10.3390/nu18020339.

Cao, Q., Shen, M., Li, R., Liu, Y., Zeng, Z., Zhou, J., Niu, D., Zhang, Q., Wang, R., Yao, J., Zhang, G., 2025. Elucidating the specific mechanisms of the gut-brain axis: the short-chain fatty acids-microglia pathway. J. Neuroinflam. 22 (1). doi:10.1186/s12974-025-03454-y.

Caradus, J.R., Chapman, D.F., Rowarth, J.S., 2024. Improving human diets and welfare through using herbivore-based foods: human and animal perspectives. Animals 14 (7). doi:10.3390/ani14071077.

Cheatham, C.L., Nieman, D.C., Neilson, A.P., Lila, M.A., 2022. Enhancing the cognitive effects of flavonoids with physical activity: is there a case for the gut microbiome? Front. Neurosci. 16. doi:10.3389/fnins.2022.833202.

Chen, S., Wang, X., Cheng, Y.u, Gao, H., Chen, X., 2023. A review of classification, biosynthesis, biological activities and potential applications of flavonoids. Molecules 28 (13), 4982. doi:10.3390/molecules28134982.

Cheng, L., Shi, C., Li, X., Matsui, T., 2024. Impact of peptide transport and memory function in the brain. Nutrients 16 (17), 2947. doi:10.3390/nu16172947.

Cheng, N., Bell, L., Lamport, D.J., Williams, C.M., 2022. Dietary flavonoids and human cognition: a meta-analysis. Mol. Nutr. Food Res. 66 (21). doi:10.1002/mnfr.202100976.

Cirkovic Velickovic, T.D., Stanic-Vucinic, D.J., 2018. The role of dietary phenolic compounds in protein digestion and processing technologies to improve their antinutritive properties. Compr. Rev. Food Sci. Food Saf. 17 (1), 82–103. doi:10.1111/1541-4337.12320.

Clerici, L., Bottari, D., Bottari, B., Microbiome, G., 2025. Diet and depression: literature review of microbiological, nutritional and neuroscientific aspects. Curr. Nutr. Rep. 14 (1). doi:10.1007/s13668-025-00619-2.

Conboy Stephenson, R., Ross, R.P., Stanton, C., 2021. Carotenoids in milk and the potential for dairy based functional foods. Foods 10 (6), 1263. doi:10.3390/foods10061263.

Coyle, D., Hood, P., Marsh, D., 2010. CLIL: Content and Language Integrated Learning. Cambridge University Press, Cambridge.

Crocetta, A., Liloia, D., Costa, T., Duca, S., Cauda, F., Manuello, J., 2024. From gut to brain: unveiling probiotic effects through a neuroimaging perspective—A systematic review of randomized controlled trials. Front. Nutr. 11. doi:10.3389/fnut.2024.1446854.

Czapski, G.A., Strosznajder, J.B., 2021. Glutamate and gaba in microglia-neuron cross-talk in alzheimer's disease. Int. J. Mol. Sci. 22 (21). doi:10.3390/ijms222111677.

Daneshzad, E., Heshmati, J., Basirat, V., Keshavarz, S.A., Qorbani, M., Larijani, B., Bellissimo, N., Azadbakht, L., 2022. The effect of the dietary approaches to stop hypertension (DASH) diet on sleep, mental health, and hormonal changes: a randomized clinical trial in women with type 2 diabetes. Front. Nutr. 9. doi:10.3389/fnut.2022.775543.

Derbyshire, E., Obeid, R., 2020. Choline, neurological development and brain function: a systematic review focusing on the first 1000 days. Nutrients 12 (6), 1731. doi:10.3390/nu12061731.

Dicks, L.M.T., 2022. Gut bacteria and neurotransmitters. Microorganisms. 10 (9), 1838. doi:10.3390/microorganisms10091838.

Dima, C., Assadpour, E., Dima, S., Jafari, S.M., 2020. Bioavailability and bioaccessibility of food bioactive compounds; overview and assessment by in vitro methods. Compr. Rev. Food Sci. Food Saf. 19 (6), 2862–2884. doi:10.1111/1541-4337.12623.

Djuric, D.M., Todorović, D., Bajić, Z., Krneta, S.M., Šobot, T., 2024. Is homocysteine a biomarker of environmental health risk and epigenetic-DNA methylation: links to cardiovascular pathogenesis and B vitamins. Adv. Biochem. Health and Dis. 30, 383–452. doi:10.1007/978-3-031-62806-1_15.

Dong, T., Zhang, T., Wang, H., Zhang, J., Abdullah, R., Sun, B., Peng, G., 2025. Microbiota–gut–brain axis and bile acids–driven neuromodulation. Neural Regen. Res. doi:10.4103/nrr.nrr-d-25-00927.

Dudley-Evans, T., St John, M.J., 1998. Developments in English for Specific Purposes: A Multi-disciplinary Approach. Cambridge University Press, Cambridge.

Elnazer, H.Y., Baldwin, D.S., 2014. Investigation of cortisol levels in patients with anxiety disorders: a structured review. Curr. Top. Behav. Neurosci. 18, 191–216. doi:10.1007/7854_2014_299.

Encardone, A., (2025). Healthy Dietary Patterns' Impact on Depressive Symptoms: A Meta-Analysis of Randomized Controlled Trials (Master's thesis, Clemson University, South Carolina, USA).

Esgunoglu, L., Liaquat, M., Gillings, R., Lazar, A., Leddy, A., Brooks, J., Penny, W., Sami, S., Hornberger, M., Stevenson, E., Jennings, A., Minihane, A.M, 2024. Acute effect of a Mediterranean-style dietary pattern (MDP) on mood, anxiety and cognition in UK adults with mild to moderate anxiety and depression: the MediMood randomised controlled trial protocol. BMJ Open 14 (12), e082935. doi:10.1136/bmjopen-2023-082935.

Fekete, M., Lehoczki, A., Tarantini, S., Fazekas-Pongor, V., Csípő, T., Csizmadia, Z., Varga, J.T., 2023. Improving cognitive function with nutritional supplements in aging: a comprehensive narrative review of clinical studies investigating the effects of vitamins, minerals, antioxidants, and other dietary supplements. Nutrients 15 (24), 5116. doi:10.3390/nu15245116.

Fernandes, A., Mateus, N., Freitas, V., 2023. Polyphenol-dietary fiber conjugates from fruits and vegetables: nature and biological fate in a food and nutrition perspective. Foods 12 (5), 1052. doi:10.3390/foods12051052.

Firth, J., Marx, W., Dash, S., Carney, R., Teasdale, S.B., Solmi, M., Stubbs, B., Schuch, F.B., Carvalho, A.F., Jacka, F., Sarris, J., 2019. The effects of dietary improvement on symptoms of depression and anxiety: a meta-analysis of randomized controlled trials. Psychosom. Med. 81 (3), 265–280. doi:10.1097/PSY.0000000000000673.

Fock, E., Parnova, R., 2023. Mechanisms of blood–brain barrier protection by microbiota-derived short-chain fatty acids. Cells 12 (4), 657. doi:10.3390/cells12040657.

Fung, T.C., 2020. The microbiota-immune axis as a central mediator of gut-brain communication. Neurobiol. Dis. 136, 104714. doi:10.1016/j.nbd.2019.104714.

Gade, A., Kumar, M.S., 2023. Gut microbial metabolites of dietary polyphenols and their potential role in human health and diseases. J. Physiol. Biochem. 79 (4), 695–718. doi:10.1007/s13105-023-00981-1.

Gao, Y., Liu, Y., Ma, T., Liang, Q., Sun, J., Wu, X., Song, Y., Nie, H., Huang, J., Mu, G., 2025. Fermented dairy products as precision modulators of gut microbiota and host health: mechanistic insights, clinical evidence, and future directions. Foods 14 (11), 1946. doi:10.3390/foods14111946.

Garzoni, L., Faure, C., Frasch, M.G., 2013. Fetal cholinergic anti-inflammatory pathway and necrotizing enterocolitis: the brain-gut connection begins in utero. Front. Integr. Neurosci. 7. doi:10.3389/fnint.2013.00057.

Ghallab, Y.K., Elassal, O.S., 2024. Biochemical and neuropharmacology of psychiatric disorders. In: Elassal, O.S., Ghallab, Y.K. (Eds.), Nutrition and Psychiatric Disorders: An Evidence-Based Approach to Understanding the Diet-Brain Connection. Springer Nature Singapore, Singapore, pp. 25–47.

González-Rodríguez, N., Vázquez-Liz, N., Rodríguez-Sampedro, A., Regal, P., Fente, C., Lamas, A., 2025. The impact of A1-and A2 β-casein on health outcomes: a comprehensive review of evidence from human studies. Appl. Sci. 15 (13), 7278.

Hafeez, Z., Benoit, S., Cakir-Kiefer, C., Dary, A., Miclo, L., 2021. Food protein-derived anxiolytic peptides: their potential role in anxiety management. Food Funct. 12 (4), 1415–1431. doi:10.1039/d0fo02432e.

Hajzer, Z.E, Alibrahem, W., Helu, N.K, Oláh, C., Prokisch, J., 2025. Functional foods in clinical trials and future research directions. Foods. 14 (15), 2675. doi:10.3390/foods14152675.

Han, A., Almeida, L., Anand, N., Salloum, I.M., Kanaan, S., Gadad, B.S., Daher, J.P.L., 2025. Exploring neuropsychiatric manifestations of vitamin B complex deficiencies. Front. Psychiatry 16. doi:10.3389/fpsyt.2025.1569826.

Havenaar, R., Bellmann, S., 2022. Results from a validated in vitro gastrointestinal model (TIM) used as input data for in silico modeling give highly predictive information for the human situation. Med. Res. Arch. 10 (9). doi:10.18103/mra.v10i9.3136.

Hayashi, A., Terasaka, S., Nukada, Y., Kameyama, A., Yamane, M., Shioi, R., Iwashita, M., Hashizume, K., Morita, O., 2022. 4″-Sulfation is the major metabolic pathway of epigallocatechin-3-gallate in humans: characterization of metabolites, enzymatic analysis, and pharmacokinetic profiling. J. Agric. Food Chem. 70 (27), 8264–8273. doi:10.1021/acs.jafc.2c02150.

Hoda, M., Hemaiswarya, S., Doble, M., 2019. Pharmacokinetics and pharmacodynamics of polyphenols. In: Hemaiswarya, S., Doble, M. (Eds.), Role of Phenolic Phytochemicals in Diabetes Management: Phenolic Phytochemicals and Diabetes. Springer Singapore, Singapore, pp. 159–173.

Ishijima, T., Nakajima, K., 2021. Inflammatory cytokines TNFα, IL-1β, and IL-6 are induced in endotoxin-stimulated microglia through different signaling cascades. Sci. Prog. 104 (4). doi:10.1177/00368504211054985.

Jenkins, T.A., Nguyen, J.C.D., Polglaze, K.E., Bertrand, P.P., 2016. Influence of tryptophan and serotonin on mood and cognition with a possible role of the gut-brain axis. Nutrients 8 (1). doi:10.3390/nu8010056.

Jovanović, V., 2015. Beyond the PANAS: incremental validity of the scale of positive and negative experience (SPANE) in relation to well-being. Person. Indiv. Differ. 86, 487–491. doi:10.1016/j.paid.2015.07.015.

Kalkman, H.O., Hersberger, M., Walitza, S., Berger, G.E., 2021. Disentangling the molecular mechanisms of the antidepressant activity of omega-3 polyunsaturated fatty acid: a comprehensive review of the literature. Int. J. Mol. Sci. 22 (9). doi:10.3390/ijms22094393.

Karbownik, M.S., Mokros, L., Dobielska, M., Kowalczyk, M., Kowalczyk, E., 2022. Association between consumption of fermented food and food-derived prebiotics with cognitive performance, depressive, and anxiety symptoms in psychiatrically healthy medical students under psychological stress: a prospective cohort study. Front. Nutr. 9. doi:10.3389/fnut.2022.850249.

Katayama, S., Nakamura, S., 2019. Emerging roles of bioactive peptides on brain health promotion. Int. J. Food Sci. Technol. 54 (6), 1949–1955. doi:10.1111/ijfs.14076.

Kezer, G., Paramithiotis, S., Khwaldia, K., Harahap, I.A., Čagalj, M., Šimat, V., Smaoui, S., Elfalleh, W., Ozogul, F., Esatbeyoglu, T., 2025. A comprehensive overview of the effects of probiotics, prebiotics and synbiotics on the gut-brain axis. Front. Microbiol. 16. doi:10.3389/fmicb.2025.1651965.

Khiroya, K., Sekyere, E., McEwen, B., Bayes, J., 2023. Nutritional considerations in major depressive disorder: current evidence and functional testing for clinical practice. Nutr. Res. Rev. 1–12.

Khorasanchi, Z., Ahmadihoseini, A., Hajhoseini, O., Zare-Feyzabadi, R., Haghighi, M., Heidari, M., Esfehani, A.J., Dehnavi, Z., Sharifan, P., Rashidianyazd, S., Latifi, M.R., Rastgooy, F., Ildarabadi, D., Bajgiran, M.M., Soflaei, S.S., Ferns, G., Mobarhan, M.G., 2022. Adherence to dietary approaches to stop hypertension (DASH) diet in relation to psychological function in recovered COVID-19 patients: a case–control study. BMC Nutr. 8 (1). doi:10.1186/s40795-022-00633-5.

Kickbusch, I., Pelikan, J., Apfel, F., Tsouros, A., 2013. Health Literacy: The Solid Facts. WHO Regional Office for Europe, Copenhagen.

Kidd, P.M., 2007. Omega-3 DHA and EPA for cognition, behavior, and mood: clinical findings and structural-functional synergies with cell membrane phospholipids. Altern. Med. Rev. 12 (3), 207–227.

Kim, C.H., 2021. Control of lymphocyte functions by gut microbiota-derived short-chain fatty acids. Cell. Mol. Immunol. 18 (5), 1161–1171. doi:10.1038/s41423-020-00625-0.

Kiran, B.S., Bashir, O., Pawase, P.A., Dash, K.K., Amin, T., Shams, R., Shaikh, A.M., Béla, K., 2025. A review of in-vitro digestibility models on diverse foods in various segments of human digestive tract. Discov. Food 5 (1). doi:10.1007/s44187-025-00454-y.

Kita, M., Obara, K., Kondo, S., Umeda, S., Ano, Y., 2018. Effect of supplementation of a whey peptide rich in tryptophan-tyrosine-related peptides on cognitive performance in healthy adults: a randomized, double-blind, placebo-controlled study. Nutrients 10 (7), 899. doi:10.3390/nu10070899.

Kolarš, B., Jovin, V.M., Živanović, N., Minaković, I., Gvozdenović, N., Kokeza, I.D, Lesjak, M., 2025. Iron deficiency and iron deficiency anemia: a comprehensive overview of established and emerging concepts. Pharmaceuticals 18 (8), 1104. doi:10.3390/ph18081104.

Krajčová, A., Schulzová, V., Lojza, J., Křížová, L., Hajšlová, J., 2010. Phytoestrogens in bovine plasma and milk - LC-MS/MS analysis. Czech J. Food Sci. 28 (4), 264–274. doi:10.17221/138/2010-cjfs.

Kruger, J., Sus, N., Moser, A., Scholz, S., Adler, G., Venturelli, S., Frank, J., 2024. Low β-carotene bioaccessibility and bioavailability from high fat, dairy-based meal. Eur. J. Nutr. 63 (6), 2261–2270. doi:10.1007/s00394-024-03423-w.

Kurhaluk, N., Kołodziejska, R., Kamiński, P., Tkaczenko, H., 2025. Integrative neuroimmune role of the parasympathetic nervous system, vagus nerve and gut microbiota in stress modulation: a narrative review. Int. J. Mol. Sci. 26 (23), 11706. doi:10.3390/ijms262311706.

Kus, M., Ibragimow, I., Piotrowska-Kempisty, H., 2023. Caco-2 cell line standardization with pharmaceutical requirements and in vitro model suitability for permeability assays. Pharmaceutics 15 (11), 2523. doi:10.3390/pharmaceutics15112523.

Lalonde, R., 2024. Anxiolytic-like effects of milk proteins. Pharmacol. Biochem. Behav. 240, 173789. doi:10.1016/j.pbb.2024.173789.

Lassale, C., Batty, G.D., Baghdadli, A., Jacka, F., Sánchez-Villegas, A., Kivimäki, M., Akbaraly, T., 2019. Healthy dietary indices and risk of depressive outcomes: a systematic review and meta-analysis of observational studies. Mol. Psychiatry 24 (7), 965–986. doi:10.1038/s41380-018-0237-8.

Li, C.X., Wang, F.R., Zhang, B., Deng, Z.Y., Li, H.Y., 2023. Stability and antioxidant activity of phenolic compounds during in vitro digestion. J. Food Sci. 88 (2), 696–716. doi:10.1111/1750-3841.16440.

Li, Y., Hu, N., Yang, D., Oxenkrug, G., Yang, Q., 2017. Regulating the balance between the kynurenine and serotonin pathways of tryptophan metabolism. FEBS J. 284 (6), 948–966. doi:10.1111/febs.14026.

Likpalimor, O.N., Zhang, S., Adagba, F.I., Fatoye, S., Adega, K.J., Njue, J., 2025. Mechanisms of microplastics sorption of antibiotics and impacts on aquatic ecosystems for sustainable development goals. Discov. Environ. 3 (1). doi:10.1007/s44274-025-00431-3.

Lobo-Silva, D., Carriche, G.M., Castro, A.G., Roque, S., Saraiva, M., 2016. Balancing the immune response in the brain: IL-10 and its regulation. J. Neuroinflam. 13 (1). doi:10.1186/s12974-016-0763-8.

Luo, Y., Li, Z., Gu, L., Zhang, K., Nevárez-Moorillón, G.V., 2023. Fermented dairy foods consumption and depressive symptoms: a meta-analysis of cohort studies. PLoS One 18 (2), e0281346. doi:10.1371/journal.pone.0281346.

Giollabhui, N.M., Ng, T.H., Ellman, L.M., Alloy, L.B., 2021. The longitudinal associations of inflammatory biomarkers and depression revisited: systematic review, meta-analysis, and meta-regression. Mol. Psychiatry 26 (7), 3302–3314. doi:10.1038/s41380-020-00867-4.

Marx, W., Suo, C., Dissanayaka, T., Maleki, S., Nguyen, L., Travica, N., Mohebbi, M., Lotfaliany, M., Yucel, M., McGuinness, A.J., Berk, M., Aslam, H., Jacka, F.N., 2026. Effects of a probiotic fermented dairy product on hippocampal metabolites, structure and function: an 8-week randomised, placebo-controlled trial in healthy women. Gut 75 (3), 511–520. doi:10.1136/gutjnl-2025-335398.

Mat, D.J.L., Le Feunteun, S., Michon, C., Souchon, I., 2016. In vitro digestion of foods using pH-stat and the INFOGEST protocol: impact of matrix structure on digestion kinetics of macronutrients, proteins and lipids. Food Res. Int. 88, 226–233. doi:10.1016/j.foodres.2015.12.002.

Morris, G., Berk, M., Carvalho, A., Caso, J.R., Sanz, Y., Walder, K., Maes, M., 2017. The role of the microbial metabolites including tryptophan catabolites and short chain fatty acids in the pathophysiology of immune-inflammatory and neuroimmune disease. Mol. Neurobiol. 54 (6), 4432–4451. doi:10.1007/s12035-016-0004-2.

Mujica-Coopman, M.F., Paules, E.M., Trujillo-Gonzalez, I., 2024. The role of prenatal choline and its impact on neurodevelopmental disorders. Front. Nutr. 11. doi:10.3389/fnut.2024.1463983.

Murdoch, B., Caulfield, T., 2018. Pragmatic clinical trials and the consent process. Res. Ethics 14 (2), 1–14. doi:10.1177/1747016117733506.

Nair, M.R.R., Frady, K., Yu, Z., Cantet, J.M., Hasan, M.S., Ríus, A.G., 2025. Resveratrol supplementation improves productivity and health in heat-stressed lactating Holstein cows. J. Dairy Sci. 108 (6), 6400–6411. doi:10.3168/jds.2024-25988.

National Academies of Sciences, Engineering, and Medicine, 2017. Communicating Science Effectively: A Research Agenda. National Academies Press, Washington, DC.

Nechifor, M., 2025. Magnesium and zinc in schizophrenia. Biomedicines. 13 (9), 2249. doi:10.3390/biomedicines13092249.

Neshatdoust, S., Saunders, C., Castle, S.M., Vauzour, D., Williams, C., Butler, L., Lovegrove, J.A., Spencer, J.P.E., 2016. High-flavonoid intake induces cognitive improvements linked to changes in serum brain-derived neurotrophic factor: two randomised, controlled trials. Nutrit. Healthy Aging 4 (1), 81–93. doi:10.3233/nha-1615.

Nolan, J.M., Meagher, K.A., Howard, A.N., Moran, R., Thurnham, D.I., Beatty, S., 2016. Lutein, zeaxanthin and meso-zeaxanthin content of eggs laid by hens supplemented with free and esterified xanthophylls. J. Nutrit. Sci. 5. doi:10.1017/jns.2015.35.

Nutbeam, D., 2000. Health literacy as a public health goal. Health Promot. Int. 15 (3), 259–267.

Oleskin, A.V., Shenderov, B.A., Rogovsky, V.S., 2017. Role of neurochemicals in the interaction between the microbiota and the immune and the nervous system of the host organism. Probiotics Antimicrob. Proteins 9 (3), 215–234. doi:10.1007/s12602-017-9262-1.

Pham, N., Morrin, M., Bublitz, M.G., 2019. Flavor halos and consumer perceptions of food healthfulness. Eur. J. Mark. 53 (4), 685–707. doi:10.1108/EJM-10-2017-0663.

Porter, K., Hoey, L., Hughes, C., Ward, M., McNulty, H., 2016. Causes, consequences and public health implications of low B-vitamin status in ageing. Nutrients 8 (11), 725. doi:10.3390/nu8110725.

Rezaei Ahvanooei, M.R., Norouzian, M.A., Hedayati, M., Ghaffari, M.H., 2023. Comprehensive review of resveratrol as a feed additive in dairy cows: exploring its potential diverse effects and implications. Vet. Res. Commun. 47 (3), 1115–1124. doi:10.1007/s11259-023-10157-3.

Ribeiro, C.M., Vieira, T.R.R., 2025. Probiotics and the gut–brain connection on microbial pathways supporting mental well-being-a systematic review. Microb. Bioact. 8 (1), 1–10.

Ríos, J.L., Schinella, G.R., Moragrega, I., 2022. Phenolics as GABAA receptor ligands: an updated review. Molecules 27 (6). doi:10.3390/molecules27061770.

Ríos-Covián, D., Ruas-Madiedo, P., Margolles, A., Gueimonde, M., De los Reyes-Gavilán, C.G., Salazar, N., 2016. Intestinal short chain fatty acids and their link with diet and human health. Front. Microbiol. 7. doi:10.3389/fmicb.2016.00185.

Robinson, S.R., Greenway, F.L., Deth, R.C., Fayet-Moore, F., 2025. Effects of different cow-milk beta-caseins on the gut–brain axis: a narrative review of preclinical, animal, and human studies. Nutr. Rev. 83 (3), e1259. doi:10.1093/nutrit/nuae099.

Ronaldson, P.T., Davis, T.P., 2020. Regulation of blood–brain barrier integrity by microglia in health and disease: a therapeutic opportunity. J. Cereb. Blood Flow Metab. 40 (1), S6. doi:10.1177/0271678X20951995.

Roth, W., Zadeh, K., Vekariya, R., Ge, Y., Mohamadzadeh, M., 2021. Tryptophan metabolism and gut-brain homeostasis. Int. J. Mol. Sci. 22 (6), 2973. doi:10.3390/ijms22062973.

Ryan, R., Booth, S., Spathis, A., Mollart, S., Clow, A., 2016. Use of salivary diurnal cortisol as an outcome measure in randomised controlled trials: a systematic review. Ann. Behav. Med. 50 (2), 210–236. doi:10.1007/s12160-015-9753-9.

Salamone, J.D., Correa, M., 2024. The Neurobiology of activational aspects of motivation: exertion of effort, effort-based decision making, and the role of dopamine. Annu. Rev. Psychol. 75, 1–32. doi:10.1146/annurev-psych-020223-012208.

Sallam, I.E., Abdelwareth, A., Attia, H., Aziz, R.K., Homsi, M.N., von Bergen, M., Farag, M.A., 2021. Effect of gut microbiota biotransformation on dietary tannins and human health implications. Microorganisms 9 (5). doi:10.3390/microorganisms9050965.

Sarkar, A., Lehto, S.M., Harty, S., Dinan, T.G., Cryan, J.F., Burnet, P.W.J., 2016. Psychobiotics and the manipulation of bacteria–gut–brain signals. Trends Neurosci. 39 (11), 763–781. doi:10.1016/j.tins.2016.09.002.

Sergin, S., Jambunathan, V., Garg, E., Rowntree, J.E., Fenton, J.I., 2022. Fatty acid and antioxidant profile of eggs from pasture-raised hens fed a corn- and soy-free diet and supplemented with grass-fed beef suet and liver. Foods 11 (21). doi:10.3390/foods11213404.

Shahidi, F., Athiyappan, K.D., 2025. Polyphenol-polysaccharide interactions: molecular mechanisms and potential applications in food systems – a comprehensive review. Food Prod. Process. Nutr. 7 (1). doi:10.1186/s43014-025-00322-3.

Sharma, R., Gupta, D., Mehrotra, R., Mago, P., 2021. Psychobiotics: the next-generation probiotics for the brain. Curr. Microbiol. 78 (2), 449–463. doi:10.1007/s00284-020-02289-5.

Silber, B.Y., Schmitt, J.A.J., 2010. Effects of tryptophan loading on human cognition, mood, and sleep. Neurosci. Biobehav. Rev. 34 (3), 387–407. doi:10.1016/j.neubiorev.2009.08.005.

Silva, Y.P., Bernardi, A., Frozza, R.L., 2020. The role of short-chain fatty acids from gut microbiota in gut-brain communication. Front. Endocrinol. 11. doi:10.3389/fendo.2020.00025.

Skawratananond, S., McCrea, G.E., Lie, P., Buxton, M.B., Daly, S.P., Vojtkofsky, N.A., Smith, S.C., Zhang, C., Hernandez, M., Hindle, A., Logsdon, A.F., Lawrence, J.J., 2025. The synergistic interplay between vitamin A, dietary fiber, and the microbiota-gut-brain axis: a potential mechanism for preventing Alzheimer's disease. Am. J. Physiol. Gastrointest. Liver Physiol. 329 (3), G484. doi:10.1152/ajpgi.00097.2025.

Smith, P.G., Morrow, R.H., Ross, D.A., 2015. Field Trials of Health Interventions: A Toolbox (3rd ed.). Oxford University Press, Oxford, UK doi:10.1093/med/9780198732860.001.0001.

Song, Z., Toy, J.Y.H., Huang, D., 2025. Absorption, distribution, metabolism, and excretion (ADME) of bioactive food constituents. In: Bioactive Food Constituents. Royal Society of Chemistry (RSC), Cambridge, pp. 72–90. doi:10.1039/9781837674046-00072.

Sørensen, K., Van den Broucke, S., Fullam, J., Doyle, G., Pelikan, J., Slonska, Z., Brand, H., 2012. Health literacy and public health: a systematic review and integration of definitions and models. BMC Public Health 12 (1). doi:10.1186/1471-2458-12-80.

Sousa, R., Recio, I., Heimo, D., Dubois, S., Moughan, P.J., Hodgkinson, S.M., Portmann, R., Egger, L., 2023. In vitro digestibility of dietary proteins and in vitro DIAAS analytical workflow based on the INFOGEST static protocol and its validation with in vivo data. Food Chem. 404. doi:10.1016/j.foodchem.2022.134720.

Stein, R.A., Riber, L., 2023. Epigenetic effects of short-chain fatty acids from the large intestine on host cells. MicroLife 4. doi:10.1093/femsml/uqad032.

Steinshamn, H., Purup, S., Thuen, E., Hansen-Møller, J., 2008. Effects of clover-grass silages and concentrate supplementation on the content of phytoestrogens in dairy cow milk. J. Dairy Sci. 91 (7), 2715–2725. doi:10.3168/jds.2007-0857.

Tan, J., Wang, C., Tomiyama, A.J., 2024. Dietary approaches to stop hypertension (DASH) diet and mental well-being: a systematic review. Nutr. Rev. 82 (1), 60–75. doi:10.1093/nutrit/nuad038.

Tan, J.K., Macia, L., Mackay, C.R., 2023. Dietary fiber and SCFAs in the regulation of mucosal immunity. J. Allergy Clin. Immunol. 151 (2), 361–370. doi:10.1016/j.jaci.2022.11.007.

Tang, Y., Liu, W., Zhang, J., Juan, B., Zhu, Y., Zhu, L., Zhao, Y., Daglia, M., Xiao, X., He, Y., 2025. Advances in intestinal-targeted release of phenolic compounds. Nutrients 17 (16), 2598. doi:10.3390/nu17162598.

Tavakoly, R., Moosburner, A., Anheyer, D., Cramer, H., 2025. Effect of different dietary patterns on patients with depressive disorders: a systematic review and meta-analysis of randomized controlled trials. Nutrients 17 (3), 563. doi:10.3390/nu17030563.

Teixeira, A.L., Scholl, J.N., Bauer, M.E., 2024. Psychoneuroimmunology of mood disorders. Methods Mol. Biol. 2868, 49–72. doi:10.1007/978-1-0716-4200-9_4.

Tribble, R.C., Dmitrieva, J., Watamura, S.E., LeBourgeois, M.K., 2015. The cortisol awakening response (CAR) in toddlers: Nap-dependent effects on the diurnal secretory pattern. Psychoneuroendocrinology 60, 46–56. doi:10.1016/j.psyneuen.2015.05.009.

Turgeon, S.L., Brisson, G., 2020. Symposium review: the dairy matrix—bioaccessibility and bioavailability of nutrients and physiological effects. J. Dairy Sci. 103 (7), 6727–6736. doi:10.3168/jds.2019-17308.

Vanlaer, Y., Minschart, C., Snauwaert, I., Myngheer, N., Maes, T., Block, C.D., Van Pottelbergh, I., Abrams, P., Vinck, W., Leuridan, L., Driessens, S., Billen, J., Matthys, C., Bogaerts, A., Laenen, A., Mathieu, C., Benhalima, K., 2025. Diabetes risk perception in women with a recent history of gestational diabetes mellitus: a secondary analysis from a Belgian randomized controlled trial (MELINDA Study). J. Clin. Med. 14 (14), 4998. doi:10.3390/jcm14144998.

Vartanian, M., Endres, K.J., Lee, Y.T., Friedrich, S., Meemken, M.T., Schamarek, I., Rohde-Zimmermann, K., Schürfeld, R., Eisenberg, L., Hilbert, A., Beyer, F., Stumvoll, M., Sacher, J., Villringer, A., Christensen, J.F., Witte, A.V., 2025. Investigating the impact of microbiome-changing interventions on food decision-making: MIFOOD study protocol. BMC Nutrit. 11 (1). doi:10.1186/s40795-024-00971-6.

Verhagen, H., Vos, E., Francl, S., Heinonen, M., van Loveren, H., 2010. Status of nutrition and health claims in Europe. Arch. Biochem. Biophys. 501 (1), 6–15. doi:10.1016/j.abb.2010.04.012.

van Vliet, S., Provenza, F.D., Kronberg, S.L., 2021. Health-promoting phytonutrients are higher in grass-fed meat and milk. Front. Sustain. Food Syst. 4. doi:10.3389/fsufs.2020.555426.

Wallace, T.C., 2018. A comprehensive review of eggs, choline, and lutein on cognition across the life-span. J. Am. Coll. Nutr. 37 (4), 269–285. doi:10.1080/07315724.2017.1423248.

Wang, H., He, Y., Zou, P., Wang, X., Peng, G., Deng, X., 2026. From the gut to the brain: microplastic-associated neurovascular dysfunction and implications for stroke risk. Adv. Sci. 13 (11). doi:10.1002/advs.202520278.

Watson, D., Clark, L.A., Tellegen, A., 1988. Development and validation of brief measures of positive and negative affect: the PANAS scales. J. Pers. Soc. Psychol. 54 (6), 1063–1070. doi:10.1037/0022-3514.54.6.1063.

Weaver, C.M, 2021. Dairy matrix: is the whole greater than the sum of the parts? Nutr. Rev. 79 (Supplement_2), 4–15. doi:10.1093/nutrit/nuab081.

World Health Organization, 2023. General principles of food hygiene. Geneva: World Health Organization. https://www.who.int/publications/i/item/9789241594690 (Accessed: 15 April 2026).

World Medical Association, 2024. WMA Declaration of Helsinki–Ethical principles for medical research involving human participants. The World Medical Association, France.

Xavier, A.A., Mariutti, L.R., 2021. Static and semi-dynamic in vitro digestion methods: state of the art and recent achievements towards standardization. Curr. Opin. Food Sci. 41, 260–273. doi:10.1016/j.cofs.2021.08.002.

Xie, Y., Xie, W., 2020. The role of sulfotransferases in liver diseases. Drug Metab. Dispos. 48 (9), 742–749. doi:10.1124/dmd.120.000074.

Yin, W., Löf, M., Chen, R., Hultman, C.M., Fang, F., Sandin, S., 2021. Mediterranean diet and depression: a population-based cohort study. Int. J. Behav. Nutr. Phys. Act. 18 (1). doi:10.1186/s12966 021 01227 3.

Yosaee, S., Clark, C.C.T., Keshtkaran, Z., Ashourpour, M., Keshani, P., Soltani, S., 2022. Zinc in depression: from development to treatment: a comparative/dose response meta-analysis of observational studies and randomized controlled trials. Gen. Hosp. Psychiatry 74, 110–117. doi:10.1016/j.genhosppsych.2020.08.001.

Zaheer, K., 2017. Hen egg carotenoids (lutein and zeaxanthin) and nutritional impacts on human health: a review. CyTA J. Food 15 (3), 474–487. doi:10.1080/19476337.2016.1266033.

Zhao, Y., Chen, Z., Yang, Y., Lin, P., Jiao, A., Zhengyu, J., 2025. The stability of different free polyphenols during multi-stage simulated digestion and their inhibitory properties on starch digestion enzymes and cellular inflammation. Int. J. Food Sci. Technol. 60 (2). doi:10.1093/ijfood/vvaf162.

Zhou, L., Xiong, J.Y., Chai, Y.Q., Huang, L., Tang, Z.Y., Zhang, X.F., Liu, B., Zhang, J.T., 2022. Possible antidepressant mechanisms of omega-3 polyunsaturated fatty acids acting on the central nervous system. Front. Psychiatry 13. doi:10.3389/fpsyt.2022.933704.

Zhu, W., Zhang, X., Wang, D., Yao, Q., Ma, G.-L., Fan, X., 2024. Simulator of the human intestinal microbial ecosystem (SHIME®): current developments, applications, and future prospects. Pharmaceuticals 17 (12), 1639. doi:10.3390/ph17121639.

Zinkow, A., Grodzicki, W., Czerwińska, M., Dziendzikowska, K., 2025. Molecular mechanisms linking omega-3 fatty acids and the gut–brain axis. Molecules 30 (1), 71. doi:10.3390/molecules30010071.

Chapter 9

Behavioral and perceptual studies on consumers

Ahmed I. Elsherbeni[a] **and Mohamed E. Abd El-Hack**[b]

[a]*Animal Production Systems Research Department, Animal Production Research Institute, Agricultural Research Center, Giza, Egypt,* [b]*Department of Industrial Pharmacy, College of Pharmaceutical Sciences and Drug Manufacturing, Misr University for Science and Technology (MUST), Giza, Egypt*

9.1 Introduction

Over the last two decades, the market for the functional foods (FFs) sector, which are conventional food items that have been modified to provide health features beyond basic nutrition, has expanded substantially (Aiello, 2022; Ghosh et al., 2024). This expansion associates livestock and poultry nutrition with potential health features for humans by utilizing food-chain procedures such as supplementing animals and poultry with botanical or herbal-based diets, along with recognized probiotic dairy products (Pant et al., 2024). For sensible commercialization, adjusting supervision, and effective public health communication, it is necessary to understand customers' attitudes and beliefs about these FFs (Pistorio et al., 2023). Consumers are suspicious of foods that obscure the distinction between healthy food and drugs or trust narrative upstream claims, such as herbal-fed-based products (e.g., meat, milk, and egg), although there is common public knowledge and acceptance of public functional components, e.g., omega-3 and pro- and prebiotics (Halder et al., 2021; Guptill et al., 2022). Psychological agents such as supposed health profits, supposed risks, creator trust, and the intelligibility of health claims touch customer acceptance; these are qualified by individual health and demographic characteristics such as motivation and health knowledge (Szymkowiak et al., 2022; Theodorou et al., 2024). Through cognitive heuristics and emotional triggers, extrinsic cues such as labeling and branding have a substantial impact on inducing consumer expectations and trust, often resulting in phenomena like the health halo effect (Schuldt and Schwarz, 2010; Wansink and Chandon, 2014).

As observations considering herbal-fed FFs are modeled by cultural identities and traditional knowledge practices, cross-cultural differences affect acceptance

Plant-Derived Feed Additives and Animal-Based Foods: Linking Animal Health, Product Quality, and Consumer Perceptions. DOI: https://doi.org/10.1016/B978-0-443-45446-2.00006-5

models even more (Goyal et al., 2023; Vlaicu et al., 2023). To shed light on how trust, sensory expectations, cultural narratives, and ethical considerations interact to affect the uptake of functional and herbal-fed food products, this chapter examines these behavioral and perceptual aspects using both qualitative and quantitative consumer research.

9.2 Consumer knowledge and attitudes

Over the past 20 years, the market for FFs—conventional food prepared to provide health benefits beyond basic nutrition—has grown very significantly. Furthermore, to well-known product classes (e.g., yogurts, probiotics, and fortified dairy products) (Aiello, 2022; Ghosh et al., 2024), a novel generation of product ideas has surfaced that connect food-chain activities (such as feeding particular botanical or herbal-based diets to animals and poultry) with messages about human public health (Pant et al., 2024). Comprehending customers' opinions about these innovations is critical for monitoring supervision, public health communication, and responsible commercialization (Pistorio et al., 2023).

Geographically and by product type, public awareness of the varied concept of FFS differs. Public awareness and acceptance of famous functional components (e.g., omega-3, vitamins, probiotics, and prebiotic) are relatively high (Halder et al., 2021; Butnariu, 2024), whereas consumers are less certain of boundary cases, such as foods that fall somewhere between "food" and "medicine," or foods types that depend on new upstream claims, for instance, animals and poultry that fed specific herbal blends to produce "safe and healthier" products (Guptill et al., 2022; Frohlich, 2023). According to numerous research sources, baseline acceptability upsurges with the claimed benefits and awareness. On the other hand, advances that are unfamiliar or technically framed typically cause more skepticism or worry (Baker et al., 2022; Bhatnagar, 2025; Davis, 2025).

There are currently few specific studies on consumer awareness of animal and poultry-origin products made from animal or poultry fed herbal-based diets. The data that are currently available point to the fact that the overall public is more familiar with traditional fortification or probiotic and prebiotic claims (Embling et al., 2024). Regarding general models of reception for technology-based food innovations, customers' responses to such claims rely mostly on perceived naturalness, the clarity of the causal claim (how feeding herbs to animals benefits human consumers), and trust in producers or authorization. To put it briefly, awareness is low to moderate and depends on the reliability of the source and the messaging (Baker et al., 2022). Furthermore, misunderstanding how an upstream practice (such as herb-supplemented animal feed) could aid in benefits to human health, confusing "natural" with "safe/effective," and overestimating or underestimating the extent of benefit recommended by marketing claims are instances of public knowledge gaps (Grinberga-Zalite et al., 2024). When technical terms are employed without strong explanations, misconceptions are augmented. Numerous studies show that clear and credible educational messages enhance reception and reduce uncertainty, while the extent of the effect varies according to prior attitudes and trust (Ropawandi et al., 2022; Buçinca et al., 2025).

Several studies have identified the following proximate psychological factors of acceptance: perceived risk and uncertainty, perceived benefit (health gains), perceived cost (price), perceived taste, and social norms (Theodorou et al., 2024; Orynbekov et al., 2025). These direct opinions are influenced by distant factors, including the institutional environment (organizational climate and credibility of the source), personal differences (narcophobia and cognitive needs), and product attributes (familiar versus unfamiliar elements). Willingness to try a product increases when the expected benefits outweigh the perceived risks and when trust is strong. On the contrary, high perceived ambiguity or poor trust (such as an unfamiliar manufacturer or ambiguous claims) diminishes willingness. The conceptual frameworks relating to consumer acceptance of technology-based food innovations are consistent with these findings (Ronteltap et al., 2007; Meijer et al., 2021; Walten, 2022).

Higher levels of health literacy and education, as well as personal health orientation (interest in healthy eating, presence of chronic health concerns), are strong predictors of willingness to try and pay more for FFs (Szymkowiak et al., 2022; Nam and Suk, 2024). In addition, people who have active health goals are more motivated to assess health claims and might use new-fangled properties if they support their goals (Braithwaite et al., 2024). Depending on the quality of the evidence presented, education can encourage acceptance or rejection of information. Higher-educated purchasers tend to request and use scientific data, although they may analyze claims further thoroughly (Menzli et al., 2022; Hanan et al., 2024).

Regarding the recent consumer surveys, the following trends are consistent across recent scoping reviews and empirical surveys conducted in Europe, North America, and other counties: (1) Consumer interest in FFs is still stable but divided into segments, with some consumers being early adopters and others remaining skeptical; (2) the strongest predictor of acceptance is perceived health benefit, especially when benefits are tangible and well-known (e.g., calcium for bone health); (3) acceptance is increased by trust in regulatory regimes, third-party certification, and medical professionals; and (4) price premiums and sensory expectations (anticipated taste) are practical barriers to actual purchase, even among willing consumers (Lucas and Brunner, 2024; Chauhan et al., 2025;Sato et al., 2025)

Essiz and Senyuz (2024) noted that survey-based research repeatedly demonstrates that opinions toward and uptake of functional meals are influenced by demographic factors such as age, income, education, and generational cohort. For instance, more than 90% of older persons (60+) indicate the consumption of FFs, particularly probiotic yogurt, eggs enriched with omega-3, and fiber-fortified bread. These individuals are concerned about health issues such

FIGURE 9.1 Understanding consumer attitudes toward FFs. *FFs*, Functional foods.

as arthritis, vascular disease, and bone health (Abdi-Moghadam et al., 2023; Joshi et al., 2024). However, when no specific product was prescribed, factors such as age, gender, education, or the presence of children in the home were not meaningfully associated with the general acceptance of the concept of FFs, according to previous studies such as Vella et al. (2013) and Guo et al. (2023). Each generation has different perspectives on healthy foods, reliable information sources, and barriers/benefits that affect their behavior, according to recent generational cohort analysis (Dangaiso et al., 2023; Voinea et al., 2025). On the other hand, large-scale nutritional studies in China indicate that younger generations are consuming more animal foods, fruits, and dairy products and less grains and vegetables, demonstrating a shift in food preferences (including acceptance of innovative or FFs) (Zhu et al., 2023; Wu et al., 2025). Fig. 9.2 presents consumer awareness and acceptance in the FFs landscape.

9.3 Psychological priming and labeling effects

Foods that promise additional health or psychological benefits in addition to nutrition are becoming increasingly popular among modern customers. Since the markets for FFs have expanded quickly, helped along by convincing advertising that highlights features such as "herbal-fed," "organic," and "natural"

(Grunert, 2020). Extrinsic signals, including labeling, color, and brand associations, have a substantial influence on consumers' assessments of these foodstuffs, according to a behavioral and perceptual study (Wansink and Chandon, 2014).

9.4 Green labeling and mood associations

A recent study by Sörqvist et al. (2019), Braca and Dondio (2023), and Rincon (2025) noted that marketing labels (such as natural, eco-friendly, and herbal-fed) act as semantic primes that activate positive affective and moral schemas in the attention of consumers. Psychological priming is the process by which experience with particular words or colors subtly shapes attitudes and actions. The green color is correlated with health features, freshness, and ecological attributes, which enhances the perceived quality of food products and decreases consumption guilt (Magnier and Schoormans, 2015; Luo et al., 2022; Su et al., 2022).

Moreover, the use of labels such as "mood-boosting" or "relaxing" elicits expectancy-driven responses that are equivalent to placebo effects. Customers believe that a psychological improvement can expand their personal perception. For example, in comparison to identical unlabeled beverages, it was found that beverages marketed as "mood-enhancing" boosted self-stated affect (Lee et al., 2021; Mehta, 2024; García-Barón et al., 2025). Similarly, herbal-fed beef products evoke ideas of purity, conventional knowledge, and a decline in spotlight to chemical elements that improve perceived taste and safety (Basha and Lal, 2019).

All these results suggest that labeling serves as a cognitive heuristic as well as a perceptual filter. Perceptual expectations modify sensory evaluations, even when real differences between products are negligible. This event, defined as the "health halo effect," appears when a favorable feature, such as natural, is broadcast to other aspects such as flavor or ecological impact (Schuldt and Schwarz, 2010; Küst, 2019; Bschaden et al., 2022). Consequently, the labeling function is both information and a psychological stimulus that subconsciously influences perception.

9.5 Branding and trust

Chiefly for FFs or nutraceuticals whose advantages are not directly apparent, branding creates cognitive anchors that touch consumer self-confidence and perceived efficacy. A powerful, well-known brand depresses perceived danger in consumption by communicating safety and credibility (Delgado-Ballester et al., 2011). For example, expectation theory explains why customers frequently consider a probiotic yogurt from a famous international brand to be more effective than an equal unbranded product (Karagiannis et al., 2022; Sindhu, 2024;Irmak et al., 2005). This result is analogous to behavioral economics' placebo-by-label phenomenon, which states that a product's perceived performance is enhanced when customers have confidence in its label or brand (Shiv et al., 2005; Ha et al., 2022; Aslan, 2023).

Regarding behavioral economics, constrained rationality governs how consumers respond to labels; heuristics and emotional cues, rather than objective assessment, impact decisions (Kahneman, 2011; Saleem and Amjad, 2023; Ranawat, 2025). According to the placebo-by-label effect, physiological or emotional effects that are compatible with prospects of advantage—whether stimulated by branding, pricing, or label—can happen (Waber et al., 2008; Rodríguez et al., 2023; Shams et al., 2025). Previous studies claimed that members who consumed a premium-branded "energy drink" outperformed those who consumed the same beverage under a nonspecific label in terms of cognitive function (Irmak et al., 2005). These results demonstrate how perceived efficacy translates into measurable behavioral variation through belief-driven consumption. Such procedures underscore the troubles of distinguishing between biological and perceived efficacy in the context of FFs. Customers' activities, such as repeated purchases or satisfaction, display armored belief systems rather than verified outcomes when they trust a brand or believe a label's authenticity (Basso et al., 2020). This is under the need for evidence-based labeling and transparency in maintaining ethical standards in health-related marketing.

9.6 Taste, trust, and wellness perception

Choosing food is a sensory experience, a social cue, and an expression of one's identity all at once. It is not only about calories or nutrition. Foods with functional claims, such as probiotics, prebiotics, herbal-fed or natural animal products; moreover, mood-modulating labels, are becoming more and more common in the modern marketplace. These claims elevate a product from a sensory item to a recognized artifact of ethics (sustainability/organic) or health (functional claims). How can these nonsensory cues alter customers' expectations about what they would taste, what they actually report tasting, and whether they will eventually embrace and stick with those products? is the question posed by behavioral scientists (Shangguan et al., 2019; Wu et al., 2021)(Fig. 9.2).

9.7 Sensory versus functional trade-offs

Although there are few direct empirical comparisons between "herbal-fed" and conventionally fed animal products, related research offers data and techniques for drawing educated guesses. Research examining the sensory impacts of animal feeding approaches, e.g., supplementing with herbs or spices and offering pasture instead of concentrated diets, has displayed that dietary choices could change intramuscular fat composition, the chemical structure of volatile composites, and consequently, aroma and flavor features (Ponnampalam et al., 2024). Depending on the quantity and postharvest practices, these changes can be beneficial, enhancing aromatic or herbaceous scents, or detrimental, resulting in unpleasant odors. Analyses of studies conducted by the Committee on Meat Science and Sensory Evaluation confirm that production variables, such as feed type, maturity duration, slaughter methods, and spice use, significantly influence

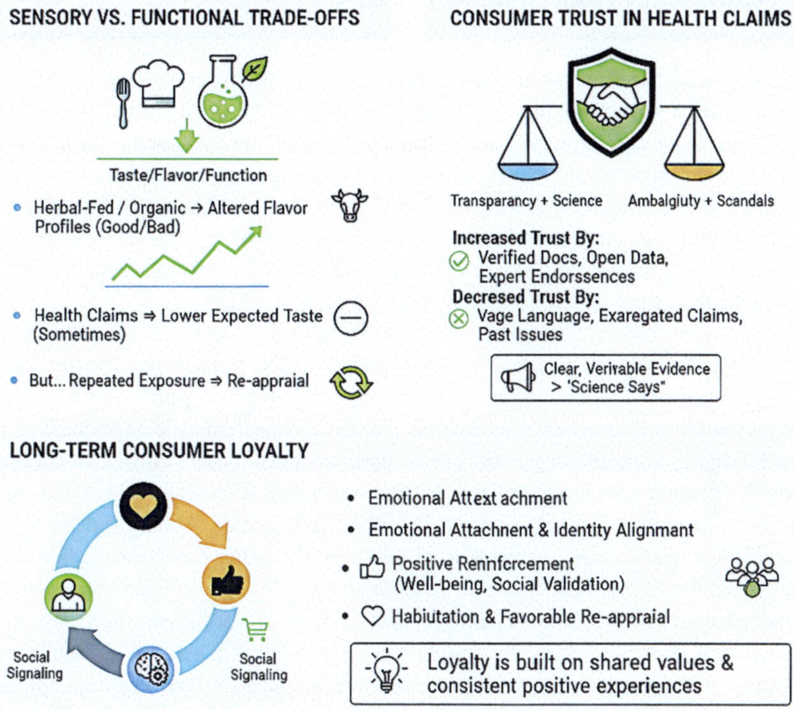

FIGURE 9.2 Taste, trust, and wellness perception: How sensory and nonsensory cues impact FFs' choice. *FFs*, Functional foods.

sensory outcomes (Saldaña et al., 2021; del Bosque et al., 2022; Yamini et al., 2025). Furthermore, cultural norms and experience with specific flavor types influence detailed consumer responses. Whether these changes are perceived as advantages or disadvantages depends on the label, familiarity, and the extent to which functional claims evoke compensatory health expectations, despite the fact that plant-based diets can quantitatively modify sensory profiles (Tura et al., 2024; Cattaneo et al., 2025).

The results of controlled trials using blind, described, and informed tasting methods are not entirely consistent. Nutritional and health claims reduce their impact on reported taste after consumption; however, they consistently enhance feelings of well-being and may lower taste expectations (Prada et al., 2021; Broeders et al., 2024). Research indicates that health claims lower consumers' expectations of a product's flavor; however, they subsequently express satisfaction with it. The authority of these claims, purchaser health awareness, and the implication of substituting flavor-improving ingredients (e.g., "low sugar" or "low fat") all affect the level of sensory reduction (Hong et al., 2022; Rramani et al., 2023).

9.8 Consumer trust in health claims

Numerous subjects influence trust in food claims, involving traceability, third-party certifications, perceived source integrity (manufacturer vs. regulator), and the availability of supporting scientific facts. Claims buoyed by verified documentations, open provenance data, or independent expert endorsements are extra trusted by customers; conversely, ambiguous language, aggressive marketing, or previous industry scandals undermine trust (Wu et al., 2021; Schrobback et al., 2023). In addition, customers believe assessments highlight the significance of perceived transparency and the social amplification of disbelief through social media (Rupprecht et al., 2020; Nguyen and Nguyen, 2025).

Behavioral evidence suggests that merely quoting "science" is insufficient; the communication of evidence is important. In FFs' marketing, credibility is increased more by independent verification, references to public registries or peer-reviewed publications, and concise, easily comprehensible explanations of research design (sample size, outcomes) than by ambiguous "clinically shown" claims (Azimont and Araujo, 2014; Martirosyan and Alvarado, 2023). Longer-term trust is supported, and perceived deception risk is decreased by transparency regarding limitations and reasonable claims. On the other hand, exaggerated or unsubstantiated neurofunctional or "mood-boosting" claims discourage uptake and breed distrust (Didenko, 2023; Vu et al., 2025). In addition, the tightening of permitted claim language by regulators also impacts corporate conduct and consumer perceptions; as a result, compliance and clarity have a direct impact on trust (Nair, 2024; Pattanaik, 2024).

9.9 Long-term consumer loyalty

Recurring loyalty and selection are distinctly expected by emotional attachment and discovery signals. Foods that are associated with a consumer's self-concept, such as acquiring ethical and locally produced beef or being health-conscious, tend to foster more significant loyalty than temporary sensory innovation (Leszczynski, 2008; Wang et al., 2019). Likewise, emotions related to value congruence, e.g., pride, belonging, and gratitude, are focal in transferring a first purchase into recurrent trade, corresponding to social science and neuroscience studies (Soscia, 2013; Yan et al., 2024). Across the development of narratives that consumers incorporate into their identities, corporate social responsibility and overt sustainability commitments can strengthen these emotional connections.

Recurring pleasurable practices, principally when paired with perceived health benefits, may result in habituation or a promising reappraisal of the sensory potentials, even in cases when there are primary sensory trade-offs, like a somewhat diverse flavor (Saha, 2018). Although early sensory worries, behavioral economics suggests that minor, unbroken positive reinforcements, for instance, social validation and perceived welfare gains, might boost the possibility of stable switching (Mobekk, 2021; Wolfenbarger, 2023). The purchase of

FFs regularly acts as a form of social signaling, conveying persons' health values, ecological concerns, or social status. When group norms advance such behavior, constant purchasing habits could develop into self-reinforcing. Marketers and public health promoters can exploit these dynamics, but should exercise caution to prevent exaggerating statements, which could result in a loss of confidence (Cao and Zhang, 2024; Anwar et al., 2025; Nazzaro et al., 2025).

9.10 Cross-cultural insights into herbal-fed foods

Alongside consumer desire for naturalness, perceived health benefits, and ethical production, attention in functional and "value-added" animal and/or poultry products (e.g., eggs, meat, milk, and cheese) produced from animal and/or poultry fed herbal feed additives has increased (Vlaicu et al., 2023; Cuchillo-Hilario et al., 2024). The grade to which herbal-fed diets are viewed as desirable, legitimate, or even acceptable, however, differs depending on cultural background, religious beliefs, historical foodways, and narrative customs.

9.11 Acceptance across regions

Giving herbs to livestock and poultry is closely linked to usual medical systems, such as Ayurveda and Traditional Chinese Medicine, in numerous Asian countries, involving China, India, and specific parts of Southeast Asia. The idea that herbs given to animals can affect the quality or healthfulness of the products made from them may be more acceptable to consumers who are familiar with the use of herbs for therapeutic purposes (Basu et al., 2023; Goyal et al., 2023; Mollahassani et al., 2025). This context exposes two vital patterns: First, old knowledge holders, such as herbalists, the aged, and veterinarians trained in ethnoveterinary practice, are often reliable; second, products that are accessible within the framework of recognized medical or culinary traditions, instead of innovative commercial claims, are more expected to be accepted. In convinced areas, manufacturers use historical narratives and local herb names to induce a feeling of security and legitimacy (Guiné et al., 2021; Pascall et al., 2022; Ndhlovu et al., 2023; Wendimu et al., 2023).

Some elements are valued in Europe, often based on the theories of artisan and terwarist farmers. Mountain grasses and Mediterranean pastures are particularly prized, and the use of grass in feeding or forage is a crucial factor in regional legitimacy (Teston, 2023; Baudar, 2023). The noticeable flavor qualities that grasses impart may appeal to customers who prefer "natural" and small-scale production. Thus rather than making direct health claims to consumers, sales campaigns in Europe tend to emphasize the sensory quality of products, animal health, and forage yield in natural pastures (Gerber et al., 2024; Abbasi et al., 2025; Ankiel et al., 2025). On the other hand, there is considerable diversity among African contexts. Practices using local grasses for animal health and growth are well-established in many rural and periurban systems. Perceived modernity, safety, and status determine whether these activities are acceptable in urban consumer markets. While some consumers view "traditional" methods as undocumented and rural, others embrace them as authentic and healthy. Scalability and trust-building for herbal-based products in broader markets are complicated by informal value chains and a lack of institutional verification (Shayanowako et al., 2021; Braamhaar et al., 2025; Niba, 2025).

Several studies have claimed that grass-fed and conventionally foraged products are primarily found in niche markets, including artisanal farms, organic markets, and specialty restaurants in North and South America. Indigenous peoples in many regions possess an ecological knowledge of how plants and animals interact, as well as how specific plants are used in grazing systems. However, elements such as certifications, proven health benefits, and brand recognition are needed to achieve overall public satisfaction (Baggio et al., 2021; Root-Bernstein et al., 2022; Török et al., 2024). Furthermore, to attract consumers who value integrity and health, manufacturers combine narratives of sustainability and small-scale farming practices with a sensory narrative that highlights the flavor profiles of their food products (Pérez, 2009).

In line with traditional beliefs, herbs used in animal feed are often associated with direct, and sometimes spiritual, benefits, including increased vitality and a marked improvement in mood balance (Abdallah et al., 2019). By assessing outcomes such as growth performance characteristics, meat fat profiles, and oxidative stability, modern advocates are calling for the measurement of these benefits and the production of herbal inputs as alternatives to synthetic materials or antibiotics. Research that reflects local knowledge and practices while producing reliable and safe information that aligns with current regulatory and market values is crucial to bridging the gap between these two methodologies (Ismail and Huda, 2024 Ponnampalam et al., 2025).

9.12 Religious and ethical considerations

Religious systems define what is acceptable in multiple ways. Plant-based food supplements are generally suitable for consumers adhering to halal or kosher dietary laws. Herbal supplements may raise concerns, however, if they affect animal welfare or the slaughtering process, or introduce ingredients prohibited under these dietary laws (Saeed, 2023; Mehsar, 2025). Transparency and certification are essential to ensure compliance. Feed and production methods must not conflict with dietary provisions or guidelines, according to religious authorities and certification bodies. Partnering with accreditation bodies and providing accurate documentation of feed ingredients and processing methods can help companies aiming to penetrate these markets overcome entry barriers (Abdallah, 2021; Conficoni et al., 2022; Johan and Plana-Casado, 2023).

Multiple studies indicate that while some white vegetarians emphasize farming methods that promote animal welfare and natural feeding habits, vegetarian consumers generally reject animal products outright (Almeida, 2024). Furthermore, highlighting improvements in animal welfare and the prohibition of growth enhancers is crucial for the successful marketing of both animal and grass-fed products. Instead of focusing solely on vegetarians, marketing efforts should target flexible omnivores and ethically holistic eaters, as the vegan market is not significantly larger than the animal market (Gioreva, 2024). Concerns about consumer awareness, environmental impacts, and animal welfare are also critical factors in the introduction of high-quality, grass-fed animal production. In addition, herbal feed also needs to be respected, principally the variances between foreign and local herbs and the quantity of land expected to grow them. Also, customers are allowed to well-versed consent, which calls for an unambiguous category of what "herbal-fed" really means; consumers who pay attention to ethics often want third-party authorization and certification of the herbal feed's full lifecycle (Sweeney et al., 2022; Hao et al., 2023; Oda et al., 2024; Phillips, 2024).

9.13 Cultural narratives and storytelling

Employing stories that connect a product to its place and the people who made it can help justify extreme prices and make it feel more authentic. For example, a story about a district herb that has been used for generations in alpine pasture grazing can make people believe the product is worth more. Also, using storytelling methods that associate the producer's identity, the established uses of herbs, and the product's sensory experiences (e.g., taste and smell) could make people ponder whether the product is worth more (Ikäheimo, 2021; Cutolo and Ferriani, 2024). But other customers respond differently to different categories of storylines. Particular people, called "traditionalists," are more likely to respond to stories that focus on tradition because they value stability with legacy. Some customers, called "health seekers," are attracted to functional claims and would require scientific verification to support them (Körmeling, 2021; Dihdashti, 2025). On the other hand, "ethical consumers" care more about sustainability and health and are more likely to respond to messages that stress these values (Lim, 2022; Kenny et al., 2023).

Higher costs may be reasonable, and a sense of accuracy is created by employing narratives that link a product to its place and the people who produced it. A story about a local herb that has been beneficial for generations in mountain pasture grazing can present the product's perceived worth (Mack et al., 2013; Kernecker, 2025). Previous research by Luthra et al. (2024) reported that the perceived value of products can be enhanced through effective narrative techniques that link the product's personality, the traditional uses of herbal plants, and the product's sensory experiences (such as taste and smell).

9.14 Focus group and survey data

Developing positive marketing strategies and evidence-based communication requires understanding how consumers perceive and evaluate foods made from

FF or herbal supplements. In parallel with Verbeke (2015), behavioral and cognitive research provides empirical support for how people learn about FF, considering attributes related to identity, trust, and sensory experience. Besides, Ares et al. (2020) stated that core groups and survey-based research are still the key instruments for understanding these dynamics.

9.15 Methodologies

Qualitative methods such as focus groups and structured interviews are utilized to explore the shared meanings, feelings, and beliefs of purchasers. They provide subtle, nuanced information regarding communication challenges and initial product acceptability (Krueger, 2015; Gordon et al., 2022; Lim, 2025). To balance the spontaneity of participants and facilitate the facilitator's role, discussions are typically conducted in a semistructured manner. Focus groups in grass-fed food research demonstrates how willingness to adopt influences perceptions of naturalness, cultural affiliation, and producer trust (Kim et al., 2023). Content analysis or discourse analysis is used to examine topics within written session data. Likert scales are used in quantitative questionnaires to assess the acceptability of certain attributes, such as sustainability claims, taste expectations, or potential health benefits. Five- to seven-point scales are often employed by scientists for comparison and clarity (DeVellis, 2017; Russo et al., 2021). Similarly, Likert-based tools make it probable to apply statistical analyses (such as regression and ANOVA) to get relationships between observation patterns and demographic data (Grunert et al., 2018; Koo and Yang, 2025).

Conjoint analysis breaks down decision-making into characteristic preferences to model buyer trade-offs. This approach evaluates how consumers assess terms such as "herbal-fed," "antibiotics-free," "organic," or "locally made" in FFS research (Oztas and Erdem, 2021; Kulshreshtha et al., 2023). This approach facilitates precise product design and communication by providing usefulness indicators that demonstrate the relative importance of each feature. Furthermore, behavioral findings are more reliable and externally verifiable when using a representative sample. Many surveys employ quota or purposive sampling to achieve significant demographic diversity, as probability-based sampling, such as stratified random sampling, is optimal but often limited by financial constraints (Hair et al., 2019). By recording both beliefs and behavioral purposes, the combination of qualitative focus groups and quantitative surveys improves authority (Creswell et al., 2018; Alordiah et al., 2023).

9.16 Outcomes and insights

Considering several global research experiments, perceived naturalness, ethical alignment, and health perception are the crucial drivers behind the approval of herbal-feed-based FFs (Ares et al., 2020; Onyango et al., 2023). Herbal-feed–based FFs labeling is related by customers to ecological responsibility, health, and clarity. Acceptance is also promoted by emotional motivators, e.g., curiosity and the desire to be seen as healthy consumers (Pieniak et al., 2016). Information from focus groups often displays mistrust when there is a lack of clarity or technical communication support. Likewise, experimental intentions could be dejected by cultural and sensory beliefs, for instance, expected taste modifications (Asioli et al., 2017). To demonstrate attitudinal modifications by segment (age, gender, and education level), survey data are regularly served in bar charts, heatmaps, and spider plots.

9.17 Recommendations for product developers

Labels must strike a balance between emotional appeal and technical reliability. Concise and reliable data, presented alongside simple, nature-inspired products, enhance understanding and comprehension of concepts (Van Loo et al., 2017; Lee, 2022). The integration of visual elements, such as images of herbs and user testimonials, also fosters a sense of authenticity. Using an interactive and educational approach, rather than overly promotional language, strengthens credibility. Consumer segments respond positively to communication that focuses on mutual benefits, such as health and sustainability, along with transparency regarding the traceable origin of herbs (Schnettler et al., 2020; Sigrist et al., 2025). Therefore developers should align their data with buyer priorities based on the results of group preference analysis. Placing the phrase "herb-fed" alongside "antibiotic-free" may have a synergistic effect. Caution is advised when making exaggerated claims, as these can jeopardize long-term trust (Wansink and Chandon, 2014; Szeggen and Oyanek, 2022; A'yuni et al., 2025).

9.18 Conclusion

A complex interplay of psychological, cultural, sensory, and informational factors influences consumer perceptions and behaviors toward functional and herbal products. This chapter has demonstrated that despite widespread recognition of functional products, misconceptions and a lack of information about early developments, such as the use of herbs in nutrition and animal products, persist. External cues, such as branding and color, serve as inferential tools that consumers use to assess authenticity, safety, and health benefits. Expectations can influence both perceived quality and emotional response, a phenomenon known as the "halo of health" or the brand-induced placebo effect. Building trust is crucial for fostering acceptance and lasting commitment. Third-party certifications, scientific verification, and clear communication contribute significantly to building consumer trust, while exaggerated or vague claims undermine credibility. Cross-cultural studies examine the adoption of herbal foods, drawing on ancient Asian medical traditions, stories of artisanal and ethical production in Europe, and emerging niche markets in Africa and the Americas, depending on the context. Perceived authenticity and ethical value are influenced by conformity, cultural narratives, and ethical transparency. The primary drivers of adoption for these products are their perceived authenticity, health awareness, and ethical conformity, which are influenced by education, financial status, and generational shifts, consistent with empirical findings from surveys and focus groups. The importance of behavioral insights is growing, and product design is guided by the organizational integration of qualitative and quantitative methodologies, including shared preference analysis, Likert scale surveys, and structured dialogues. The implications for product developers and policymakers are twofold: The success of innovations in functional and herbal foods depends on clear, empirically supported claims that are both emotionally and scientifically compelling. Furthermore, authenticity, credible sensory experiences, and alignment with the customer's identity and cultural values are crucial for building long-term loyalty. To effectively link technological innovation with customer trust in the evolving functional and herbal food landscape, it is essential to integrate behavioral research into regulatory frameworks and business communication.

References

Abbasi, A., Sarabi-Aghdam, V., Fathi, M., Abbaszadeh, S., 2025. Non-dairy fermented probiotic beverages: a critical review on the production techniques, health benefits, safety considerations and market trends. Food Rev. Int. 1–56. doi:10.1080/87559129.2025.2550596.

Abdallah, A., 2021. Has the lack of a unified halal standard led to a rise in organised crime in the halal certification sector? Forensic Sci. 1 (3), 181–193. doi:10.3390/forensicsci1030016.

Abdallah, A., Zhang, P., Zhong, Q., Sun, Z., 2019. Application of traditional Chinese herbal medicine by-products as dietary feed supplements and antibiotic replacements in animal production. Curr. Drug Metab. 20 (1), 54–64. doi:10.2174/1389200219666180523102920.

Abdi-Moghadam, Z., Darroudi, M., Mahmoudzadeh, M., Mohtashami, M., Jamal, A.M., Shamloo, E., Rezaei, Z., 2023. Functional yogurt, enriched and probiotic: a focus on human health. Clin. Nutrit. ESPEN 57, 575–586. doi:10.1016/j.clnesp.2023.08.005.

Aiello, P., 2022. Dietary habits, lifestyle and self-medication in an Italian and Spanish university population. Universidad Católica San Antonio de Murcia (UCAM, Murcia, Spain.

Almeida, A., Rodrigues, I., 2024. Animal-based food consumption among future teachers: a survey to discuss more sustainable options. Research and Innovation, Seville, Spain,IATED Academy, Valencia, Spain, pp. 1540–1545.

Alordiah, C.O., Ossai, J.N., 2023. Enhancing questionnaire design: theoretical perspectives on capturing attitudes and beliefs in social studies research. Int. J. Innov. Sci. Res. Technol. 8 (10), 603–614.

Ankiel, M., Halagarda, M., Piekara, A., Sady, S., Żmijowska, P., Popek, S., Pachołek, B., Jefmański, B., Kucia, M., Krzywonos, M., 2025. Role of certifications and labelling in ensuring authenticity and sustainability of fermented milk products. Sustainability 17 (18), 8398. doi:10.3390/su17188398.

Anwar, M.U., Abrar, M., Kiran, F., 2025. Packaging information cues and willingness to pay premium for private-label functional foods: the mediating role of brand attachment. J. Dev. Soc. Sci. 6 (3), 647–663.

Ares, G., Antúnez, L., Vidal, L., Giménez, A., Curutchet, M.R., 2020. Consumer behavior toward functional foods: Insights from focus groups and surveys. Food Qual. Prefer. 84, 103954.

Asioli, D., Næs, T., Granli, B.S., Almli, V.L., 2017. Consumer preferences for functional foods and ingredients: a cross-cultural study. Appetite 112, 131–140.

Aslan, H., 2023. The influence of halal awareness, halal certificate, subjective norms, perceived behavioral control, attitude and trust on purchase intention of culinary products among Muslim customers in Turkey. Int. J. Gastronomy Food Sci. 32, 100726. doi:10.1016/j.ijgfs.2023.100726.

A'yuni, Q., Soontrunnarudrungsri, A., Pirak, T., 2025. Consumer preferences on key Kombucha attributes among Thais and Indonesians: a conjoint analysis study for advanced food product development. ASEAN J. Sci. Technol. Rep. 28 (3), e255936. doi:10.55164/ajstr.v28i3.255936.

Azimont, F., Araujo, L., 2014. Credible Qualifications: The Case of Functional Foods. Edward Elgar Publishing, Cheltenham, UK doi:10.4337/9781782549758.00013.

Baggio, R., Overbeck, G.E., Durigan, G., Pillar, V.D., 2021. To graze or not to graze: a core question for conservation and sustainable use of grassy ecosystems in Brazil. Perspect. Ecol. Conserv. 19 (3), 256–266. doi:10.1016/j.pecon.2021.06.002.

Baker, M.T., Lu, P., Parrella, J.A., Leggette, H.R., 2022. Consumer acceptance toward functional foods: a scoping review. Int. J. Environ. Res. Public Health 19 (3). doi:10.3390/ijerph19031217.

Basha, S.A., Lal, D., Basha, S.A., Lal, D., 2019. Consumer perception and attitude towards organic food products: a study of consumers in southern India. Int. J. Recent Technol. Eng. 8 (4), 6989–6993.

Basso, F., Robert-Demontrond, P., Hayek, M., Anton, J., 2020. The placebo effect in marketing: conceptualization, mechanisms, and consequences. Psychol. Mark. 37 (5), 713–726.

Basu, R., Dasgupta, S., Babu, S.N., Noor, A., 2023. Medicinal plants in the Indian traditional medicine and current practices. In: Bioprospecting of Tropical Medicinal Plants. CRC Press, India, pp. 253–286. doi:10.1007/978-3-031-28780-0_9.

Baudar, P., 2023. The New Wildcrafted Cuisine: Exploring the Exotic Gastronomy of Local Terroir. Chelsea Green Publishing.

Bhatnagar, R., 2025. In: Social acceptance interventions framework.Wendy Deliverables, D4.1, WP 4, T. Marin Energi Testsenter AS, Haugesund, Norway, p. 4.1.

Braamhaar, D.J.M., van der Lee, J., Bebe, B.O., Oosting, S.J., 2025. From rural to urban: exploring livestock farming practices in urbanizing landscapes. Agric. Syst. 225. doi:10.1016/j.agsy.2025.104297.

Braca, A., Dondio, P., 2023. Developing persuasive systems for marketing: the interplay of persuasion techniques, customer traits and persuasive message design. Ital. J. Mark 2023 (3), 369–412. doi:10.1007/s43039-023-00077-0.

Braithwaite, J., Churruca, K., Ellis, L.A., Leask, E., Long, J.C., Sarkies, M., Zurynski, Y., Clay-Williams, R., 2024. Resilient health care performance in the real world: fixing problems that never happened. BMC Health Serv. Res. 24 (1). doi:10.1186/s12913-024-11639-z.

Broeders, I., Folkvord, F., Croes, E., 2024. Making healthy food taste better: the effect of healthy versus hedonic food product claims on consumers' taste perception and purchase. intention.Appetite 202, 107636. https://doi.org/10.1016/j.appet.2024.107636.

Bschaden, A., Schulz, J., Stroebele-Benschop, N., 2022. The sustainability halo effect: does the provision of sustainability information of a snack affect sensory and health perception, and willingness to pay? Future Foods 5, 100143. doi:10.1016/j.fufo.2022.100143.

Buçinca, Z., Swaroop, S., Paluch, A.E., Doshi-Velez, F., Gajos, K.Z., 2025. Contrastive explanations that anticipate human misconceptions can improve human decision-making skills. In: Proceedings of the 2025 CHI Conference on Human Factors in Computing Systems, pp. 1–25. doi:10.1145/3706598.3713229.

Butnariu, M., 2024. Functional foods and nutraceuticals in health maintenance. In: Nutrition and Diet in Health: Principles and Applications. CRC Press, Romania, pp. 71–88. doi:10.1201/9781003361497-8.

Cao, C., Zhang, X., 2024. What triggers consumers to purchase eco-friendly food? The impact of micro signals, macro signals and perceived value. Br. Food J. 126 (5), 2204–2226. doi:10.1108/bfj-08-2023-0697.

Cattaneo, C., Spinelli, S., Dinnella, C., Proserpio, C., Monteleone, E., Pagliarini, E., Laureati, M., 2025. Patterns of sensory and hedonic responses for salty and umami tastes and their impact on food familiarity, consumption, and nutritional status: a gender-based analysis from a large population sample. Curr. Res. Food Sci. 10, 100970. doi:10.1016/j.crfs.2025.100970.

Chauhan, A., Goyal, M.R., Aswal, J.S., 2025. Consumer perception and acceptance: navigating the challenges. In: Biotechnological Innovations in Food Processing. Springer, Cham, pp. 177–205.

Conficoni, D., Zaghi, M., Rossin, T., Brscic, M., Valerio, G., 2022. Meeting religious requirements and food safety during ritual slaughter: a case study on how Italian authorities handle the issue. Anim. Front. 12 (1), 25–34. doi:10.1093/af/vfac011.

Creswell, J.W., Plano Clark, V.L., 2018. Designing and Conducting Mixed Methods Research, 3rd ed. SAGE, CA, USA.

Cuchillo-Hilario, M., Fournier-Ramírez, M.I., Díaz Martínez, M., Benavides, S.M.ño, Calvo-Carrillo, M.C., Carrillo Domínguez, S., Carranco-Jáuregui, M.E., Hernández-Rodríguez, E., Mora-Pérez, P., Cruz-Martínez, Y.R., Delgadillo-Puga, C., 2024. Animal food products to support human nutrition and to boost human health: the potential of feedstuffs resources and their metabolites as health-promoters. Metabolites. 14 (9). doi:10.3390/metabo14090496.

Cutolo, D., Ferriani, S., 2024. Now it makes more sense: how narratives can help atypical actors increase market appeal. J. Manag. 50 (5), 1599–1642. doi:10.1177/01492063231151637.

Dangaiso, P., Makudza, F., Tshuma, S., Hogo, H., Mpondwe, N., Masona, C., Sakarombe, U., Nedure, T., Muchowe, R., Nyathi, G., Jonasi, K., Towo, T., Manhando, T., Tagwirei, D., 2023. Can social marketing undo the COVID-19 infodemic? Predicting consumer preventive health behavior in the marginalized communities in Zimbabwe. Cogent Public Health 10 (1). doi:10.1080/27707571.2023.2234599.

Davis, T., 2025. Distributed Ledger Technology Adoption in Healthcare: The Effect of Organizational Attitude Towards Patient Relations. Jacksonville University, Jacksonville, FL, USA.

del Bosque, C.I.E., Grahl, S., Nolte, T., Mörlein, D., 2022. Meat quality parameters, sensory properties and consumer acceptance of chicken meat from dual-purpose crossbreeds fed with regional faba beans. Foods 11 (8), 1074. doi:10.3390/foods11081074.

Delgado-Ballester, E., Munuera-Alemán, J.L., 2011. Brand trust in the context of consumer loyalty. Eur. J. Mark. 45 (3), 390–418.

DeVellis, R.F., 2017. Scale development: Theory and Applications, 4th ed. SAGE, Thousand Oaks, CA, USA.

Didenko, A.N., 2023. Australia's consumer data right and its implications for consumer trust. Forthcoming 50 (1), 20–23.

Dihdashti, M., 2025. Biohacking gastronomy–Volume 01: unlocking the science of taste, health & culinary innovation.Available at SSRN. https://doi.org/10.2139/ssrn.5384720.

Embling, R., Neilson, L., Mellor, C., Durodola, M., Rouse, N., Haselgrove, A., Shipley, K., Tales, A., Wilkinson, L., 2024. Exploring consumer beliefs about novel fortified foods: a focus group study with UK-based older and younger adult consumers. Appetite 193, 107139. doi:10.1016/j.appet.2023.107139.

Essiz, O., Senyuz, A., 2024. Predicting the value-based determinants of sustainable luxury consumption: a multi-analytical approach and pathway to sustainable development in the luxury industry. Bus. Strat. Environ. 33 (3), 1721–1758. doi:10.1002/bse.3569.

Frohlich, X., 2023. From Label to Table: Regulating Food in America in the Information Age. University of California Press, Oakland, CA, USA doi:10.1525/9780520970816.

García-Barón, S.E., Carmona-Escutia, R.P., Herrera-López, E.J., Leyva-Trinidad, D.A., Gschaedler-Mathis, A., 2025. Consumers' drivers of perception and preference of fermented food products and beverages: a systematic review. Foods 14 (5). doi:10.3390/foods14050713.

Georgieva, K., 2024. Ethics on the grill: Exploring vegan consumer reactions to the mainstreaming of veganism, marketing of meat alternatives, and plant-based brand identities [Master's thesis]. York St. John University (United Kingdom).

Gerber, S., Dix, S.R., Cash, S.B., 2024. Marketing plant-based versus animal-sourced foods in online grocery stores: a comparative content analysis of sustainability and other product claims in the United States. Bus. Strat. Environ. 33 (5), 4958–4973. doi:10.1002/bse.3717.

Ghosh, S., Sarkar, T., Chakraborty, R., 2024. Functional food and food innovation toward quality and safety regulations. In: Food Coatings and Preservation Technologies. Wiley, Hoboken, NJ, pp. 585–609. doi:10.1002/9781394237623.ch17.

Gordon, W., Langmaid, R., Mills, C., 2022. Qualitative Market Research: A Practitioner's and Buyer's Guide. Routledge, London doi:10.4324/9781315245553.

Goyal, S., Thirumal, D., Algin Yapar, E., Sönmez Gürer, E., Kumar, A., Babu, M.A, Sindhu, R.K, 2023. Asian veterinary medicines: from the past to the future. J. Res. Pharm. (27(4)) 1313–1328. doi:10.29228/jrp.419.

Grinberga-Zalite, G., Zvirbule, A., Hernik, J., 2024. Fostering a link between creativity and consumer acceptance: essential factors for advancing innovations in food industry. Creat. Stud. 17 (1), 309–322. doi:10.3846/cs.2024.19789.

Grunert, K.G., 2020. Consumer behavior with regard to food innovation. J. Sci. Food Agric. 100 (10), 3884–3891.

Grunert, K.G., Sonntag, W.I., Glanz-Chanos, V., Forum, E., 2018. Consumer interest in sustainable food production: insights from a European study. J. Food Prod. Mark. 24 (3), 308–326.

Guiné, R.P.F., Florença, S.G., Barroca, M.J., Anjos, O., 2021. The duality of innovation and food development versus purely traditional foods. Trends Food Sci. Technol. 109, 16–24. doi:10.1016/j.tifs.2021.01.010.

Guo, L., Huang, F., Liu, M., Zhang, Y., Zhang, J., Zhang, B., Wang, H., 2023. Generational differences in food consumption among Chinese adults of different ages. Nutrients 15 (20), 4451. doi:10.3390/nu15204451.

Guptill, A.E., Copelton, D.A., Lucal, B., 2022. Food & Society: Principles and Paradoxes. John Wiley & Sons, Hoboken.

Ha, M.T., Ngan, V.T.K., Nguyen, P.N.D., 2022. Greenwash and green brand equity: the mediating role of green brand image, green satisfaction and green trust and the moderating role of information and knowledge. Bus. Ethics Environ. Responsib. 31 (4), 904–922. doi:10.1111/beer.12462.

Hair, J.F., Black, W.C., Babin, B.J., Anderson, R.E., 2019. Multivariate Data Analysis, 8th ed. Cengage, Andover.

Halder, T., Mehta, P., Acharya, N., 2021. Trends in the Functional Food Market and Nutraceutical Product Developmen Nutraceuticals for Aging and Anti-Aging: Basic Understanding and Clinical Evidence. CRC Press, India, pp. 1–26. doi:10.1201/9781003110866-1.

Hanan, F., Karim, S., Aziz, Y., Ishak, F., Sumarjan, N., 2024. Consumer's cultured meat perception and acceptance determinants: a systematic review and future research agenda. Int. J. Consum. Stud. 48 (5). doi:10.1111/ijcs.13088.

Hao, D.C., Li, C.X., Xiao, P.G., Xie, H.T., Bao, X.L., Wang, L.F., 2023. Conservation tillage in medicinal plant cultivation in China: what, why, and how. Agronomy 13 (7). doi:10.3390/agronomy13071890.

Hong, X., Li, C., Wang, L., Gao, Z., Wang, M., Zhang, H., Monahan, F.J., 2022. The effects of nutrition and health claim information on consumers' sensory preferences and willingness to pay. Foods 11 (21), 3460. doi:10.3390/foods11213460.

Ikäheimo, J.P., 2021. Arctic narratives: brewing a brand with neolocalism. J. Brand Manag. 28 (4), 374–387. doi:10.1057/s41262-021-00232-y.

Irmak, C., Block, L.G., Fitzsimons, G.J., 2005. The placebo effect in marketing: sometimes you just have to want it to work. J. Mark. Res. 42 (4), 406–409. doi:10.1509/jmkr.2005.42.4.406.

Ismail, I., Huda, N., 2024. Current techniques and technologies of meat quality evaluation. In: Hand Book of Processed Functional Meat Products. Springer Nature, Malaysia, pp. 437–512. doi:10.1007/978-3-031-69868-2_17.

Johan, E., Plana-Casado, M.J., 2023. Harmonizing halal in ASEAN: analysis of halal food guidelines under the ASEAN way approach. J. ASEAN Stud. 11 (1), 43–67. doi:10.21512/jas.v11i1.9682.

Joshi, K., Singh, R., Singh, M., 2024. Functional foods for geriatrics. In: The Functional Foods: Nutrient and Health Benefits. Apple Academic Press, India, pp. 43–75.

Kahneman, D., 2011. Thinking, Fast and Slow. Penguin Books, London Farrar, Straus and Giroux.

Karagiannis, D., Hatzithomas, L., Fotiadis, T., Gasteratos, A., 2022. The impact of brand awareness and country of origin in the advertising effectiveness of Greek food products in the United Kingdom: the case of Greek yogurt. Foods 11 (24), 4019. doi:10.3390/foods11244019.

Kenny, T.A., Woodside, J.V., Perry, I.J., Harrington, J.M., 2023. Consumer attitudes and behaviors toward more sustainable diets: a scoping review. Nutr. Rev. 81 (12), 1665–1679. doi:10.1093/nutrit/nuad033.

Kernecker, M.L., 2025. Between Grazing and Gathering: plant knowledges, belonging and becoming in the Swiss Alps. Plant Perspect doi:10.3197/WHPPP.63876246815913.

Kim, J., Lee, S., Kim, Y., 2023. Consumer perceptions of herbal-fed poultry: focus group analysis in South Korea. Br. Food J. 125 (5), 1624–1642.

Koo, M., Yang, S.W., 2025. Likert-type scale. Encyclopedia 5 (1). doi:10.3390/encyclopedia5010018.

Körmeling, M.M., 2021. Understanding the attraction towards complementary and alternative medicine in the Netherlands–a gendered analysis. Master's thesis. Radboud University, Nijmegen, The Netherlands.

Krueger, R.A., Casey, M.A., 2015. Focus Groups: A Practical Guide for Applied Research, 5th ed. SAGE, Thousand Oaks, CA.

Kulshreshtha, K., Sharma, G., Bajpai, N., 2023. Conjoint analysis: the assumptions, applications, concerns, remedies and future research direction. Int. J. Qual. Reliab. Manag. 40 (2), 607–627. doi:10.1108/IJQRM-07-2021-0199.

Küst, P., 2019. The impact of the organic label halo effect on consumers' quality perceptions, value-in-use and well-being. J. Manag. Sci. (JUMS) 4 (2), 241–264.

Lee, D., 2022. Design thinking in the classroom: Easy-to-use teaching tools to foster creativity, encourage innovation and unleash potential in every student. Simon & Schuster, New York.

Lee, S., Park, H.J., Kim, Y., 2021. Mood and health expectation effects of functional beverage labeling. Food Qual. Preference 93, 104272.

Leszczynski, L., 2008. A/maturity in art: on instructing the midlife novice painter.Master's thesis. Simon Fraser University.

Lim, W.M., 2022. The sustainability pyramid: a hierarchical approach to greater sustainability and the United Nations sustainable development goals with implications for marketing theory, practice, and public policy. Australas. Mark. J. 30 (2), 142–150. doi:10.1177/18393349211069152.

Lim, W.M., 2025. What is qualitative research? An overview and guidelines. Australas. Mark. J. 33 (2), 199–229. doi:10.1177/14413582241264619.

Lucas, B.F., Brunner, T.A., 2024. Attitudes and perceptions towards microalgae as an alternative food: a consumer segmentation in Switzerland. Algal Res. 78. doi:10.1016/j.algal.2023.103386.

Luo, X., Zaitoon, A., Lim, L.T., 2022. A review on colorimetric indicators for monitoring product freshness in intelligent food packaging: indicator dyes, preparation methods, and applications. Compr. Rev. Food Sci. Food Saf. 21 (3), 2489–2519. doi:10.1111/1541-4337.12942.

Luthra, A., Dixit, S., Singh, A., Garg, S., Khattar, S., 2024. Sensory storytelling: crafting brand narratives through sight, sound, smell, touch, and taste. In: Sensible Selling Through Sensory Neuromarketing. IGI Global, India, pp. 230–253. doi:10.4018/979-8-3693-4236-7.ch012.

Mack, G., Walter, T., Flury, C., 2013. Seasonal alpine grazing trends in Switzerland: economic importance and impact on biotic communities. Environ. Sci. Policy 32, 48–57. doi:10.1016/j.envsci.2013.01.019.

Magnier, L., Schoormans, J., 2015. Consumer reactions to sustainable packaging: the interplay of visual appearance, verbal claim and environmental concern. J. Environ. Psychol. 44, 53–62. doi:10.1016/j.jenvp.2015.09.005.

Mahesar, S.A., Shah, S.T.H., Ramadan, M.F., Panhwar, W.N., 2025. Introduction to halal food authentication. In: Authentication of Halal Products Using FTIR Spectroscopy. Springer Nature Switzerland, Cham, pp. 1–18.

Martirosyan, D., Alvarado, A., 2023. Functional foods regulation system: proposed regulatory paradigm by functional food center. Funct. Food Sci. 3 (11), 275–287. doi:10.31989/ffs.v3i11.1265.

Mehta, A., 2024. Effects of extrinsic characteristics, mental simulation and repetitive exposure on the hedonic, emotional and physiological responses of consumers: A study of food choices and purchasing behaviours: A thesis submitted in partial fulfilment of the requirements for the Degree of Doctor of Philosophy at Lincoln University. Lincoln University.

Meijer, G.W., Lähteenmäki, L., Stadler, R.H., Weiss, J., 2021. Issues surrounding consumer trust and acceptance of existing and emerging food processing technologies. Crit. Rev. Food Sci. Nutr. 61 (1), 97–115. doi:10.1080/10408398.2020.1718597.

Menzli, L.J., Smirani, L.K., Boulahia, J.A., Hadjouni, M., 2022. Investigation of open educational resources adoption in higher education using Rogers' diffusion of innovation theory. Heliyon 8 (7). doi:10.1016/j.heliyon.2022.e09885.

Mobekk, H., 2021. Influencing choice behavior and improving well-being by applying behavior analysis and behavioral economics.Ph.D. Department of Behavioural Sciences, Faculty of Health Sciences, OsloMet – Oslo Metropolitan University, Oslo, Norway.

Mollahassani, A., Cheng, A., Zhang, W., Wu, Y., 2025. The transmission of herbal medicine to China via the Silk Road in Medieval ages. Chin. Med. Cult. 8 (1), 13–23. doi:10.1097/mc9.0000000000000130.

Nair, K., 2024. Regulatory policies and consumer confusion: assessing the impact and mitigation strategies across industries. Int. J. Mark. Bus. Commun. 13 (2), 1–5.

Nam, S.J., Suk, J., 2024. Influence of health food literacy on willingness to pay for healthier foods: focus on food insecurity. Int. J. Equity in Health 23 (1). doi:10.1186/s12939-024-02135-1.

Nazzaro, C., Uliano, A., Lerro, M., Stanco, M., 2025. From claims to choices: how health information shapes consumer decisions in the functional food market. Foods 14 (4), 699. doi:10.3390/foods14040699.

Ndhlovu, P.T., Asong, J.A., Omotayo, A.O., Otang-Mbeng, W., Aremu, A.O., 2023. Ethnobotanical survey of medicinal plants used by indigenous knowledge holders to manage healthcare needs in children. PLoS One 18 (3). doi:10.1371/journal.pone.0282113.

Nguyen, M.H., Nguyen, D.H., 2025. How do transparency and traceability enhance purchasing behaviors via consumer trust? Insights for food supply chains. J. Consum. Mark doi:10.1108/JCM-07-2024-7005.

Niba, A.T., 2025. Emerging peri-urban and urban animal production in the Western Highlands of Cameroon: The search for novel strategies for improving sustainability and productivity. Trop. Agric. 102 (1), 124–140. https://journals.sta.uwi.edu/ojs/index.php/ta/article/download/8409/7750.

Oda, B.K., Lulekal, E., Warkineh, B., Asfaw, Z., Debella, A., 2024. Ethnoveterinary medicinal plants and their utilization by indigenous and local communities of Dugda District, Central Rift Valley, Ethiopia. J. Ethnobiol. Ethnomed. 20 (1). doi:10.1186/s13002-024-00665-0.

Onyango, B., Govindasamy, R., Nayga, R.M., 2023. Factors influencing consumer adoption of functional foods in emerging markets. J. Consum. Mark. 40 (2), 181–195.

Orynbekov, D., Kalibekkyzy, Z., Nurgazezova, A., Nurymkhan, G., Kassenov, A., Yermekov, Y., 2025. Consumer perception of irradiated food products in the Abai region of Kazakhstan. Foods 14 (9), 1625. doi:10.3390/foods14091625.

Oztas, G.Z., Erdem, S., 2021. Framework selection for developing optimization algorithms: assessing preferences by conjoint analysis and best–worst method. Soft Comput. 25 (5), 3831–3848. doi:10.1007/s00500-020-05411-8.

Pant, H., Swaroop, D., Singh, M.K., Verma, J., Pandey, P.R., 2024. Environment and agriculture: new technological applications. Society of Biological Sciences and Rural Development (SBSRD). Prayagraj, India.

Pascall, M.A., DeAngelo, K., Richards, J., Arensberg, M.B., 2022. Role and importance of functional food packaging in specialized products for vulnerable populations: implications for innovation and policy development for sustainability. Foods 11 (19). doi:10.3390/foods11193043.

Pattanaik, D., 2024. Public perception of corporate environmental commitments: how communication shapes trust. J. Jurivox 1 (2), 63–74.

Pérez, R.L., 2009. Tasting culture: Food, family and flavor in Greater Mexico. New York University.

Phillips, C.J.C., 2024. Farm animal welfare—from the farmers' perspective. Animals 14 (5), 671. doi:10.3390/ani14050671.

Pieniak, Z., Verbeke, W., Vanhonacker, F., Guerrero, L., Hersleth, M., 2016. Association between traditional food consumption and motives for food choice in six European countries. Appetite 87, 336–344.

Pistorio, E., Chinnici, G., Zarbà, C., Bellia, C., Pappalardo, G., 2023. The revolution of functional food: a market analysis. In: International Multidisciplinary Scientific GeoConference Surveying Geology and Mining Ecology Management, SGEM, 6. Sofia, Bulgaria, pp. 433–441. doi:10.5593/sgem2023V/6.2/s25.53.

Ponnampalam, E.N., Jairath, G., Gadzama, I.U., Li, L., Santhiravel, S., Ma, C., Flores, M., Priyashantha, H., 2025. Production systems and feeding strategies in the aromatic fingerprinting of animal-derived foods: invited review. Foods 14 (19). doi:10.3390/foods14193400.

Ponnampalam, E.N., Priyashantha, H., Vidanarachchi, J.K., Kiani, A., Holman, B.W.B., 2024. Effects of nutritional factors on fat content, fatty acid composition, and sensorial properties of meat and milk from domesticated ruminants: an overview. Animals 14 (6). doi:10.3390/ani14060840.

Prada, M., Saraiva, M., Sério, A., Coelho, S., Godinho, C.A., Garrido, M.V., 2021. The impact of sugar-related claims on perceived healthfulness, caloric value and expected taste of food products. Food Qual. Prefer. 94. doi:10.1016/j.foodqual.2021.104331.

Ranawat, M., 2025. Redefining consumer decision-making: the neuropsychological foundation of behavioral economics in modern marketing. Voice Creat. Res. 7 (2), 121–134. doi:10.53032/tvcr/2025.v7n2.15.

Rincon, J.M.C., 2025. Strategic framing in communication to promote sustainable consumer choices.Ph.D. dissertation. Aarhus BSS. Aarhus University, Aarhus, Denmark.

Rodríguez, V.J.C., Antonovica, A., Martín, D.L.S., 2023. Consumer neuroscience on branding and packaging: a review and future research agenda. Int. J. Consum. Stud. 47 (6), 2790–2815. doi:10.1111/ijcs.12936.

Ronteltap, A., van Trijp, J.C.M., Renes, R.J., Frewer, L.J., 2007. Consumer acceptance of technology-based food innovations: lessons for the future of nutrigenomics. Appetite 49 (1), 1–17. doi:10.1016/j.appet.2007.02.002.

Root-Bernstein, M., Vargas, B.H., Bondoux, A., Guerrero-Gatica, M., Zorondo-Rodríguez, F., Huerta, M., Valenzuela, R., Bello, Á.V., 2022. Silvopastoralism, local ecological knowledge and woodland trajectories in a category V- type management area. Biodivers. Conserv. 31 (2), 543–564. doi:10.1007/s10531-021-02349-7.

Ropawandi, D., Halim, L., Husnin, H., 2022. Augmented reality (AR) technology-based learning: the effect on physics learning during the COVID-19 pandemic. Int. J. Inf. Educ. Technol. 12 (2), 132–140. doi:10.18178/ijiet.2022.12.2.1596.

Rramani, Q., Barakat, Y., Jacob, G., Ohla, K., Lim, S.X.L., Schicker, D., Freiherr, J., Saruco, E., Pleger, B., Weber, B., Schultz, J., 2023. Nutrition claims influence expectations about food attributes, attenuate activity in reward-associated brain regions during tasting, but do not impact pleasantness. Brain Behav. 13 (1). doi:10.1002/brb3.2828.

Rupprecht, C.D.D., Fujiyoshi, L., McGreevy, S.R., Tayasu, I., 2020. Trust me? Consumer trust in expert information on food product labels. Food Chem. Toxicol. 137. doi:10.1016/j.fct.2020.111170.

Russo, G.M., Tomei, P.A., Serra, B., Mello, S., 2021. Differences in the use of 5-or 7-point Likert scale: an application in food safety culture. Organ. Cult. 21 (2), 1.

Saeed, T.A., 2023. Food additives and their applications in the production of halal and kosher foods. Halal and Kosher Food: Integration of Quality and Safety for Global Market Trends 247–259. doi:10.1007/978-3-031-41459-6_19.

Saha, D.P., 2018. A study of methods in computational psychophysiology for incorporating implicit affective feedback in intelligent environments.Ph.D. dissertation. Virginia Polytechnic Institute and State University (Virginia Tech, Blacksburg, VA, USA.

Saldaña, E., Merlo, T.C., Patinho, I., Rios-Mera, J.D., Contreras-Castillo, C.J., Selani, M.M., 2021. Use of sensory science for the development of healthier processed meat products: a critical opinion. Curr. Opin. Food Sci. 40, 13–19. doi:10.1016/j.cofs.2020.04.012.

Saleem, H., Amjad, A., Saleem, H., Amjad, A., 2023. Psychological influences on economic decision-making: analyzing biases, heuristics, and consumer behavior. Insights J. Life Soc. Sci. 1 (2), 34–38.

Sato, K., Kodama, K., Sengoku, S., 2025. The co-evolution of markets and regulation in the Japanese functional food industry: balancing risk and benefit. Foods 14 (9), 1581. doi:10.3390/foods14091581.

Schnettler, B., Miranda-Zapata, E., Lobos, G., Schnettler, B., Miranda-Zapata, E., Lobos, G., 2020. Trust and food labeling in Latin American consumers. Food Res. Int. 131, 108988.

Schrobback, P., Zhang, A., Loechel, B., Ricketts, K., Ingham, A., 2023. Food credence attributes: a conceptual framework of supply chain stakeholders, their motives, and mechanisms to address information asymmetry. Foods 12 (3), 538. doi:10.3390/foods12030538.

Schuldt, J.P., Schwarz, N., 2010. The "organic" path to obesity? organic claims influence calorie judgments and exercise recommendations. Judgm. Decis. Mak. 5 (3), 144–150. doi:10.1017/s1930297500001017.

Sezgin, E., Uyanık, B., 2022. Priorities of consumers for restaurant preferences: a conjoint analysis study on generation Z. J. Tour. Leis. Hosp. 4 (2), 6.

Shams, S.R., Brown, D.M., Hardcastle, K., 2025. Pricing strategy for people, planet, and profit. In: Sustainable Marketing: Strategic Marketing for People, Planet and Profit. Springer Nature Switzerland, Cham, pp. 169–225.

Shangguan, S., Afshin, A., Shulkin, M., Ma, W., Marsden, D., Smith, J., Saheb-Kashaf, M., Shi, P., Micha, R., Imamura, F., Mozaffarian, D., 2019. A meta-analysis of food labeling effects on consumer diet behaviors and industry practices. Am. J. Prev. Med. 56 (2), 300–314. doi:10.1016/j.amepre.2018.09.024.

Shayanowako, A., Morrissey, O., Tanzi, A., Muchuweti, M., Mendiondo, G., Mayes, S., Modi, A., Mabhaudhi, T., 2021. African leafy vegetables for improved human nutrition and food System resilience in Southern Africa: a scoping review. Sustainability 13 (5), 2896. doi:10.3390/su13052896.

Shiv, B., Carmon, Z., Ariely, D., 2005. Placebo effects of marketing actions: consumers may get what they pay for. J. Mark. Res. 42 (4), 383–393. doi:10.1509/jmkr.2005.42.4.383.

Sigrist, A., Richardet, E., Senn, N., Schwarz, J., 2025. Discussing human and environmental health co-benefits related to diet and mobility behaviours in the primary care setting: a qualitative exploratory study. Int. J. Environ. Res. Public Health 22 (10), 1503. doi:10.3390/ijerph22101503.

Sindhu, T.A., 2024. A comparative study on the consumer preference towards pddp milk and other milk brands.MBA dissertation. Department of Management Studies, PSG Institute of Management, Coimbatore, India.

Sörqvist, P., Langeborg, L., 2019. Why people harm the environment although they try to treat it well: an evolutionary–cognitive perspective on the environmental belief–behavior gap. Front. Psychol. 10, 348.

Soscia, I., 2013. Emotions and Consumption Behaviour. Edward Elgar Publishing, Cheltenham doi:10.4337/9780857937971.

Su, Y., Khaskheli, A., Raza, S.A., Yousufi, S.Q., 2022. How health consciousness and social consciousness affect young consumers purchase intention towards organic foods. Manag. Environ. Qual. Int. J. 33 (5), 1249–1270. doi:10.1108/MEQ-12-2021-0279.

Sweeney, S., Regan, Á, McKernan, C., Benson, T., Hanlon, A., Dean, M., 2022. Current consumer perceptions of animal welfare across different farming sectors on the island of Ireland. Animals 12 (2), 185. doi:10.3390/ani12020185.

Szymkowiak, A., Antoniak, M.A., Borusiak, B., 2022. The role of health orientation in determining purchase intention and behaviour. Br. Food J. 124 (13), 559–577. doi:10.1108/BFJ-12-2021-1272.

Teston, M., 2023. Strategies to generate added value for mountain livestock farming systems.Ph.D. dissertation. University of Padova, Padova, Italy.

Theodorou, A., Alessandro, M., Federica, D., Mei, X., Marino, B., 2024. The social-psychological ante-cedents of risk perception of natural hazards: a meta-analytic approach. Int. J. Psychol. 59 (Suppl. 1), 289.

Török, P., Lindborg, R., Eldridge, D., Pakeman, R., 2024. Grazing effects on vegetation: biodiversity, management, and restoration. Appl. Veg. Sci. 27 (3). doi:10.1111/avsc.12794.

Tura, M., Gagliano, M.A., Valli, E., Petracci, M., Gallina Toschi, T., 2024. A methodological review in sensory analyses of chicken meat. Poult. Sci. 103 (11). doi:10.1016/j.psj.2024.104083.

Van Loo, E.J., Hoefkens, C., Verbeke, W., van Loo, E.J., Hoefkens, C., Verbeke, W., 2017. Healthy, sustainable and plant-based eating: perceived benefits and barriers. Appetite 108, 137–149. doi:10.1016/j.foodpol.2017.03.001.

Vella, M.N., Stratton, L.M., Sheeshka, J., Duncan, A.M., 2013. Exploration of functional food consumption in older adults in relation to food matrices, bioactive ingredients, and health. J. Nutr. Gerontol. Geriatr. 32 (2), 122–144. doi:10.1080/21551197.2013.781419.

Verbeke, W., 2015. Profiling consumers who are ready to adopt insects as a meat substitute in a Western society. Food Qual. Prefer. 39, 147–155. doi:10.1016/j.foodqual.2014.07.008.

Vlaicu, P.A., Untea, A.E., Varzaru, I., Saracila, M., Oancea, A.G., 2023. Designing nutrition for health—incorporating dietary by-products into poultry feeds to create functional foods with insights into health benefits, risks, bioactive compounds, food component functionality and safety regulations. Foods. 12 (21). doi:10.3390/foods12214001.

Voinea, L., Badea, A.M., Dina, R., Popescu, D.V., Bucur, M., Negrea, T.M., 2025. Perspectives in the scientific literature on the barriers and benefits of the transition to a plant-based diet: a bibliometric analysis. Foods 14 (17). doi:10.3390/foods14172942.

Vu, N., Ghadge, A., Bourlakis, M., 2025. The impact of Blockchain adoption on supply chain performance: evidence from food industry. Int. J. Prod. Res. 63 (14), 5402–5427. doi:10.1080/00207543.2024.2414375.

Waber, R.L., Shiv, B., Carmon, Z., Ariely, D., 2008. Commercial features of placebo and therapeutic efficacy. JAMA 299 (9), 1016–1017. doi:10.1001/jama.299.9.1016.

Walten, L., 2022. Consumer acceptance of technology-based food innovations: towards an integrated framework of the consumer acceptance process and effective communication approaches.Ph.D. dissertation. Leibniz University Hannover, Hannover, Germany.

Wang, S., Lehto, X., Cai, L.A., Behnke, C., 2019. Encountering exotic cuisine on foreign lands: narratives from American travelers. TTRA 2018 International Conference: Encouraging Innovation in the Universe of Tourism Research. Miami/Coral Gables, FL.

Wansink, B., Chandon, P., 2014. Slim by design: Redirecting the accidental drivers of mindless overeating. J. Consum. Psychol. 24 (3), 413–431. doi:10.1016/j.jcps.2014.03.006.

Wendimu, A., Bojago, E., Abrham, Y., Tekalign, W., 2023. Practices of ethnoveterinary medicine and ethnobotanical knowledge of plants used to treat livestock diseases, Wolaita zone, southern Ethiopia. Cogent Food Agric. 9 (1). doi:10.1080/23311932.2023.2248691.

Winslow, D., 2012. Food for thought: Sustainability, Community-Engaged Teaching and Research, and Critical Food Literacy. Syracuse University, Syracuse, NY.

Wolfenbarger, C.L. (2023). Gratitude impact on organizational citizenship behavior, employee well-being, and three generational cohorts: generation X, millennials, generation Z.Ph.D. dissertation. Liberty University, Lynchburg, VA, USA

Wu, H., Zhang, Z., Xu, J., Song, J., Han, J., Zhang, J., Mei, Q., Cheng, F., Zhuang, H., Li, S., 2025. Food consumption away from home had divergent impacts on diet nutrition quality across urban and rural China. Food Secur. 17 (1), 41–56. doi:10.1007/s12571-024-01514-4.

Wu, W., Zhang, A., van Klinken, R.D., Schrobback, P., Muller, J.M., 2021. Consumer trust in food and the food system: a critical review. Foods 10 (10). doi:10.3390/foods10102490.

Yamini, S., Paswan, V.K., Hmar, B.Z., Verma, D.K., Baruah, S., 2025. Quality, nutrition, and functionality of meat products. In: Trends in Animal-Based Foods. Springer Science+Business Media, India, pp. 149–179. doi:10.1007/978-3-032-00575-5_8.

Yan, L., Keh, H.T., Murray, K.B., 2024. Feeling the values: how pride and awe differentially enhance consumers' sustainable behavioral intentions. J. Acad. Mark. Sci. 52 (1), 75–96. doi:10.1007/s11747-023-00928-4.

Zhu, Y., Wang, Z., Zhu, X., 2023. New reflections on food security and land use strategies based on the evolution of Chinese dietary patterns. Land Use Policy 126, 106520. doi:10.1016/j.landusepol.2022.106520.

Chapter 10

Ethics, environment, and sustainable practices

Ahmed I. Elsherbeni[a] and Mohamed E. Abd El-Hack[b]
[a]*Animal Production Systems Research Department, Animal Production Research Institute, Agricultural Research Center, Giza, Egypt,* [b]*Department of Industrial Pharmacy, College of Pharmaceutical Sciences and Drug Manufacturing, Misr University for Science and Technology (MUST), Giza, Egypt*

10.1 Introduction

Most animal production systems are subject to increasing and ongoing scrutiny, not only regarding productivity but also their environmental impact, ethical standards, and contribution to public health. While most conventional systems, which rely heavily on antibiotics, support intensive animal production, they are also strongly associated with greenhouse gas emissions, reduced biodiversity, depletion of available resources, and rising rates of antimicrobial resistance (AMR) (Ahmed et al., 2023;Adegbeye et al., 2025). These steady pressures have prompted an intense search for substitutes that improve animal performance, environmental safety, and public expectations.

Herbal and plant-based dietary systems have emerged as promising strategies in this new shift. Bioactive compounds derived from plants, such as polyphenols, tannins, essential oils, and saponins, have the potential to regulate gut microbiota, enhance nutrient utilization, reduce methane emissions, and decrease reliance on synthetic antimicrobials (Hamdi et al., 2026). Unlike antibiotics, which exert selective pressure on a particular set of bacterial pathways, plant-derived compounds exert their effect through multitargeted processes, significantly reducing the risk of bacterial resistance developing (AL-Azzawi et al, Hasan and Barrak, 2024). In addition to controlling animal productivity, these systems (herbal-based) are mostly included in several rangelands and regenerative outlines that improve soil carbon sequestration, nutrient cycling processes, and ecosystem services (Malisch et al., 2024).

Growing universal concern about AMR, an acute public health threat matching the WHO and the FAO, has resulted in an increased requirement to lower the use of antibiotics in the animal and poultry sectors. Likewise, purchaser

Plant-Derived Feed Additives and Animal-Based Foods: Linking Animal Health, Product Quality, and Consumer Perceptions. DOI: https://doi.org/10.1016/B978-0-443-45446-2.00013-2

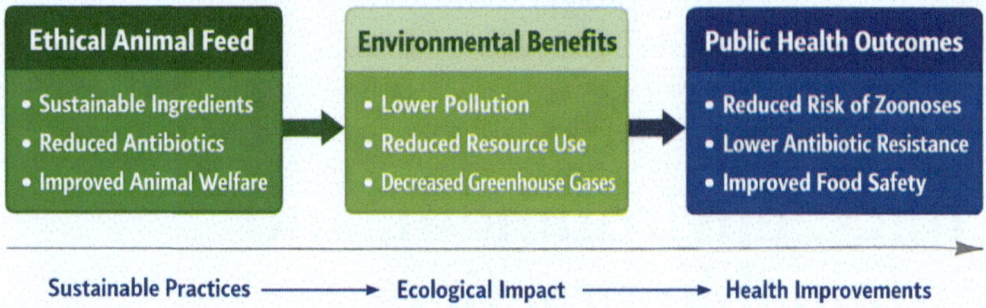

FIGURE 10.1 Conceptual model linking ethical feed → environmental benefit → public health outcome.

requirements for antibiotic-free and ethically produced animal products have emerged, establishing a deep link among animal welfare, environmental sustainability, and food safety in purchasing choices (Stygar et al., 2022). This confluence of trends highlights the significance of the "One Health" framework, which foci the interconnectedness of animals, humans, and environmental health. Although promising evidence, the regular shift toward herbal-based feeding systems requires strong regulatory organization, long-term field verification, and multidisciplinary evaluation employing life cycle and socioeconomic methods, see Fig. 10.1. This chapter attempts to exhibit the environmental, ethical, and public health dimensions of herbal-based feeding systems, placing them as a broad strategy for supporting sustainable animal production in coordination with One Health attitudes.

10.2 Environmental benefits of herbal feed systems

The livestock sector contributes approximately 24% of total global intestinal CH_4 emissions, making it a central focus for climate change mitigation work (Malyugina et al., 2025). Simultaneously, several traditional livestock sectors are under intense scrutiny regarding antibiotic use, water and land consumption, and their impact on biological diversity. In recent years, herbal-based diets have emerged as a potential avenue for lowering the environmental footprint of livestock production while maintaining and enhancing productivity (Adegbeye et al., 2025). Herbal-based diet systems rely on two main aspects: The provision of bioactive compounds that modify rumen fermentation processes and the creation of dietary mixtures that provide nutritional and environmental benefits throughout the production cycle (Ahmed et al., 2024).

10.3 Reduction in greenhouse gas emissions

10.3.1 Impact of herbal supplementation on enteric fermentation

Intestinal CH_4 gas in ruminants is produced as a result of bacterial fermentation procedures in the rumen, where CH_4-producing archaea convert hydrogen and carbon dioxide into CH_4 via specific metabolic pathways

(Ungerfeld et al., 2023). In this context, herbal-based feed additives are an indicated option for inhibiting these processes via several mechanisms, providing sustainable alternatives to other artificial inhibitors. New in vitro studies have scientifically assessed the capacity of various natural feed additives to decrease CH_4 production, collecting them according to their mechanisms of action (Hodge et al., 2024).

Rumen microbiome modifiers, e.g., garlic (oil and powder), allicin, *Yucca schidigera* extract, and essential oil mixtures, could decrease the growth of CH_4-producing archaea, either through direct antimicrobial effects or by modifying the rumen microbial balance (Niu et al., 2026). In contrast, hydrogen sinks, such as calcareous seaweed and its magnesium oxide (MgO)-enriched formulations, decrease the availability of free hydrogen, a key for CH_4 production (Cui et al., 2024). In addition, enzyme inhibitors, e.g., the red alga *Asparagopsis taxiformis*, directly target methyl coenzyme M reductase (MCR), a crucial enzyme in the CH_4 metabolic pathway formation (Indugu et al., 2024; Li et al., 2025).

The greatest declines in CH_4 emissions were attained via the synergistic use of additives with complementary mechanisms. For example, merging garlic powder with a buffered solution of calcareous seaweed resulted in about 64.5% reduction in relative CH_4 emissions ($P < .05$), together with about 37% decrease in overall gas production. It was also observed that several of these treatments enhanced rumen fermentation efficiency. Calcareous seaweed supplemented with magnesium oxide increased total volatile fatty acids by 39.8%, whereas garlic oil and yucca extract improved their production by 24.4% and 22.6%, respectively (Hodge et al., 2025). Results recommend a double benefit of reducing greenhouse gas emissions while improving energy availability for the animal, addressing traditional concerns about the impact of CH_4 reduction strategies on productive performance. Besides, an essential oil blend involving coriander seed oil, eugenol, geraniol acetate, and geraniol verified antimicrobial activities against rumen bacteria (Gram-positive) and protozoa, contributing to an average 12.9% decrease in CH_4 emissions in cattle (Caroprese et al., 2023; Enikuomehin, 2025). Nevertheless, numerous pieces of evidence indicate that microbial adaptation may emerge during extended feeding periods, underscoring the urgent need to create integrated strategies that combine diverse additives with diverse mechanisms of action to maintain long-term effectiveness (Flores et al., 2024). In addition, field studies in multispecies rangeland systems have also revealed additional potential mechanisms for reducing CH_4 emissions. For example, alfalfa (*Lotus corniculatus* L.) produces tannins, which have been shown to reduce intestinal methane emissions in cattle by approximately 13% (Verma et al., 2024).

10.3.2 Comparative carbon footprints of herbal versus conventional systems

Life cycle assessment is a well-established methodological tool for comparing the environmental effects of herbal-based diet systems versus conventional systems. A comprehensive assessment of poultry diets fortified with bioactive plant compounds verified that these systems are linked to significantly lower environmental burdens compared with conventional feeding programs that are based on antibiotics and vitamins (Vlaicu et al., 2023). The study revealed that replacing both antibiotics and vitamins with bioactive plant extracts (raw or encapsulated form) is a sustainable option for lowering the environmental footprint of various production systems.

The environmental advantages extend direct emissions reductions far; herbal-based diet systems can reform the greenhouse gas balance through combined pathways. First, lowering belief in pharmaceutical inputs contributes to depressing emissions correlated with antibiotic and vitamin manufacturing chains. Second, developed feed conversion efficiency, as reported in studies indicating that herbal extracts improve growth performance and feed consumption (Ashour et al., 2025a), is reflected in lowered emissions per unit of animal product. Third, some forage types play an obvious role in decreasing soil-borne emissions; for instance, plantain (*Plantago lanceolata*) creates nitrification inhibitors, hence restricting nitrous oxide (N_2O) emissions (Chisholm, 2024). More broadly, comparative analyses of animal feed systems imply that efficiently managed grazing systems can attain better environmental performance even compared to some modern plant-based meat alternatives. Cradle-to-distribution lifecycle analyses have exhibited that plant-based meat alternatives, while having a lower footprint than beef raised in intensive fattening systems, can surpass the footprint of beef produced in well-managed pastures (Caradus et al., 2024). These outcomes emphasize the critical role of pasture management quality, involving variety and herb composition, in deciding the sustainability of the production systems.

10.4 Water and land use efficiency

10.4.1 Resource savings via reduced antibiotic-related processing

Replacing antibiotics with herbal-based bioactive compounds leads to significant resource efficiency across the animal production chains. The production of standard antibiotics requires high water and energy consumption through the stages of industrial fermentation, purification, and chemical synthesis (Zhgun, 2023). While studies of the water footprint of synthetic drugs (e.g., antibiotics and vitamins) production contrasted with plant extracts are limited, the environmental consequence of adding bioactive compounds is remarkably less than that of conventional additives across a variety of sustainability indicators (Bekavac et al., 2025). These resource-efficient profits outspread to on-farm treatment applications, where plant-based veterinary applications offer low-priced, locally available alternatives to numerous chemical drugs, which are mostly vital in pastoral systems in arid regions and environments with limited resources (Zaroug, 2010; Venkataraman et al., 2023; Abd El-Hack et al., 2025). From a sustainability perspective, reorganized herbal healthcare systems lead to reduced costs of transportation and packaging, which are requirements for the veterinary drug supply chain, while also supporting the cultivation of native plant and herb species and preserving traditional knowledge (Obahiagbon and Ogwu, 2024). Also, the widespread use of oral administration shows the prospect of incorporating these treatments directly into animal feeding programs without the need for particular equipment or intensive additional labor, which improves efficiency and facilitates their application in various production systems (Rehman et al., 2025).

10.4.2 Agroecological herb source

The environmental benefits of herbal-based feed systems depend seriously on sourcing and production systems. Assuming agroecological methods for herb and plant cultivation, which emphasize improving biological diversity, enhancing soil fertility, and supporting farmers' livelihoods, contribute to the overall sustainability of herb supplements (Oyenuga et al., 2025).

Integrated supply chain initiatives also display the ability to accomplish sustainable herb sourcing on a commercial scale. For instance, many programs (e.g., the TEROVA) worked directly with spice and herb farmers to establish transparent, traceable, and sustainability-certified supply chains. These methods incorporate providing training about sustainable practices, such as decreasing the reliance on chemical fertilizers, supporting organic composting, and augmenting soil fertility, along with using measures to enhance ingredient safety and quality and lessen waste (Hart, 2021; Manea et al., 2024). On the other hand, the economic viability of incorporating herbs into crop rotation systems depends on numerous related factors, such as the availability of fertilizer and the costs of transportation (Al-Musawi et al., 2025). Yet, in the occurrence of limited fertilizer accessibility or an absence of alternatives, herb farming can become extra economically competitive even at minor prices compared to unfertilized grain farming, improving its application as a sustainable long-term agricultural decision (Ibrahim et al., 2023).

10.5 Biodiversity and soil health

10.5.1 Cultivation of native herbs and ecosystem services

Enhancing plant diversity in rangelands is a cornerstone of rebuilding sustainable agricultural ecosystems. Integrating multispecies herblands into agricultural landscapes presents an ecologically sound substitute for monoculture farming, which is linked to biological diversity loss and lowered environmental services (Malisch et al., 2024). New evidence proposes that multi-species rangelands, combining herbs, legumes, and flowering grasses, expand habitat structure and increase the food resources availability for pollinators, in that way improving the stability of insect communities and their ecological roles. This is related to the temporal and spatial differences in flowering and the functional complementarity between plant species, which offer better continuity of nectar and pollen accessibility compared to a monoculture system (Baker, 2025; Khan et al., 2026).

A recent study suggested that plant type diversity in rangelands directly influences plant–pollinator interaction networks by enhancing the structural complexity of ecosystems and increasing their resilience to environmental stresses, e.g., climate change or the use of intensive pesticides (Rakosy et al., 2022). Multispecies systems are illustrated by intensified functional overlap and ecological resilience, depressing the probability of immediate pollination network failure and advertising the stability of long-term ecosystem services. Plant variation furthermore aids in facing the adverse effects of agricultural disturbances through supporting both specific and general pollinator populations, thus raising and sustaining agricultural production (Diyaolu and Folarin, 2024).

Likewise, recent studies suggest that rangeland diversity boosts ecological multifunctionality, involving forage production, carbon sequestration, improvement of soil properties, water control, and biodiversity support (Waterhouse et al., 2024). This is correlated to increased plant functional diversity, which leads to developed resource use efficiency and diminished variability in production under climate change conditions. These attributes are remarkably focal in the context of climate change, as ecological models have shown that species-rich plant populations are more flexible to heat stress and drought than simple systems (Singh et al., 2024). Improved and seminatural rangelands should not be viewed as competing systems, but rather as systems that provide integrated ecosystem services. Improved rangelands provide important resources for both pollinators and food production, while semi-natural rangelands are crucial for the protection of specialized and endangered species (Simba et al., 2024). Overall, the results show that multispecies rangelands not only support pollinators but also form the cornerstone of enhancing the resilience of agricultural systems, improving production stability, and reducing reliance on chemical inputs. However, further long-term studies are still needed to understand the complex interactions between plant diversity, microbial communities, and biochemical processes, particularly in different agricultural environments (Baker, 2025).

10.5.2 Integration into regenerative agriculture

Multispecies pasture-based feeding systems align with the principles of regenerative agriculture by promoting carbon and nitrogen cycling within the farming system and reducing reliance on external inputs, which supports climate change mitigation and enhances the resilience of livestock production. At the mechanistic levels, these systems improve soil carbon fixation by improving root inputs and enhancing organic carbon stability, markedly in deeper, less degraded areas (Wróbel et al., 2023; Roderick and Jones, 2025).

The root structure of multispecies pastures varies qualitatively from monoculture systems based on species, e.g., *Lolium perenne*, which create homogeneous, shallow roots. Conversely, multispecies mixtures show integrated rooting strategies. Legumes (e.g., *Trifolium pratense*) promote biofixation of nitrogen and decrease the use of synthetic fertilizers, which is associated with carbon emissions (Yang et al., 2025). Also, dandelion (*Taraxacum officinale*) loosens hardened layers and increases carbon deposition in deeper soils, while *Plantago lanceolata* displays nitrification inhibition, lowering nitrogen loss and nitrous oxide (N_2O) emissions (Ding et al., 2025). This functional integration significantly contributes to improved soil structure, reduced water infiltration, and increased organic carbon stability. Furthermore, some plant-based forage materials contain bioactive compounds that play a crucial role in mitigating animal emissions. For example, the tannins found in *Onobrychis viciifolia* and *Lotus corniculatus* modify rumen fermentation processes and reduce intestinal CH_4 production rates, thereby significantly lowering emission intensity. Therefore multispecies pastures offer a comprehensive approach to mitigating the effects of climate change (Zhao et al., 2025).

10.6 Ethical production and humane treatment

10.6.1 Consumer expectations and animal welfare

Buyer expectations are effectively driving increased demand for ethical animal products, as a broad segment of consumers chooses foods that align with general animal welfare values (Schröder and McEachern, 2004). Many studies indicate that consumer interest goes beyond animal treatment practices alone, as they seek products with certifications or claims related to so-called humane and cruelty-free production (Thaemert, 2023). A recent study revealed that most consumers believe that claims related to animal welfare, such as "humane breeding," should be based on a clear and measurable set of standards, while on the other hand, another segment of consumers believes that such claims should not be allowed without independent verification (Dilley, 2025). Demands for transparency are also evident in the European market, where animal welfare is consistently ranked among sustainability indicators that influence consumer purchasing decisions, particularly for red meat and dairy products (Stygar et al., 2022). On the other hand, other studies discover a considerable perception gap between purchaser expectations and actual market practices (Kouas and Akrout, 2025). In addition, consumer attitudes toward natural production systems, remarkably pasture-based systems, emphasize the growing importance of animal welfare in shaping purchasing behavior. Pasture grazing is often identified as an indicator of improved quality of life and as allowing animals to express their natural behaviors, thus improving the ethical image of the product (Rivero and Lee, 2022). Also, this trend is associated with the general perception that natural feeding systems support wider sustainability objectives, involving protecting biodiversity and lowering the environmental effects of the livestock sector (Kühl et al., 2022).

The demand for the term "natural" is growing in animal and poultry feeding practices, as consumers in regions such as Europe and North America show increasing interest in products derived from animals raised without the use of artificial additives, such as antibiotics or synthetic growth promoters. This mirrors a substantial shift toward production systems that spotlight sustainability, ethics, and health (Peter, 2025). Taken together, this evidence demonstrates that the concept of ethical animal production is no longer merely a theoretical orientation but has become a set of tangible consumer expectations that emphasize humane treatment, natural environmental conditions, and high transparency in animal production.

10.6.2 Industry best practices

Responding to growing consumer awareness of ethical and sustainability issues, as well as increasing demand for ethically produced animal products, the agricultural sector has witnessed a gradual shift in recent decades toward adopting strategies that promote animal welfare. This includes adopting diverse environmental enrichment practices and humane treatment of animals and birds, and a range of effective tools for improving the behavioral and physiological aspects of modern production systems, which often lack sufficient environmental incentives. Producers recognize that improving animal welfare is not only an ethical obligation but is also related to improved production performance, lower stress, and boosted quality and safety of animal products (Taylor et al., 2023). In this context, environmental enrichment practices are a promising, low-cost approach that promotes natural behaviors and reduces undesirable behaviors, such as those associated with stress. A recent study revealed that providing simple welfare tools, such as large brushes, in the cattle fattening sector led to significant improvements in behavioral welfare indicators. Animals receiving these resources showed a significant increase in positive behaviors, such as play and social interaction, accompanied by a decrease in aggressive behavior, promoting improved mental health and reduced stress. Furthermore, this reflects a marked improvement in herd management, reduced worker exposure to hazards, improved product quality, and a positive interplay between welfare and sustainable production (Pilarczyk et al., 2025). In addition to adopting these practices, independent verification mechanisms have become essential to enhance consumer confidence. This has led to the emergence of a range of third-party certification programs, which serve as reliable tools for assessing animal welfare standards and product differentiation in global markets. A prime example is the USDA's Organic Produce Program (Dimitri and Nehring, 2022; Miranda-de La Lama and Estévez-Moreno, 2025).

On the other hand, more rigorous programs, such as the "Certified in Animal Welfare" certification offered by the "Green World" organization, embody a more holistic approach. These programs require the provision of real grazing areas and the improvement of the indoor environment to promote natural animal behaviors and focus on reducing painful practices and improving sustainable management methods (Ankiel et al., 2025). Furthermore, the Italian Animal Welfare Quality System, as a clear example of government initiatives, demonstrates ongoing efforts to harmonize national standards and establish formal accreditation that emphasizes transparency, strict control, and effective communication with consumer groups regarding sustainable production practices (Aprillian and Leucci, 2025). Although these developments have led to an excess of certification programs, their altered standards can confuse customers and undermine their ability to distinguish between several welfare levels. Consequently, there is a necessity to develop clear regulatory frameworks and mandatory, evidence-based classification systems to confirm consistency and transparency in labeling (Broojerdi et al., 2024). Table 10.1 provides specific, evidence-based support for the claims received about the benefits of herbal feed on animal welfare.

10.6.3 Antibiotic-free meat as a wellness strategy

The shift toward producing meat free from antibiotics and their by-products is a strategic approach that goes beyond traditional food safety considerations to encompass broader dimensions of public health and well-being. These shifts are compounded by the growing global challenge of AMR, which the World Health Organization (WHO) has identified as one of the top ten threats to global health (Aminudin et al., 2025). The overuse of conventional antibiotics in animal and poultry production systems, especially as a growth promoter and for the prevention of common diseases, leads to selective pressure that accelerates the development and spread of resistant bacterial strains through a series of interactions between the three aspects: Animals, humans, and the environment (Ahmad et al., 2023). These resistant strains and their genes could be transmitted through either direct contact, the food supply chain, or environmental contamination, and this reinforces the great importance of the "One Health" concept, which emphasizes the interconnectedness of human, animal, and ecosystem health (Al-Khalaifah et al., 2025). The continued overuse of antibiotics in intensive production systems increases concerns about the declining effectiveness of clinically important antibiotics, which may influence treatment choices in both human and veterinary medicine. Based on the above, antibiotic-free animal production systems are now being promoted as effective preventive interventions aimed at reducing selective pressure and strengthening international efforts to contain AMR (Muteeb et al., 2023; Hussain et al., 2026).

In the same vein, herbal-based feed additives have emerged as viable alternatives to antibiotics used in the livestock and poultry sectors (such as essential oils, tannins, saponins, and flavonoids). These additives possess antibacterial, antioxidant, and anti-inflammatory properties, in addition to their potential to regulate the intestinal microbiome and improve intestinal barrier integrity

TABLE 10.1 Enhanced animal welfare indicators through herbal supplementation.

Welfare indicator	Indicator	Basic mechanics of herbs action	Herbs/bioactive components	Observation	References
Physiological stress	Cortisol levels and stress hormones	Adaptogenic, antioxidant, and antiinflammatory impacts decline the hypothalamic–pituitary–adrenal axis activation	Turmeric, ginseng, ashwagandha, oregano	Lowered levels of plasma cortisol; improved stress resilience through heat stress, transport, and crowding	Alagawany et al. (2022), Ntsongota et al. (2025), Sejian et al. (2025)
Oxidative status	Levels of MDA and antioxidant enzymes (e.g., CAT, SOD, GPx)	Polyphenols and flavonoids lower the influence of reactive oxygen species (ROS) and boost endogenous antioxidant defense	Rosemary, green thyme, tea, curcumin	Enhanced antioxidant status; lowered oxidative damage; good resilience to various environmental stressors	Surai et al. (2019), Papatsiros et al. (2022), Abd El-Ghany (2025), Gu et al. (2025), Fathi et al. (2025)
Immunity	Immunoglobulins (e.g., IgG, IgA) and cytokines	Immunomodulatory and anti-inflammatory activities boost the balance of immune and mucosal protection	Garlic, thyme, echinacea, saponins	Improved both humoral and cellular immunity; lowered inflammatory biomarkers	Ashour et al. (2025), Deeb et al. (2025), Budiastuti et al. (2026), Chung et al. (2026)
Gut health	Intestinal morphology, microbiota balance	Adjustment of gut microbiome, developed barrier integrity, and digestion of nutrients	Tannins, essential oils	Increased height of villus; lowered pathogenic load; superior gut integrity	Windisch et al. (2008), Elbaz et al. (2025), Oni and Oke (2025)

(*continued on next page*)

Heat stress tolerance	Heat shock proteins, temperature of rectal	Enhanced cellular stress tolerance, as well as ant oxidant protection	Curcumin, resveratrol, plant polyphenols	Reduced body temperature and respiratory rates; enhanced heat stress adaptation	Habashy et al. (2022), Eder et al. (2025), Fathi et al. (2025), Gu et al. (2025)
Inflammation—Pain	Proinflammatory cytokines (e.g., TNF-α and IL-6)	Anti-inflammatory and analgesic phytochemicals modify inflammatory pathways	Boswellia, turmeric, willow bark	Reduced inflammatory responses and improved recovery from injuries	Papatsiros et al. (2022), Abdelnour et al. (2025), Rehman et al. (2025)
Integument and Health of Skin	Condition of the feather and coat, lesions	Antimicrobial and antioxidative influences enhance integument health	Essential oils, neem, and aloe vera	Diminished dermatitis and lesions; developed plumage and skin conditions	Ntsongota et al. (2025), Ravi et al. (2025), Saad et al. (2025)
Locomotion and lameness	Gait score, joint inflammation	Anti-inflammatory and cartilage-protective features	Curcumin, omega-rich plant oils	Enhanced locomotion and decreased lameness prevalence	Alagawany et al. (2022), Fathi et al. (2025), Li (2025)
Reproductive welfare	Fertility rates, embryonic viability	Hormonal balance, antioxidant, and anti-stress impacts	Fenugreek, maca, ginseng	Enhanced reproductive performance and maternal welfare	Budiastuti et al. (2026), Cao et al. (2026), Sisodiya et al. (2026)
Mortality rates and morbidity	Disease frequency, survival	Antimicrobial and immune-improving features	Essential oils, thyme, and garlic	Decreased disease and mortality rates	Banday et al. (2025), Nhan et al. (2025), Windisch et al. (2008)
General welfare score	Composite welfare indicators	Integrated impacts on physiology, immunity, metabolism, and behavior	Multiherb phytogenic blends	Enhanced welfare, productivity, and resilience	Papatsiros et al. (2022), Fedynyshyn (2025), Ntsongota et al. (2025), Alian et al. (2026)

(Hamdi et al., 2026). Some studies have shown that certain herbal formulations can reduce the intestinal microbial load, improve rumen fermentation processes, and increase nutrient utilization efficiency, thereby enhancing production performance and reducing antibiotic use (Bąkowski and Kiczorowska, 2021). Unlike antibiotics, which target specific bacterial strains, these plant compounds act through multiple, simultaneous mechanisms, thus limiting the potential for bacterial strains to develop resistance. However, there are still many challenges in aspects of standardizing chemical specifications, determining dosages, and assessing the safety of long-term use, which are critical aspects for ensuring sustainable efficacy and preventing unintended adverse environmental or microbial effects (AL-Azzawi et al., 2024). Choosing "clean" foods is also seen as part of preventative health strategies, as reducing exposure to antibiotic residues and resistant organisms is believed to support immunity and gastrointestinal health by maintaining a balanced human microbiome (Endale et al., 2023). Although there is limited direct clinical evidence to suggest tangible physiological effects on both immunity and intestinal function, reducing exposure rates to resistant bacteria is in itself a preventive measure that promotes overall health and also reduces the long-term risks that may be associated with AMR (Ifedinezi et al., 2024).

Overall, antibiotic-free meat production represents a dynamic intersection of agricultural innovation, sensible management of AMR, and customer-led health trends. The success of this model depends on adopting integrated strategies that include endorsing the responsibility of antibiotic use, emerging evidence-based plant-based and organic alternatives, improving label transparency, and applying effective observation and follow-up systems (García Herrero et al., 2024).

10.7 Sustainability narratives in consumer psychology

10.7.1 Branding and storytelling

In the herbal nutrition market, branding and storytelling have emerged as fundamental, actual tools for customers, not only through the browser's use of sustainability labels on packaging. This has led to a growing number of brands not merely describing their products but also creating compelling narratives for urban customers around concepts of origin and animal responsibility (Jain and Singh, 2024). Consequently, research indicates that consumers do not always have a strong understanding of technical sustainability standards; however, they respond positively to authentic and impactful stories that humanize the supply chain (Madden, 2025). A study conducted on a range of seaweed-based beverages revealed that more information about nutritional value and consumption positively enhanced participants' sensory experiences and emotional responses, leading to an increased sense of security and comfort. This suggests that information conveyed through the brand and its message influences purchaser expectations, subsequent higher quality, and greater value for the products (Moss et al., 2024).

10.7.2 Greenwashing concerns

However, the power of influential discourse in the areas of sustainability can achieve substantial results that go beyond environmental responsibility, as it can extend beyond marketing performance to ethical performance and positive product impacts. The credibility of the herbal nutrition sector depends principally on assurance based on all recommendations for proof and verifiability, as complex molecules and market associations may occur as a result of the entanglement of information on labels and the variances associated with them (You et al., 2024). The danger is remarkably on the commercial side; this is clearly obvious in study that has tested protein-rich foods, which are also categorized by a variety of products bearing inaccurate labels, and indeed by contamination with toxic ingredients, e.g., mercury and banned pesticides, which is totally at odds with the natural and pure pictures that will often appear in their marketing (Lin et al., 2025). In this context, licensing authorities play a critical role in allowing the rights to withdraw funds and supporting claims of sustainability and quality. Such certifications provide consumers with reliable assurance that a product meets specific and rigorous standards (Kosoe et al., 2024). Effective certification programs rely not only on formal reviews but also on on-site audits of good manufacturing practices. For example, ISURA certifications involve testing for more than 800 contaminants to ensure the accuracy of product data and the absence of any contaminants. This system is characterized by its accuracy and independence and has become a leading system in the field of product consumption, especially with the increasing demand for organic, traceable, and standardized products (Sindhu and Tewary, 2025). Fig. 10.2 illustrates a visual comparison of the level of consumer trust associated with different types of labels, reinforcing the argument for rigorous, verifiable standards.

10.8 Aligning public health and ecological goals

10.8.1 One Health framework

The shift from antibiotic-dependent animal production systems to grass-based feed solutions reflects a growing commitment to the "One Health" concept, which links animal, human, and environmental health and is supported by international organizations such as the WHO and the Food and Agriculture Organization (FAO). Studies show that AMR is a serious global problem, exacerbated by the use of these drugs in the animal production sector (Tang et al., 2023). In conventional production systems, particularly poultry, antibiotics are used to promote growth and control disease; however, evidence links these practices to the emergence and spread of resistant bacterial strains through food chains and occupational exposure ((Mak et al., 2022; Amato, 2022). Therefore reducing the reliance of animal and poultry production systems on antibiotics has become a key objective in international public health efforts to combat AMR (Tang et al., 2023; Alhassan et al., 2025).

FIGURE 10.2 Infographic comparing consumer trust in different sustainability labels.

Herbal nutrition presents a promising nexus strategy by amalgamating animal health administration with environmental and public health objectives. Herbal-based feed additives, such as plant extracts and polyphenolic compounds, may contribute to altering the gut microbiome, enhancing the immune system, and improving oxidative status in livestock and poultry production systems (Elsherbeni et al., 2025). These bioactive molecules have diverse effects, including, for example, eliminating pathogenic bacteria and reducing inflammation. Such activity contributes to maintaining gut health and improving its resistance to various disorders (Benameur et al., 2023). Their multitarget approaches may substantially reduce selective pressure for resistance in contrast to usual antibiotics. Also, herbal-based systems are usually correlated to alternative and pasture-based animal production models that stimulate biodiversity, raise nutrient cycling, and minimize levels of environmental pollution (Yang et al., 2024). These combined results show herbal nutrition as a preventive and systems-oriented approach allied with One Health principles. Fig. 10.3 visualizes the core principle of the One Health concept.

10.8.2 Policy and research integration

A sustainable transition to herbal-fed livestock and poultry systems requires coordinated policy structures and specialized research addressing the complexities of antimicrobial management. Currently, the regulatory framework for this transition remains incomplete, and there is significant variation in how plant

FIGURE 10.3 One Health triangle: Animal, human, and environment interconnected via herbal feed use.

additives are assessed and approved across different countries (Thakkar et al., 2020). For example, the European Food Safety Authority (EFSA) has established procedures for assessing safety and efficacy. In contrast, many developing countries and Third World countries lack consistent regulatory approaches. To ensure the quality of animal products, consumer confidence, and the flourishing of international trade, enhanced international cooperation and the promotion of regulatory convergence are essential (Gerardi, 2023). Engaging diverse stakeholders, such as policymakers, scientists, veterinarians, and producers, is crucial for developing evidence-based general recommendations that balance sustainability, safety, and innovation (Shrivastava et al., 2025).

Subsequent studies should focus on scientific research rather than solely responding to short-term performance indicators. Longitudinal studies that assess bacterial environment, emissions, resistance mechanisms, and production efficiency are crucial for livestock production systems. Combining life cycle estimation, socioeconomic modeling, and micronutrient technologies will contribute to a better understanding of the sustainability and scalability of herbal-based systems (Li et al., 2024). Government investments in antibiotic-free and residue-free livestock sectors, along with incentive and licensing programs, can effectively accelerate this process (Hussain et al., 2026). Investment in knowledge transfer, extension services, and farmer-focused innovation platforms will also enhance implementation across most livestock production systems. Overall, aligning regulatory, research, and funding objectives with the One Health concept will strengthen livestock and poultry production sectors to provide safe, sustainable, and ethically produced animal protein globally (Hosseini et al., 2026).

10.9 Conclusion

Herb-based feeding systems represent a promising new approach to sustainable livestock and poultry production, addressing a variety of challenges related to environmental impacts, antibiotic resistance, and consumer ethical requirements. These herbal-based supplements and their derivatives demonstrate numerous tangible benefits, such as reducing methane emissions, improving nutrient utilization, enhancing animal welfare indicators, and decreasing antibiotic dependence. This aligns with the "One Health" framework by simultaneously promoting animal and human health and environmental sustainability through reduced chemical inputs, improved biodiversity, and more efficient use of available resources. Despite this promising evidence, several challenges remain, including standardizing bioactive compounds, modifying gut bacteria through long-term feeding systems, and the fragmentation of regulatory frameworks across different sectors. Therefore realizing the potential of herbal-based feeding systems requires a systematic and coordinated effort in scientific research, policy, and industry practices, including standardized approval procedures, comprehensive verification studies, and transparent certification mechanisms. As the livestock sector faces climate change and antibiotic resistance, herbal-based feeding systems offer a practical and viable path toward sustainable, resilient, and ethically responsible food production.

References

Abd El-Ghany, W.A., 2025. The impacts of using rosemary (Rosmarinus officinalis Linn.) on oxidative stress, performance, serum biochemistry, microbial community, and immunity of poultry. CABI Rev. 20 (1). doi:10.1079/cabireviews.2025.0085.

Abd El-Hack, M.E., Aldhalmi, A.K., Ashour, E.A., Elsherbeni, A.I., 2025. An updated review on dandelion (Taraxacum officinale) polysaccharides as a natural substitute for antibiotics in poultry production: benefits and applications. Worlds Poult. Sci. J. 81 (4), 1461–1479. doi:10.1080/00439339.2025.2522903.

Abdelnour, S.A., Abdelaal, M., Sindi, R.A., Alfattah, M.A., Khalil, W.A., Bahgat, L.B., Sheiha, A.M., 2025. Physio-metabolic response, immune function, epigenetic markers, and reproductive performance of rabbits under environmental stress: the mitigating role of boswellia essential oil nanoemulsion. BMC Vet. Res. 21 (1). doi:10.1186/s12917-025-04587-1.

Adegbeye, M.J., Widiawati, Y., Christopher, T., Ojugbele, A.O., Ogunnusi, O.J., Hernández-Ruiz, P.E., Elghandour, M.M.M.Y., Salem, A.Z.M., Alvarado-Ramírez, E.R., 2025. Nanotechnology in livestock production: enhancing efficiency, health, and climate resilience. In:

Handbook of Climate Change Mitigation and Adaptation. Springer Science+Business Media, Indonesia, pp. 3121–3154. doi:10.1007/978-3-031-84483-6_210.

Ahmad, N., Joji, R.M., Shahid, M., 2023. Evolution and implementation of One Health to control the dissemination of antibiotic-resistant bacteria and resistance genes: a review. Front. Cell. Infect. Microbiol. 12. doi:10.3389/fcimb.2022.1065796.

Ahmed, M.G., Elwakeel, E.A., El-Zarkouny, S.Z., Al-Sagheer, A.A., 2024. Environmental impact of phytobiotic additives on greenhouse gas emission reduction, rumen fermentation manipulation, and performance in ruminants: an updated review. Environ. Sci. Pollut. Res. 31 (26), 37943–37962. doi:10.1007/s11356-024-33664-5.

Alagawany, M., Elnesr, S.S., Farag, M.R., El-Sabrout, K., Alqaisi, O., Dawood, M.A.O., Soomro, H., Abdelnour, S.A., 2022. Nutritional significance and health benefits of omega-3, -6 and -9 fatty acids in animals. Anim. Biotechnol. 33 (7), 1678–1690. doi:10.1080/10495398.2020.1869562.

AL-Azzawi, M.K, Hasan, N.A, Barrak, M.M., 2024. A review of the development of an understanding of antibiotic interactions, from mechanisms of action to novel resistance and the search for natural alternatives. Eur. J. Med. Genet. Clin. Biol. 1 (6), 78–102. doi:10.61796/jmgcb.v1i6.585.

Alhassan, M.Y, Kabara, M.K, Ahmad, A.A, Abdulsalam, J., Habib, H.I, 2025. Revisiting antibiotic stewardship: veterinary contributions to combating antimicrobial resistance globally. Bull. Natl. Res. Cent. 49 (1), 2522–8307. doi:10.1186/s42269-025-01317-3.

Alian, H.A., Abdellatif, M.M., Abdelmonem, R., El-Hack, A., M., E., 2026. Herbal supplementation in livestock: impacts on food quality. In: Functional Foods from Animal Sources: Properties, Human Health Benefits, and Applications. Springer Nature Switzerland, Cham, pp. 179–215.

Al-Khalaifah, H., Rahman, M.H., Al-Surrayai, T., Al-Dhumair, A., Al-Hasan, M., 2025. A One-Health perspective of antimicrobial resistance (AMR): human, animals and environmental health. Life 15 (10). doi:10.3390/life15101598.

Al-Musawi, Z.K., Vona, V., Kulmány, I.M., 2025. Utilizing different crop rotation systems for agricultural and environmental sustainability: a review. Agronomy 15 (8). doi:10.3390/agronomy15081966.

Amato, H.K., (2022). Environmental Determinants of Community-Acquired Antibiotic Resistance in Low-and Middle-Income Countries (Doctoral dissertation, University of California, Berkeley).

Aminudin, M.H., Amalina, F., Ab Hamid, M.R., Sulaiman, S., Azfa, N., Abd Razak, A.S., 2025. Antibiotic residues in poultry products: public health risks, antimicrobial resistance, and sustainable alternatives. Challenge 12 (55), 64.

Ankiel, M., Halagarda, M., Piekara, A., Sady, S., Żmijowska, P., Popek, S., Pacholck, B., Jefmański, B., Kucia, M., Krzywonos, M., 2025. Role of certifications and labelling in ensuring authenticity and sustainability of fermented milk products. Sustainability 17 (18), 8398. doi:10.3390/su17188398.

Aprillian, R., Leucci, F., 2025. Animal welfare claims in Indonesia: comparative study of EU standards and proposed regulatory framework. Trop. Anim. Sci. J. 48 (6), 564–571. doi:10.5398/tasj.2025.48.6.564.

Ashour, E.A., Al-Ardhi, S.A., Abd El-Hack, M.E., Elsherbeni, A.I., Elolimy, A.A., Madkour, M., Tufarelli, V., Swelum, A.A., 2025a. Effects of lactic acid and herbal blend as antibiotic alternatives on growth, carcass traits, blood chemistry, and microbial load in broiler chickens. Poult. Sci. 104 (5). doi:10.1016/j.psj.2025.105050.

Ashour, E.A., Aldhalmi, A.K., Ismail, I.E., Kamal, M., Elolimy, A.A., Swelum, A.A., Abd El-Hack, M.E., 2025b. The effect of using Echinacea extract as an immune system stimulant and antioxidant on blood indicators, growth efficiency, and carcass characteristics in broiler chickens to produce a healthy product. Poult. Sci., 104 doi:10.1016/j.psj.2024.104392.

Baker, S., Moloney, T., 2025. Designing and developing multispecies swards for future grassland systems. In: Smith, L.J., Brown, P.R. (Eds.), Advances in Temperate Grassland Science and Management. Burleigh Dodds Science Publishing, Cambridge, pp. 383–398.

Bąkowski, M., Kiczorowska, B., 2021. Probiotic microorganisms and herbs in ruminant nutrition as natural modulators of health and production efficiency–a review. Ann. Anim. Sci. 21 (1), 3–28. doi:10.2478/aoas-2020-0081.

Banday, M.T., Wani, M., Alqahtani, F.M., Adriani, L., Alhomrani, M., Adil, S., Alsanie, W.F., Alamri, A.S., Abdulaziz, O., 2025. Enhancing meat-type chicken performance through Thymus vulgaris leaf powder supplementation by affecting serum lipid profile, stress physiology, immunity, antioxidants, cecal microbiology, and jejunal histomorphology. Arch. Anim. Breed. 68 (2), 311–323. doi:10.5194/aab-68-311-2025.

Bekavac, N., Krog, K., Stanić, A., Šamec, D., Šalić, A., Benković, M., Jurina, T., Gajdoš Kljusurić, J., Valinger, D., Jurinjak Tušek, A., 2025. Valorization of food waste: extracting bioactive compounds for sustainable health and environmental solutions. Antioxidants 14 (6), 714. doi:10.3390/antiox14060714.

Benameur, T., Porro, C., Twfieg, M.E., Benameur, N., Panaro, M.A., Filannino, F.M., Hasan, A., 2023. Emerging paradigms in inflammatory disease management: exploring bioactive compounds and the gut microbiota. Brain Sci. 13 (8). doi:10.3390/brainsci13081226.

Broojerdi, A.K., Salvati, A.L., Abdelfattah, M.R., Dehaghi, R.O.A., Sillo, H.B., Gaspar, R., 2024. WHO-listed authorities (WLA) framework: transparent evidence-based approach for promoting regulatory reliance towards increased access to quality-assured medical products. Front. Med. 11. doi:10.3389/fmed.2024.1467229.

Budiastuti, B., Khairullah, A.R., Moses, I.B., Effendi, M.H., Pratama, B.P., Tyasningsih, W., Tang, J.Y.H., Kurniasih, D.A.A., Ahmad, R.Z., Yulianti, W., Rehman, S., Ma'ruf, I.F., Utomo, D.L.H., Martindah, E., Wardhani, B.W.K., Yuri, A.J.B., 2026. Potential use of phytobiotics as an alternative to antibiotics in livestock. J. Adv. Vet. Res. 16 (2), 301–312.

Cao, H., Qiu, Z., He, X., Bai, Q., Ge, S., Shi, X., Liu, Y., Xin, R., 2026. Dietary fenugreek leaf powder improves growth performance, immune function, antioxidant status, and intestinal health in yellow-feathered broilers. Poult. Sci. 105 (3), 106325. doi:10.1016/j.psj.2025.106325.

Caradus, J.R., Chapman, D.F., Rowarth, J.S., 2024. Improving human diets and welfare through using herbivore-based foods: environmental consequences and mitigations. Animals 14 (9), 1353.

Caroprese, M., Ciliberti, M.G., Marino, R., Santillo, A., Sevi, A., Albenzio, M., 2023. Essential oil supplementation in small ruminants: a review on their possible role in rumen fermentation, microbiota, and animal production. Dairy 4 (3), 497–508. doi:10.3390/dairy4030033.

Chisholm, C., (2024). The Abundance, Activity, and Community Composition of Comammox Nitrospira and Canonical Ammonia Oxidisers in New Zealand Soils: A Thesis Submitted in Partial Fulfilment of the Requirements for the Degree of Doctor of Philosophy at Lincoln University (Doctoral dissertation, Lincoln University, Lincoln, Canterbury).

Chung, E.L.T., Kassim, N.A., Zheng, A.L.T., Jesse, F.F.A., Loh, T.C., Dunshea, F.R., 2026. Phytogenic effects of Brachiaria decumbens saponin extract on growth performance, immune response, and telomere dynamics in broiler chickens under tropical conditions. Trop. Anim. Health Prod. 58 (2). doi:10.1007/s11250-025-04840-7.

Cui, Y., Zeng, Y., Hu, H., Zhang, Y., Wang, D., Feng, D., 2024. Biochar, phosphate, and magnesium oxide in seaweed and cornstarch dregs co-composting: enhancing organic matter degradation, humification, and nitrogen retention. Waste Manage. 187, 207–217. doi:10.1016/j.wasman.2024.07.024.

Deeb, S.E., Ashour, E.A., Youssef, I.M., Alshehry, G., Abuljadayel, D.A., Aljahdali, N., Al-baqami, N.M., Bahnas, M.M., Abd El-Hack, M.E., 2025. Enhancing broiler growth, carcass quality, blood parameters and intestinal microbial load: the role of dietary garlic powder as a natural growth promoter. Ann. Anim. Sci. 25 (4), 1467–1478. doi:10.2478/aoas-2025-0024.

Dilley, A.J., (2025). Consumer Knowledge, Attitude, and Practice for Animal-Raising and Environment-Related Labeling Claims (Master's thesis, University of Arkansas, Fayetteville, USA)

Dimitri, C., Nehring, R., 2022. Thirty years of organic dairy in the United States: the influences of farms, the market and the organic regulation. Renewable Agric. Food Syst. 37 (6), 588–602. doi:10.1017/s1742170522000254.

Ding, K., Luo, J., Welten, B., de Klein, C.A.M., 2025. The effect of plantain (Plantago lanceolata L.) in pasture swards on gaseous nitrogen emissions from nitrogen-enriched urine patches. Plant Soil 509 (1-2), 363–376. doi:10.1007/s11104-024-06865-7.

Diyaolu, C.O, Folarin, I.O, 2024. The role of biodiversity in agricultural resilience: protecting ecosystem services for sustainable food production. Int. J. Res. Publ. Rev. 5 (10), 1560–1573. doi:10.55248/gengpi.5.1024.2741.

Eder, K., Ringseis, R., Gessner, D.K., 2025. Effects of grape by-products on oxidative stress and inflammation in farm animals: an overview of studies performed in pigs, chickens, and cattle. Animals 15 (11). doi:10.3390/ani15111536.

Elbaz, A.M., Ashmawy, E.S., Mourad, D.M., Amin, S.A., Khalfallah, E.K.M., Mohamed, Z.S., 2025. Effect of oregano essential oils and probiotics supplementation on growth performance, immunity, antioxidant status, intestinal microbiota, and gene expression in broilers experimentally infected with Eimeria. Livest. Sci. 291. doi:10.1016/j.livsci.2024.105622.

Elsherbeni, A.I., Aldhalmi, A.K., Youssef, I.M., Kamal, M., Ashour, E.A., Moustafa, M., Al-kahtani, M.A., Al-Shehri, M., Abd El-Hack, M.E., 2025. The role of trimethylglycine (betaine) as an anti-heat stress agent in sustainable poultry production: enhancing growth, stress resilience, and nutrient utilization. J. Therm. Biol 130. doi:10.1016/j.jtherbio.2025.104152.

Endale, H., Mathewos, M., Abdeta, D., 2023. Potential causes of spread of antimicrobial resistance and preventive measures in one health perspective-a review. Infect. Drug Resist. 16, 7515–7545. doi:10.2147/idr.s428837.

Enikuomehin, J.M., (2025). Investigating the Antibacterial Properties of Terpene and Phenylpropanoid Essential Oils: A Study on Growth, Survival, and Genetic Alterations in Rumen-Gut Bacteria (Master's thesis, North Carolina Agricultural and Technical State University, North Carolina, USA). (2025).

Fathi, M., Saeedian, S., Darvishi, F., 2025. Impact of curcumin nanoparticles supplementation on antioxidant status, growth performance, immune function, gut microbiota, and hematobiochemical parameters in heat-stressed broiler chickens. Ital. J. Anim. Sci. 24 (1), 2109–2118. doi:10.1080/1828051x.2025.2563312.

Fedynyshyn, N., 2025. Bioeconomic modelling of reproductive additives in rabbit nutrition: roi-based kit production analysis. Int. J. Educ. Bus. Econ. Res. 05 (05), 14–25. doi:10.59822/ijeber.2025.5502.

Flores, J.N., Lubin, J.B., Silverman, M.A., 2024. The case for microbial intervention at weaning. Gut Microbes 16 (1). doi:10.1080/19490976.2024.2414798.

García Herrero, L., Casonato, C., Pérez Cornago, A., Sanyé-Mengual, E., Sarasa Renedo, A., Bakogianni, I., Listorti, G., 2024. Overview and analysis of sustainable product procurement criteria in the EU food sector. Publications Office of the European Union, Luxembourg https://doi.org/10.2760/1286793.

Gerardi, A., 2023. Global Food Safety Initiative (GFSI): Underpinning the Safety of the Global Food Chain, Facilitating Regulatory Compliance, Trade, and Consumer Trust. Elsevier BV, Amsterdam, pp. 1089–1098. doi:10.1016/b978-0-12-819470-6.00058-5.

Gu, Z., Xu, W., Tian, Y., Lu, L., Chen, G., 2025. Effects of green tea dust on the biochemical parameters, antioxidant capacity, and intestinal microbiota composition in goose. Front. Vet. Sci. 12. doi:10.3389/fvets.2025.1694350.

Hamdi, Z., Ashour, E.A., Nour-Eldeen, A.E.H.M., Sabike, I., Osman, A.O., Mahgoub, S.A., Elsherbeni, A.I., El-Mekkawy, M.M., Khan, M.M.H., Alharbi, H.M., Alwutayd, K.M., Abd El-Hack, M.E., 2026. Effects of Echinacea purpurea extract in drinking water on growth performance, carcass characteristics, blood biochemistry, meat quality, and cecal microbiota in broiler chickens. J. Appl. Poult. Res. 35 (2). doi:10.1016/j.japr.2026.100681.

Hart, S.L., 2021. Griffith Foods: Nourishing the World Pioneering Family Firms' Sustainable Development Strategies. Edward Elgar Publishing Ltd., United States, pp. 203–233. doi:10.4337/9781789904420.00015.

Hodge, I., Quille, P., O'Connell, S., 2024. A review of potential feed additives intended for carbon footprint reduction through methane abatement in dairy cattle. Animals 14 (4), 568. doi:10.3390/ani14040568.

Hosseini, E., Tsegay, Z.T., Smaoui, S., Elfalleh, W., Antoniadou, M., Varzakas, T., Caraher, M., 2026. One health approaches to ethical, secure, and sustainable food systems and ecosystems: plant-based diets and livestock in the African context. Foods 15 (1). doi:10.3390/foods15010085.

Hussain, S.A., Abbas, S.R., Lee, S.W., 2026. Consumer perspectives on antibiotic-free animal products: a systematic review identifying critical gaps in non-pharmaceutical intervention research. Animals 16 (1). doi:10.3390/ani16010070.

Ibrahim, L.A., Abu-Hashim, M., Shaghaleh, H., Elsadek, E., Hamad, A.A.A., Hamoud, Y.A., 2023. A comprehensive review of the multiple uses of water in aquaculture-integrated agriculture based on international and national experiences. Water (Switzerland) 15 (2). doi:10.3390/w15020367.

Ifedinezi, O.V., Nnaji, N.D., Anumudu, C.K., Ekwueme, C.T., Uhegwu, C.C., Ihenetu, F.C., Obioha, P., Simon, B.O., Ezechukwu, P.S., Onyeaka, H., 2024. Environmental Antimicrobial resistance: implications for food safety and public health. Antibiotics 13 (11). doi:10.3390/antibiotics13111087.

Indugu, N., Narayan, K., Stefenoni, H.A., Hennessy, M.L., Vecchiarelli, B., Bender, J.S., Shah, R., Dai, G., Garapati, S., Yarish, C., Welchez, S.C., Räisänen, S.E., Wasson, D., Lage, C., Melgar, A., Hristov, A.N., Pitta, D.W., 2024. Microbiome-informed study of the mechanistic basis of methane inhibition by Asparagopsis taxiformis in dairy cattle. mBio 15 (8). doi:10.1128/mbio.00782-24.

Jain, S., Singh, R., 2024. Exploring the dynamics of packaging and labelling: a comprehensive review. Indian J. Agric. 38, 98–116.

Khan, I.N, Shewale, V.S, Khairmode, P.V, 2026. Integrated perspectives on insect–animal interactions in agricultural and natural ecosystems: a review. Uttar Pradesh J. Zool. 47 (3), 26–45. doi:10.56557/upjoz/2026/v47i35484.

Kosoe, E.A., Achana, G.T., Ogwu, M.C., 2024. Regulations and policies for herbal medicine and practitioners', in Herbal Medicine Phytochemistry: Applications and Trends. Springer International Publishing, Cham, pp. 1–23.

Kouas, I., Akrout, F., 2025. The expected-real gap: an exploration of psychological factors. J. Consum. Satisf. Dissatisf. Complain. Behav. 38 (2), 177–196. https://jcsdcb.com/index.php/JCSDCB/issue/view/57.

Kühl, N., Müller, P.M., Rung, T., 2022. Adjoint complement to the universal momentum law of the wall. Flow Turbul. Combust. 108 (2), 329–351. doi:10.1007/s10494-021-00286-7.

Li, S. (2025). Evaluation of the anti-inflammatory effects of different types and concentrations of omega-3 polyunsaturated fatty acids (PUFA) in pullet and laying hen diets (Doctoral dissertation, University of Manitoba, Winnipeg, Canada). https://mspace.lib.umanitoba.ca/server/api/core/bitstreams/c1269a3d-6ae0-49fa-b65c-5661f89e50f0/content.

Li, S., Sun, Y., Cao, S., Guo, T., Tong, X., Zhang, Z., Sun, J., Yang, Y., Wang, Q., Li, D., Min, L., 2025. Asparagopsis taxiformis mitigates ruminant methane emissions via microbial modulation and inhibition of methyl-coenzyme M reductase. Front. Microbiol. 16. doi:10.3389/fmicb.2025.1586456.

Li, S., Zhu, Y., Zhong, G., Huang, Y., Jones, K.C., 2024. Comprehensive assessment of environmental emissions, fate, and risks of veterinary antibiotics in China: an environmental fate modeling approach. Environ. Sci. Technol. 58 (12), 5534–5547. doi:10.1021/acs.est.4c00993.

Lin, P., Tretola, M., Zenneg, M., Raemy, M., Peiry, S., Pinotti, L., Lerch, S., 2025. Characterization of packaging remnants and chemical contaminants in former foodstuff products for animal nutrition. In: Innovating Feed–Shaping the Future. University of Novi Sad, Novi Sad, p. 76.

Madden, A., 2025. Telling the right story: how brand and product narratives shape trust and intention in sustainable fashion: exploring the mediating roles of authenticity, brand value, and emotional connection, and the moderating effect. of narrative transportation. University of Leeds, Leeds, UK.

Mak, P.H.W., Rehman, M.A., Kiarie, E.G., Topp, E., Diarra, M.S., 2022. Production systems and important antimicrobial resistant-pathogenic bacteria in poultry: a review. J. Anim. Sci. Biotechnol. 13 (1). doi:10.1186/s40104-022-00786-0.

Malisch, C.S., Finn, J.A., Eriksen, J., Loges, R., Brophy, C., Huguenin-Elie, O., 2024. The importance of multi-species grassland leys to enhance ecosystem services in crop rotations. Grass and Forage Sci. 79 (2), 120–134. doi:10.1111/gfs.12670.

Malyugina, S., Holik, S., Horky, P., 2025. Mitigation strategies for methane emissions in ruminant livestock: a comprehensive review of current approaches and future perspectives. Front. Anim. Sci. 6. doi:10.3389/fanim.2025.1610376.

Manea, E.E, Bumbac, C., Dinu, L.R, Bumbac, M., Nicolescu, C.M, 2024. Composting as a sustainable solution for organic solid waste management: current practices and potential improvements. Sustainability 16 (15), 6329. doi:10.3390/su16156329.

Miranda-de La Lama, G.C., Estévez-Moreno, L.X., 2025. Introduction: international livestock animal welfare policies and standards: an overview. Natl. Int. Anim. Welf. Audit. 1–28. doi:10.1163/9789004750593_2.

Moss, R., Stright, A., Nicolle, L., Richelle, E., Baxter, L., Frampton, K., Dabas, T., Gorman, M., McSweeney, M.B., 2024. Impact of information about nutritional benefits, sustainability and consumption on consumer acceptance and emotional response to smoothies containing brown seaweed, Ascophyllum nodosum. Food Humanit. 3. doi:10.1016/j.foohum.2024.100373.

Muteeb, G., Rehman, M.T., Shahwan, M., Aatif, M., 2023. Origin of antibiotics and antibiotic resistance, and their impacts on drug development: a narrative review. Pharmaceuticals 16 (11). doi:10.3390/ph16111615.

Nhan, P., Chuc, N.T., Tri, N.M., Mo, T.T.H., Phuong, N.T.M., 2025. Dose-response effects of aqueous garlic extract supplementation via drinking water on growth, hematology, immunity, and coccidiosis in Japanese Quails (Coturnix Japonica). J. Indones. Trop. Anim. Agric. 50 (4), 293–303. doi:10.14710/jitaa.50.4.293-303.

Niu, K., Wang, L., Lu, Y., Zhang, Z., Sheng, P., Zhao, Z., 2026. Potential complementary modulation of rumen fermentation and lipid metabolism in sheep: a hypothesis framework for fermented Chinese chive juice and black soldier fly-derived fatty acids. Vet. Sci. 13 (2), 173. doi:10.3390/vetsci13020173.

Ntsongota, Z., Ikusika, O., Jaja, I.F., 2025. The role of phytogenic feed additives in growth and immune response in livestock production: a global systematic review. Front. Anim. Sci. 6. doi:10.3389/fanim.2025.1703112.

Obahiagbon, E.G., Ogwu, M.C., 2024. Sustainable supply chain management in the herbal medicine industry. In: Herbal Medicine Phytochemistry: Applications and Trends. Springer International Publishing, Cham, pp. 1–29. doi:10.1007/978-3-031-43199-9_64.

Oni, A.I., Oke, O.E., 2025. Gut health modulation through phytogenics in poultry: mechanisms, benefits, and applications. Front. Vet. Sci. 12. doi:10.3389/fvets.2025.1616734.

Oyenuga, A.O., Sam-Bulya, N.J., Attah, R.U., 2025. Enhancing climate resilience and profitability in medicinal herb farming systems. Int. J. Multidiscip. Res. Growth Eval. 6 (1), 989–994. doi:10.54660/.ijmrge.2025.6.1.989-994.

Papatsiros, V.G., Katsogiannou, E.G., Papakonstantinou, G.I., Michel, A., Petrotos, K., Athanasiou, L.V., 2022. Effects of phenolic phytogenic feed additives on certain oxidative damage biomarkers and the performance of primiparous sows exposed to heat stress under field conditions. Antioxidants 11 (3). doi:10.3390/antiox11030593.

Peter, S. M., 2025. International business strategies for natural pet foods: Responding to consumer demand, ethical sourcing, and regulatory challenges. In: Natural Pet Food Markets and Global Strategies. Springer International Publishing, Cham, Switzerland, pp. 1–20.

Pilarczyk, B., Pilarczyk, R., Bąkowska, M., Tomza-Marciniak, A., Seremak, B., Kwita, E., Juszczak-Czasnojć, M., Matusevičius, P., Mišeikienė, R., 2025. The welfare of cattle in different housing systems. Animals 15 (13), 1972. doi:10.3390/ani15131972.

Rakosy, D., Motivans, E., Ştefan, V., Nowak, A., Świerszcz, S., Feldmann, R., Kühn, E., Geppert, C., Venkataraman, N., Sobieraj-Betlińska, A., Grossmann, A., Rojek, W., Pochrząst, K., Cielniak, M., Gathof, A.K, Baumann, K., Knight, T.M, Lázaro, A., 2022. Intensive grazing alters the diversity, composition and structure of plant-pollinator interaction networks in Central European grasslands. PLoS One 17 (3), e0263576. doi:10.1371/journal.pone.0263576.

Ravi, S., Biswas, A., Kolluri, G., Pipaliya, G., Deo, C., Rokade, J., Goyari, K., Parte, P., 2025. Effect of dietary supplementation of herbal immuno-modulator on growth performance, immune response and serum biochemical indices of broiler chickens. Sci. Rep. 15 (1). doi:10.1038/s41598-025-23484-z.

Rehman, A.U., Buzdar, J.A., Arain, M.A., Fazlani, S.A., Arslan, M., Zhou, C., 2025. An in-depth overview of the nutritional advantages of medicinal plant supplementation in poultry feed. Worlds Poul. Sci. J. 81 (2), 569–604. doi:10.1080/00439339.2025.2468696.

Rivero, M.J., Lee, M.R.F., 2022. A perspective on animal welfare of grazing ruminants and its relationship with sustainability. Anim. Prod. Sci. 62 (18), 1739–1748. doi:10.1071/AN21516.

Roderick, S., Jones, H., 2025. Optimizing soils, pasture and grassland management for organic dairy farming. In: Smith, L.J., Brown, P.R. (Eds.), Advances in Organic Dairy Cattle Farming. Dodds Science Publishing, Cambridge: Burleigh, pp. 35–55.

Saad, H., Khalil, R., Aboushouk, A.A., Saad, A.H., Elhussieny, O., Rohiem, A.H., Elgazzar, D., El-Maksoud, E.M.A., Younis, N.A., Abdelhiee, E.Y., 2025. Ameliorative impacts of neem leaf powder on cadmium toxicity in nile tilapia: Growth performance, haemato-biochemical, oxidative and pathological insights. Egypt. J. Vet. Sci. 0 (0), 1–22. doi:10.21608/ejvs.2025.405387.2975.

Schröder, M.J.A., McEachern, M.G., 2004. Consumer value conflicts surrounding ethical food purchase decisions: a focus on animal welfare. Int. J. Consum. Stud. 28 (2), 168–177. doi:10.1111/j.1470-6431.2003.00357.x.

Sejian, V., Devaraj, C., Shashank, C.G., Silpa, M.V., Sahoo, A., Bhatta, R., 2025. The role of prebiotic and herbal supplementation in enhancing welfare and resilience of Kenguri sheep subjected to transportation stress. Vet. Sci. 12 (5). doi:10.3390/vetsci12050442.

Shrivastava, P., Sahoo, S.K., Venkatesh, K.M., 2025. Role of Veterinarians in Public Health Epidemiology and Environmental Hygiene in Veterinary Public Health. Wiley, India, pp. 287–298. doi:10.1002/9781394208180.ch24.

Simba, L.D., te Beest, M., Hawkins, H.J., Larson, K.W., Palmer, A.R., Sandström, C., Smart, K.G., Kerley, G.I.H., Cromsigt, J.P.G.M., 2024. Wilder rangelands as a natural climate opportunity: Linking climate action to biodiversity conservation and social transformation. Ambio 53 (5), 678–696. doi:10.1007/s13280-023-01976-4.

Sindhu, M., Tewary, A., 2025. Advanced liquid chromatography techniques for detecting and analyzing food adulteration: ensuring safety and quality in food products. In: Advanced Techniques against Food Adulteration. Springer Science+Business Media, India, pp. 225–252. doi:10.1007/978-3-031-99454-8_8.

Singh, H., Kumar, N., Kumar, A., 2024. Enhancing resource use efficiency in crops through plant functional traits. In: Plant Functional Traits for Improving Productivity. Springer Nature, India, pp. 97–117. doi:10.1007/978-981-97-1510-7_6.

Sisodiya, N., Dwivedi, S., Avi, R., Khabiya, R., 2026. Immunomodulators in poultry: natural pathways to enhanced health and productivity. Worlds Poult. Sci. J. 1–43. doi:10.1080/00439339.2026.2616710.

Stygar, A.H., Krampe, C., Llonch, P., Niemi, J.K., 2022. How far are we from data-driven and animal-based welfare assessment? a critical analysis of European quality schemes. Front. Anim. Sci. 3. doi:10.3389/fanim.2022.874260.

Surai, P.F., Kochish, I.I., Fisinin, V.I., Kidd, M.T., 2019. Antioxidant defence systems and oxidative stress in poultry biology: an update. Antioxidants 8 (7). doi:10.3390/antiox8070235.

Tang, K.W.K., Millar, B.C., Moore, J.E., 2023. Antimicrobial resistance (AMR). Br. J. Biomed. Sci. 80. doi:10.3389/bjbs.2023.11387.

Taylor, P.S., Schrobback, P., Verdon, M., Lee, C., 2023. An effective environmental enrichment framework for the continual improvement of production animal welfare. Anim. Welfare 32. doi:10.1017/awf.2023.5.

Thaemert, J., 2023. Animal welfare consumer protection litigation: challenges and possibilities for bringing about more "humane" labeling practices. Animal L. 29, 215.

Thakkar, S., Anklam, E., Xu, A., Ulberth, F., Li, J., Li, B.o, Hugas, M., Sarma, N., Crerar, S., Swift, S., Hakamatsuka, T., Curtui, V., Yan, W., Geng, X., Slikker, W., Tong, W., 2020. Regulatory landscape of dietary supplements and herbal medicines from a global perspective. Regul. Toxicol. Pharm. 114, 104647. doi:10.1016/j.yrtph.2020.104647.

Ungerfeld, E.M., Cancino-Padilla, N., Vera-Aguilera, N., 2023. Fermentation in the rumen Microbial Fermentations in Nature and as Designed Processes. Wiley Blackwell, Chile, pp. 133–165. doi:10.1002/9781119850007.ch4.

Venkataraman, S., Khan, I., Habibi, P., Le, M., Lippert, R., Hefferon, K., 2023. Recent advances in expression and purification strategies for plant made vaccines. Front. Plant Sci. 14. doi:10.3389/fpls.2023.1273958.

Verma, S., Akpensuen, T.T., Wolffram, S., Salminen, J.-P., Taube, F., Blank, R., Kluß, C., Malisch, C.S., 2024. Investigating the efficacy of purified tannin extracts from underutilized temperate forages in reducing enteric methane emissions in vitro. Sci. Rep. 14 (1). doi:10.1038/s41598-024-63434-9.

Vlaicu, P.A., Untea, A.E., Varzaru, I., Saracila, M., Oancea, A.G., 2023. Designing nutrition for health—incorporating dietary by-products into poultry feeds to create functional foods with insights into health benefits, risks, bioactive compounds, food component functionality and safety regulations. Foods 12 (21). doi:10.3390/foods12214001.

Waterhouse, H., Aburto, F., Rees, G., Griffin-LaHue, D.E., Salls, W.B., Rippner, D.A., Tian, Z., Scow, K., O'Geen, A.T., 2024. Diversified vegetation types on rangelands promote multiple soil-based ecosystem services. Land Degrad. Dev. 35 (3), 1011–1028. doi:10.1002/ldr.4967.

Windisch, W., Schedle, K., Plitzner, C., Kroismayr, A., 2008. Use of phytogenic products as feed additives for swine and poultry. J. Anim. Sci. 86 (14), E140. doi:10.2527/jas.2007-0459.

Wróbel, B., Zielewicz, W., Staniak, M., 2023. Challenges of pasture feeding systems—opportunities and constraints. Agriculture 13 (5), 974. doi:10.3390/agriculture13050974.

Yang, Y., Lei, D., Zhang, Y., Zhao, Z., Xie, H., Deng, F., Bao, X., Zhang, X., He, H., 2025. Effects of pig slurry coupled with straw mulching on soil nitrogen dynamics and maize Growth. Agronomy 15 (5), 1062. doi:10.3390/agronomy15051062.

Yang, Z., Deng, X., Yang, Z., Han, M., Ismail, N., Chan, K.W, Razis, A.F., Esa, N.M, Ho, K.L, Bakar, M.Z.A, 2024. Polygonum minus: a tropical medicinal herb with vast applications in food, agricultural, and medicinal fields. Food Biosci. 60, 104511. doi:10.1016/j.fbio.2024.104511.

You, H., Abraham, E.J., Mulligan, J., Zhou, Y., Montoya, M., Willig, J., Chen, B.K., Wang, C.K., Wang, L.S., Dong, A., Shamtsyan, M., Nguyen, H., Wong, A., Wallace, T.C., 2024. Label compliance for ingredient verification: regulations, approaches, and trends for testing botanical products marketed for "immune health" in the United States. Crit. Rev. Food Sci. Nutr. 64 (9), 2441–2460. doi:10.1080/10408398.2022.2124230.

Zaroug, M.G., Mirreh, M.M., 2010. Range livestock production systems in the Near East. Range Anim. Sci. Resour. Manag. 1, 92–135.

Zhao, X., Zhang, S., Zhang, Y., 2025. Tannin-based strategies for mitigating greenhouse gas emissions through nitrogen and carbon metabolism in ruminants. Agriculture 15 (21), 2234. doi:10.3390/agriculture15212234.

Zhgun, A.A., 2023. Industrial production of antibiotics in fungi: current state, deciphering the molecular basis of classical strain improvement and increasing the production of high-yielding strains by the addition of low-molecular weight inducers. Fermentation 9 (12), 1027. doi:10.3390/fermentation9121027.

Chapter 11

Regulatory frameworks and market opportunities for botanicals in animal nutrition

Mohammad Mehedi Hasan Khan [a], Mohamed E. Abd El-Hack [b] and Nazifa Rahman [a]

[a] *Department of Biochemistry and Chemistry, Sylhet Agricultural University, Sylhet, Bangladesh,*
[b] *Department of Industrial Pharmacy, College of Pharmaceutical Sciences and Drug Manufacturing, Misr University for Science and Technology (MUST), Giza, Egypt*

11.1 Introduction

All animals need a balanced diet, which usually consists of cereal grains and protein sources, to maintain optimal health and productivity. The nutritional value of feed, which is influenced by several variables such as the presence of antinutritional components, microbial contamination, digestibility, palatability, and its impact on intestinal health, is mostly determined by feed additives. A small amount of feed additives is present in animal diets; however, they help improve the quality of food and feed as well as the health and performance of the animals (Shahbakht et al., 2024). Animal feeds frequently contain modest concentrations of antibiotic growth promoters (AGPs) to boost growth rates and enhance animal performance (Fernández Miyakawa et al., 2024). The therapeutic efficiency of both veterinary and human antimicrobial medicines may be threatened by the emergence and dissemination of antibiotic-resistant genes caused by these growth promoters (Steiner and Syed, 2015). Farmers are now utilizing phytogens as feed additives as a result of the EU's ban on antibiotic use and growing consumer awareness of the possible risk of microbial resistance. These plant-based items, which comprise a wide range of herbs, spices, and essential oils, offer advantageous qualities when ingested in the right amounts and combinations (Gadde et al., 2017).

Phytogenic feed additives (PFAs), also referred to as phytoadditives, are alternative feed additives derived from spices or aromatic plants. These additives are categorized according to their composition, processing techniques, and botanical origin. It contains oleoresins extracted by solvent extraction, aromatic spices,

Plant-Derived Feed Additives and Animal-Based Foods: Linking Animal Health, Product Quality, and Consumer Perceptions. DOI: https://doi.org/10.1016/B978-0-443-45446-2.00003-X

essential oils derived from different plant components, and medicinal herbs and nonwoody plants (Babinszky et al., 2021). Farm animals are being selectively bred for increased efficiency because of the rising demand for meat. However, several factors, including nutrition and genetics, affect animal output. PFAs have become a viable substitute for antibiotics in animal feed in recent years. PFAs include a variety of substances such as herbs, spices, and essential oils. They can be categorized according to how they are used in animal feed. For instance, adding pigment-rich plants to chicken feed to produce a more vivid orange yolk color is an example of how sensory enhancers improve the flavor and aroma of animal foods. Animal health and feed quality are enhanced by technological additions (Pandey et al., 2019).

According to the European Food Safety Authority (EFSA) (2010), clear regulatory frameworks are necessary to guarantee the safety, effectiveness, and consistent quality of PFAs, which are becoming more and more important in sustainable animal nutrition. Regulatory clarity makes it possible to evaluate PFAs consistently, avoiding the use of contaminated or unverified products and safeguarding the health of consumers and animals. In addition, ensuring that foods derived from animals are free of dangerous toxins and residues, it fosters consumer confidence (Greathead, 2003). Furthermore, clear laws promote innovation by providing industrialists with the assurance to invest in research and production of novel phytogenic formulations (Windisch et al., 2008). As PFAs gain popularity as natural substitutes for AGPs, harmonized international rules are especially crucial for promoting international trade and lowering market barriers. Overall, the safe, efficient, and sustainable use of PFAs in animal production is supported by regulatory clarity.

In response to growing consumer demand for "natural," antibiotic-free, and more sustainable animal products, the global market for PFAs is steadily expanding. Because of AGP bans, rising meat consumption in Asia-Pacific, and increased R&D in botanical formulation, market estimates cluster around a global market of about USD 1.0–1.3 billion in 2024–2025 with projected compound annual growth rates of ~5%–8%, depending on the report and expected to reach approximately USD 1.5–2.1 billion by the early 2030s (market CAGRs commonly quoted between ~6% and 8% depending on the analyst) (Grand View Research, 2024;Future Market Insights, 2025). BlueWave expects the Global PFAs Market size to expand at a compound yearly growth rate (CAGR) of 8.20%, reaching a value of USD 1587.0 million by 2030. The market is projected to grow from USD 39.80 billion in 2025 to USD 53.49 billion by 2032, exhibiting a CAGR of 4.31% during the forecast period, as estimated by Fortune Business Insight (2024/2025). Industry analyses and trade coverage frequently point to the demand for natural-origin additives (such as essential oils, oleoresins, and flavonoids) as the main factor driving the adoption of PFAs. Consumers are increasingly demanding "clean-label" and antibiotic-free animal products and are putting pressure on retailers and producers for transparency on farming practices and feed inputs. According to Feed Phytogenic Market Size & Share Analysis-Growth Trends And Forecast (2025–2030), Regional dynamics are important: Europe leads on regulatory scrutiny and documented efficiency expectations, driving manufacturers to invest in standardized extracts and delivery technologies to satisfy both regulatory and consumer requirements, while Asia-Pacific exhibits the fastest uptake because of rising protein demand and price sensitivity. Overall, the market is now experiencing strong but dispersed expansion; several market research firms indicate a growing market size and positive projections, but the precise numbers differ depending on the methodology, highlighting the necessity of consulting primary market studies for accurate planning.

11.2 Global regulatory landscape

Botanical-derived ingredients (phytogenic, botanicals, and plant extracts) occupy a complex regulatory space because they can be defined as feed ingredients, feed additives, nutraceuticals, or—if therapeutic claims are made—veterinary medicinal products (VMPs). Each jurisdiction's premarket data requirements, safety and efficiency evaluation standards, permissible labeling and claims, and regulatory route to market are all determined by this classification (EFSA, 2017).

11.2.1 Strategies for international direction and harmonization

11.2.1.1 Codex Alimentarius (FAO/WHO)

Baseline definitions and basic guidelines for evaluating additives and ingredients used in food and feed are provided by international organizations and bodies, most notably Codex Alimentarius (FAO/WHO) and associated guidelines. Although Codex standards are not directly binding, they serve as national authorities' primary international reference and, when incorporated into national legislation, promote trade harmonization. Under the Codex regime, instead of having their own "botanicals" category, botanicals used as feed ingredients or additives are classified under "feed ingredients/additives." This implies that they must adhere to the same manufacturing, storage, labeling, safety, and contamination-control requirements as other feed additives. The risk assessment criteria (from the Risk Assessment Guidelines) should be used for botanical additives: possible residues in animal-derived food items, raw material quality variability, and hazards from pollutants (such as pesticide residues, heavy metals, and mycotoxins). The regulatory oversight of Codex does not directly address therapeutic claims or specific efficacy for botanicals because it concentrates on food safety (i.e., safety for humans consuming animal products); it leaves "functional claims," "nutraceuticals," and "medicinal claims" to national-level regulatory frameworks. Codex standards offer a baseline safety and quality framework for nations that adopt them as national regulations; however, local feed and additive rules must still be followed for botanicals to be approved as functional or therapeutic additives.

11.2.1.2 European Union (EFSA)—Conservative, science-evidence driven

The European Union (EU) Feed Additives Regulation requires the EFSA FEEDAP Panel to assess botanicals meant for use as feed additives. The EU Feed Additives Regulation requires the EFSA FEEDAP Panel to assess botanicals meant for use as feed additives. EFSA mandates through compositional characterization, testing for contaminants, toxicity and residue evaluations (where applicable), and proof of efficacy based on stated functionalities. The rigorous applicant requirements for botanicals and botanical preparations are outlined in EFSA's official guidance materials, and the assessment process is highly structured, leading to high consumer confidence but prolonged approval times as well as substantial data costs for applicants (EFSA, 2017).

11.2.1.3 United States-FDA/AAFCO: Generally recognized as safe routes and ingredients definitions

In the United States, for animal foods and ingredients, the FDA's Center for Veterinary Medicine (CVM) implements regulatory monitoring in accordance with the Federal Food, Drug, and Cosmetic Act (FD&C Act). According to the FDA, many botanical compounds can be introduced as traditional feed additives or through a Generally Recognized as Safe (GRAS) assessment for the intended purpose; when there is a consensus regarding safety, GRAS status can permit market entry without a formal food-additive petition. Any product that claims to be healthy or therapeutic for animals, however, might be governed as a veterinary medication and needs far more proof and permission. State-level enforcement and labeling requirements are significantly influenced by AAFCO's ingredient definitions and model laws.

11.2.2 Comparison between regulatory categories

Products used in animal feeding are classified differently in different jurisdictions, especially botanical preparations. Botanical ingredients used in animal nutrition are categorized into three broad regulatory categories depending on their composition, intended use, and the claims made: Feed additives (including technical additives and sensory agents), nutraceuticals/feed supplements, and veterinary medicines (animal drugs) by the EFSA. The regulatory requirements for safety testing, efficacy validation, labeling, claims, and market authorization are determined by this classification (Windisch et al., 2008). Companies seeking to sell phytogenic products internationally must comprehend the differences.

11.2.2.1 Feed additives

Feed additives are compounds introduced into animal feed that boost feed quality, health, or performance. Usually. Botanical-based additions are classified as "sensory additives," "digestive enhancers," or "zootechnical additives". Regulatory evaluation by EFSA typically puts emphasis on product efficacy, customer safety, ecological security, and target species safety (EFSA Panel on Additives and Products or Substances used in Animal Feed (FEEDAP) et al., 2021). Several botanicals in the United States are eligible for GRAS designation, which facilitates a more straightforward route to market. However, only functional or nutritional claims are followed under GRAS; therapeutic claims are not (Hallagan et al., 2020).

11.2.2.2 Nutraceuticals

Combining the words "nutrition" with "pharmaceuticals," "nutraceuticals" refers to products made from herbal sources, dietary supplements (such as nutrients), specific diets, and processed foods with both therapeutic and nutritional value, such as cereals, soups, and drinks (Kalra, 2003). Nutraceutical products include substances that offer physiological benefits or prevent chronic disease. These products can prolong life, prevent chronic illnesses, slow down the aging process, maintain the body's structure and function, and improve health (Chauhan et al., 2013). Nutraceuticals are a broad category of compounds that can be utilized as feed additives, feed materials, or pharmaceuticals. Because multiple procedures must be followed to gain the right marketing authorization, this has a significant impact on the regulatory side of marketing authorization in the EU. When nutraceuticals are used as feed additives for animal nutrition, they must adhere to Regulation No. 1831/2003 on additives for use in animal nutrition because particular regulations do not apply to them. Nutraceuticals must abide by Commission Regulation (EU) No. 68/2013 even though they are used as feed components. Nutraceuticals must adhere to Directive 2001/82/EC if they are used with medical claims or if they have a pharmacological effect. The EU's legislation on this subject is quite intricate and complicated. However, it permits the acquisition of a marketing authorization with a broad safety margin, protecting the environment, human health, and animal health. The European Commission uses the views produced by the Scientific Committees and panels of the EFSA to enact laws pertaining to animal nutrition. The European Medicines Agency (EMA) or the appropriate national agencies of the Member States oversee granting authorization for veterinary medicinal goods (Tedesco and Cagnardi, 2019).

11.2.2.3 Veterinary medicines

As an alternative and complementary approach to animal healthcare, herbal veterinary medicines (HVMs) have drawn a lot of interest. It explores the categorization of medicinal plants, their active ingredients, and the pharmacological processes that enable them to be used to cure a variety of animal ailments (Trivedi et al., 2025). In most jurisdiction, HVMs, or medicinal products for animals whose active ingredients are plant materials, extracts, or botanically derived mixtures, are regulated either (1) under the nation's VMP framework (treated as veterinary drugs) or (2) under feed/food/functional-product frameworks if they are supplied as minor dietary constituents with no therapeutic claims (reported by U.S. Food & Drug Administration, n.d.). Products that make therapeutic, preventative, or disease- related claims are often categorized as veterinary medicines and are subject to the entire veterinary-drug regulatory pathway. According to the EMA (European Medicines Agency, 2020) (Guideline on the quality of herbal medicinal products), veterinary medicines are controlled by strict regulations and require thorough scientific research to prove their safety, pharmacokinetics, dose accuracy, and therapeutic efficacy. Veterinary medicines are regulated to preserve the environment, animal welfare, and public health (including food safety and antimicrobial resistance issues) while ensuring the safety, effectiveness, and quality of medications used in animals. World Organization for Animal Health (WOAH) reported that Authorization, production, distribution, labeling, pharmacovigilance, and postmarket surveillance are all governed by national and regional agencies, while international groups assist with standardization, capacity building, and standards creation. According to Regulation (EU) 2019/6 and EMA implementing guidance, herbal veterinary products are typically approved in the EU under the same regulatory framework as conventional VMPs. The FDA separates foods and dietary supplements from botanical pharmaceuticals (which are subject to drug development/IND/NDA pathways). Products claiming to diagnose, treat, prevent, or mitigate disease for veterinary uses are subject to FDA/CVM regulations and must meet safety and efficacy standards; botanical origin does not grant regulatory leniency. FDA/CVM precedents and the FDA's botanical drug advice (human CDER) offer helpful comparisons for creating herbal medicines with medicinal claims. While traditional herbal medicines are recognized by active regulatory systems in major Asian markets, their treatment in veterinary usage varies. Although there are national pharmacopeial monographs for some herbs and a documented history of HVM in India, formal registration as a VMP requires adherence to national VMP regulations where therapeutic claims are made (Rastogi et al., 2015) (Table 11.1).

11.3 Regional frameworks

11.3.1 European Union-EFSA/EU regulatory frameworks

The Feed Additives Regulation (Regulation (EC) No 1831/2003) governs the EU's treatment of botanicals as feed additives. Regulation (EC) No 767/2009 (labeling/marketing) further regulates the usage and marketing of feeds. Documents that comply with EFSA FEEDAP Panel guidelines on identity, characterization,

TABLE 11.1 Comparison among feed additives, nutraceuticals, and veterinary medicines.

Topic	Feed additives	Nutraceuticals	Veterinary medicines
Definition	Substances that are added to feed without making any therapeutic claims to enhance nutrition, digestion, gut health, growth efficiency, or feed quality (Windisch et al., 2008)	Products that promote immunity, digestion, overall health, and physiological processes are frequently derived from plants (Chauhan et al., 2013)	Products, such as herbal veterinary medicinal products (HVMPs), that are meant to diagnose, treat, cure, or prevent disease (Trivedi et al., 2025)
Safety requirements	Should demonstrate that the target consumers, animals, and environment are safe (Rychen et al., 2017)	Products should be Generally Recognized As Safe (GRAS). In the United States, the manufacturer must confirm that the products are safe before marketing. FDA acts against unsafe products after marketing. Premarket toxicology test is not required (U.S. Food and Drug Administration, 2024)	Requires extensive safety data, both preclinical and clinical, to ensure safety for the target animals, the consumer, and the environment (European Medicines Agency, 2024)

(*continued on next page*)

TABLE 11.1 Comparison among feed additives, nutraceuticals, and veterinary medicines—cont'd

Topic	Feed additives	Nutraceuticals	Veterinary medicines
Efficacy evidence requirements	Required for the claim of zootechnical, technological, physiological effect under specified use condition (EFSA Panel on Additives and Products or Substances Used in Animal Feed (FEEDAP) et al., 2023)	Generally, less important than feed additives and drugs. Efficacy data recommended but often not legally mandated unless claims imply performance enhancement (U.S. Food and Drug Administration, 2024)	Therapeutic efficacy must be demonstrated through controlled field and laboratory trials (European Medicines Agency, 2024)
Labeling requirements	Manufacturer identity, authorized uses, inclusion rate, conditions of use, warnings, storage, user safety, and, in many jurisdictions, residues/withdrawal for medicated feeds. Specific wording required by authorization (EFSA Panel on Additives and Products or Substances Used in Animal Feed (FEEDAP) et al., 2023)	In the United States, for dietary supplements: Statement of identity, net quantity, ingredients list, contact of manufacturer/distributers, claims and disclaimers, and batch/lot info are required (U.S. Food and Drug Administration, 2024)	Detailed product name, active substances, concentration, indication, dosage and administration, route, contraindications, adverse reaction, withdrawal period, prescription status, and manufacturer information are mandatory (European Medicines Agency, 2024)

efficacy, and safety, including consumer safety and environmental risk, must be submitted by applicants requesting authorization for a botanical-derived additive. Documents for botanicals are frequently more complicated and expensive than those for individual additives because of three major obstacles: (1) Robust compositional characterization of complex botanical mixtures, (2) clear evidence of efficacy for the claimed function in target species, (3) toxicology/safety data for target animals, consumers, and the environment. Regulation 767/2009 governs the labeling of feed and feed additives, and the FEEDAP findings outline any usage requirements and labeling declarations to guarantee safe use. Chemical fingerprinting, batch-to-batch specifications, ADME/toxicology or history of safe use evidence, and carefully written claims that correspond to approved functional categories are required by the manufacturers (EFSA Panel on Additives and Products or Substances used in Animal Feed (FEEDAP) et al., 2023).

11.3.2 United States

The FDA CVM, state regulators, and AAFCO Association of American Feed Control Officials (AAFCO) product classifications all play a major role in shaping the regulatory environment in the United States. For botanical compounds used as animal feed, there are two typical routes: (1) An approved food additive rare for botanicals, or 2) a GRAS determination for the intended purpose. The FDA has released guidelines outlining its enforcement policy toward AAFCO-defined ingredients and how manufacturers can promote animal food containing such ingredients. In addition, several plant-derived components are included in AAFCO ingredients definitions. AAFCO inclusion can allow for ongoing market use while the FDA assesses safety and policy, according to a recent FDA guideline. However, manufacturers should be ready to supply safety and identity data and, when applicable, adhere to AAFCO name and labeling conventions. Major obstacles include proving a GRAS basis (published literature, experience-based use, or expert consensus), ensuring label claims do not classify the product as an unapproved medicine, and aligning with AAFCO ingredient names (AAFCO, 2024; U.S. Food and Drug Administration, Center for Veterinary Medicine, 2024).

11.3.3 Asia-Pacific

11.3.3.1 China

China has been continuously modifying feed/additive regulations and has published guidelines, especially on applications for plant-extracted feed additives in 2023. China maintains feed additives and ingredients catalogues and needs registration and approval, or declaration in accordance with the catalog system. Ministry of Agriculture and Rural Affairs (U.S. Department of Agriculture, Foreign Agricultural Service, 2023) guidance seeks to create a clearer approach for plant extracts, but it still requires compositional data, safety/efficacy evidence, and compliance with catalog entries- so the regulatory door is opening, but documentation requirements are stiff.

11.3.3.2 Japan

Japan has a long-standing framework in which feed additive designation has been granted by the Ministry of Agriculture, Forestry and Fisheries (MAFF) after evaluation by the Agricultural Materials Council, and there are comprehensive Ministerial Ordinances and Handbooks that address feed additive digestion, safety, and labeling. Japan maintains a strong emphasis on consumer and animal safety and mandates monitoring efficacy/safety dossiers (MAFF).

11.3.3.3 India

In India, the regulatory framework governing animal feed is changing. In an effort to harmonize the feed standards (including pet foods and commercial feeds), the Food Safety and Standards Authority of India (FSSAI) and the Department of Animal Husbandry and Dairying (DADF) have released guidelines and consolidated lists of feed additives in recent years, along with draft and regulatory modifications. For botanical companies, regulatory uncertainty and changing regulations present both potential and challenges. Firms need to keep an eye on FSSAI/DADF alerts and be ready for country-specific testing and labeling needs (Food Safety and Standards Authority of India, 2019; Food Safety and Standards Authority of India, 2021; DAHD, 2023).

11.3.4 Middle East and Africa

Middle East & Africa have diverse regulatory systems. While national agencies carry out registration and labeling, some Gulf nations (GCC members) use regional norms and controls for imports. Other markets' sustainability depends on imports of food additives and premixes, and they are improving import-control and registration programs (Cooperation Council for the Arab States of the Gulf). Although many African nations still struggle with gaps in laboratory capacity, regulatory clarity, and enforcement, continental initiatives (African Union – Interafrican Bureau for Animal Resources (AU-IBAR), Resilient African Feed and Fodder Systems (RAFFS), Pan-African regulatory collaboration) point to efforts to build capacity and harmonize veterinary products/feed regulations. This opens opportunities for exporters but also risks, which include long approval timelines, changing import rules (Food and Agriculture Organization of the United Nations & African Union, 2007). Companies can anticipate varying documentation requirements, a focus on halal, traceability in the Gulf, and participation in regional and regulatory capacity-building initiatives in Africa.

11.3.5 Latin America

11.3.5.1 Brazil

In Brazil, the Ministry of Agriculture, Livestock and Food Supply, also known as Ministério da Agricultura, Pecuária e Abastecimento (MAPA), administers feed and feed additives. The normative instructions, such as the Noemative Instruction Series like NI 54/2018 and Decree frameworks, specify dossiers, technical requirements, and registration for zootechnical additives, mycotoxin absorbents, and plant-derived products. Brazil is a major consumer and producer in the region and has a reasonably established market with clear dossier requirements such as technical reports, safety/efficacy data (Ministério da Agricultura, Pecuária e Abastecimento (MAPA), Ministry of Agriculture, Livestock and Food Supply (MAPA, 2019).

11.3.5.2 Argentina

Argentina has an active feed additives sector and uses a national food/code framework (Argentina Food Code, SENASA oversight for animal health). Administrative processes vary by agency, but regulatory approaches are similar in concept of safety/identity/efficacy evidence, import controls (U.S. Department of Agriculture, Foreign Agricultural Service, 2018). In comparison with several small Latin American Markets, Brazil and Argentina are regional leaders with more structured submission routes.

11.4 Challenges in regulation

11.4.1 Lack of standardized definitions for botanicals in animal nutrition

One of the major obstacles to the regulation of botanical ingredients for animal feed is the lack of a standardized global definition. Botanicals are classified differently by regulatory organizations; some treat them as feed additives, while others classify them as sensory additives, zootechnical additives or even veterinary supplements (EFSA, 2017). For instance, the European Union (EU) mandates that the botanical preparation be categorized according to primary function; however, this is challenging because many botanicals have numerous bioactive qualities (European Commission, 2003).

Similar to this, however, there are no standard definitions for botanicals in animal feed in the United States; instead, approval is determined by GRAS status, AAFCO definitions, or Food Additives Petitions, which causes inconsistencies for industry stakeholders (AAFCO, 2024). In addition, emerging markets such as China, Brazil, and India frequently classify botanicals under general terms such as natural additives or plant-based feed ingredients due to a lack of precise terminology (MAPA, 2019;DAHD, 2023).

Regulatory ambiguity brought up by this lack of definitional clarity causes inconsistent recognition across jurisdictions and delays in product approval (FAO and AU, 2007). Manufacturers find it difficult to conduct cross-border trade in the absence of standardized classifications, and it becomes challenging to assess scientific evidence for efficacy across the market.

11.4.2 Variability in quality control, safety testing, and dosage guidelines

Botanical feed additives have a significant degree of natural variability due to variations in plant species, growing conditions, harvesting time, extraction techniques, and storage (EFSA FEEDAP Panel, 2024). Therefore, regulators often emphasize that botanical products need to show consistency in active substances. Nonetheless, phytochemical analytical standards are still developing (European Medicines Agency, 2024).

In the EU, EFSA mandates thorough characterization of botanicals, including phytochemical fingerprinting, containment analysis, and proof of batch-to-batch uniformity, but numerous small producers lack the technology necessary for such assessments (EFSA Panel on Additives and Products or Substances used in Animal Feed (FEEDAP) et al., 2023). Although safety evidence is required for FDA regulation through the CVM in the United States, standardized testing standards for multicomponent plant extracts are not well established (U.S. Food and Drug Administration, Center for Veterinary Medicine, 2024). The developing countries are facing greater problems such as weak laboratory capacity, poor GMP enforcement, and inadequate national standards for additives. This results in varying dosage guidelines for animal species, inconsistent quality, and possible adulteration issues. Safe use is mostly dependent on manufacturer knowledge rather than a standardized regulatory threshold due to the absence of harmonized maximum inclusion rates and dosage guidelines.

11.4.3 Complex approval timelines and costs for small/medium enterprises

Botanical feed additive regulatory approval procedures can often be lengthy, expensive, and data intensive. For instance, many small- and medium-sized enterprises (SMEs) cannot afford the substantial toxicological, genotoxicity and effectiveness data requirements of feed additive applicants in the EU, which can cause several hundred thousand euros (EFSA FEEDAP Panel, 2024). Because of the intricacy of the dossier and the repeated demands for material, the examination process frequently takes longer than a year or two.

In the United States, acquiring a new AAFCO definition or a Feed Additive Petition requires extensive administrative review and scientific explanation, which creates obstacles for smaller companies with limited R&D budgets

(AAFCO, 2024). Because of the new FDA enforcement regulations on AAFCO-defined substances, the SMEs are now facing additional regulatory compliance challenges (U.S. Food and Drug Administration, Center for Veterinary Medicine, 2024).

Developing regions experience additional delay because of large reliance on imported botanicals, fragmented regulatory systems, and insufficient institutional capacity. In some countries, regulatory approval is slow or uneven due to outdated legislation or a lack of technical expertise, which increases uncertainty for the innovators (African Union-InterAfrican Bureau for Animal Resources).

Reduced innovation, fewer market entrants, and a slower adoption of botanical alternatives in animal production systems are all the consequences of high regulatory costs, prolonged approval processes, and complicated data requirements.

11.5 Linking regulation to market opportunities

11.5.1 Regulatory clarity encourages innovation and investment

Clear and reliable regulatory frameworks reduce uncertainty for both entrepreneurs and investors by specifying the data requirements, deadlines, and permissible claims that the applicants must fulfill. Companies can budget R&D, design studies to meet regulator expectations, and lower the risk of rework or rejection when agencies publish transparent guidance that includes dossier templates, study design expectations, and standard analytical methods. This encourages private investment and helps SMEs to create commercialization plans (IFIF (International Feed Industry Federation), 2013;Making Better Policies for Food Systems, 2021). Because companies that adopt a well-defined regime in one market can more readily map their dossier to nearby markets, regulatory clarity also lowers transaction costs in cross-border markets IFIF (International Feed Industry Federation, 2013). On the other hand, fragmented or unclear regulations discourage smaller companies and increase investor risk, which concentrates innovation in well-capitalized firms and slows sectoral growth (Making Better Policies for Food Systems, 2021). Policy analyses show that jurisdictions with harmonized approval processes and clear guidance tend to see more submissions and wider industry engagement because firms can predict costs and time to market. EFSA's publication of topic-specific guidance and review procedures for food additives is an example of how thorough regulatory guidance reduces procedural ambiguity for applicants.

11.5.2 Consumer trust and willingness to pay for certified botanical-based products

Authorizations, official labeling, and third-party certification based on regulatory standards are examples of regulatory signals that serve as credence cues to lessen customer doubt regarding the safety and quality of products. Consumers' willingness to pay (WTP) for goods they believe to be safer, more natural, or more sustainable is increased by credible certification labels and institutional trust, according to research conducted across the food industry (Ali et al., 2021; Wu et al., 2021). The relationship is indirect but significant for inputs derived from botanicals used in animal production: Certified feed and feed additives (or retailer claims that livestock were fed on a certified botanical-based diet) can be marketed as value-adding attributes (animal welfare, reduced synthetic inputs), allowing producers and retailers to command higher prices if customers believe the claims. WTP is consistently favorable for labeled features (organic, reduced synthetic inputs, animal welfare claims), according to empirical studies; nevertheless, the premium is heavily dependent on label familiarity and credibility or regulatory/third-party support (Ali et al., 2021; Wu et al., 2021). The market can create distinctive, higher-value goods that compensate upstream expenditures in botanical feed advances when authorities offer strict evaluation and clear labeling regulations.

11.5.3 Case study

11.5.3.1 European Union (EFSA+FEEDAP Guidance)

Under Regulation (EC) No 1831/2003, the EFSA is in charge of one of the most organized and scientifically based regulating systems for botanical feed additives in the world. Over the years, the EU's EFSA and the FEEDAP Panel have released comprehensive guidelines on safety and efficacy testing, dossier content, and techniques for characterizing complicated botanical mixes. Although the EU pathway is demanding, structured investment has been made possible by the clarity of EFSA's requirements: applicants are aware of what toxicological, residue and effectiveness studies are required and conduct those studies appropriately. When a dossier satisfies the EFSA requirements, authorization (with specific conditions of use and labeling) creates a strong market signal, enabling companies to market authorized botanical additives across the whole EU single market. For example, after assessing target species safety, consumer exposure, and environmental effects, FFSA recently approved coriander oil (*Coriandrum sativum L.*) as a sensory additive (EFSA, 2017). Similarly, according to published guidance guidelines and harmonized EU evaluation methods, essential oils like thyme, oregano, caraway, and clove oils have been evaluated and approved as flavoring ingredients in feed (EFSA, 2017).

11.5.3.2 China

In recent years, through a strict "'catalog+registration" system, the Ministry of Agriculture and Rural Affairs (MARA) has strengthened regulations on feed and feed additives. Only additives specified in a country's Feed Additive Catalog (or positively registered via application) can be utilized in animal feeds (CIRS Group, 2022). Demand for botanical-based functional additives, such as

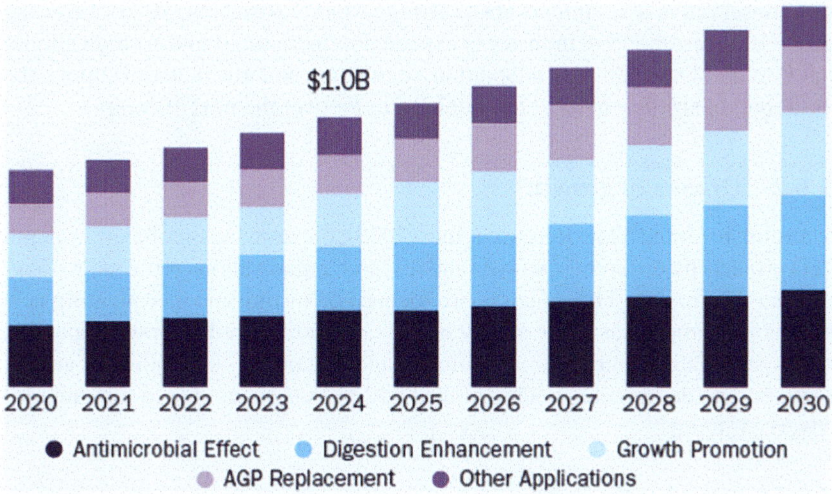

FIGURE 11.1 Phytogenic feed additives market size, by application, 2020–2030 (USD Billion). (*Data from Grand View Research (2025)*).

essential oils, herbal extracts, and phytogenic compounds, has increased after the 2022 prohibition on antibiotic growth boosters, which has led to changes in the market and regulations (Zhou, 2023). A dossier containing compositional information (botanical source, extraction technique), safety tests (toxicity and residues when necessary), stability, and quality control data must be submitted in accordance with MARA regulations (CIRS Group, 2024). As a result, several feed additives have entered the Chinese market, filling the void left by the antibiotic ban and gaining acceptance among livestock producers looking for natural substitutes. This indicates that even in a less standardized regulatory framework, more precise regulation combined with a supportive policy (the antibiotic ban) can create real opportunities for botanical feed solutions.

11.6 Market dynamics and growth potential

11.6.1 Global market size and forecast for phytogenic feed additives

The global PFAs market was estimated to be worth USD 1047.9 million in 2024 and is expected to reach USD 1484.7 million by 2030 at a CAGR of roughly 6.04%, according to a recent market analysis by Future Market Insights (2025);Grand View Research, Inc. (2025), projects a slightly better forecast, from USD 1.11 billion in 2025 to about USD 2.00 billion by 2035, or a long-term CAGR of almost 6.0% (Fig. 11.1).

Some industry analyst projections are even more optimistic, with a projected value of USD 54.19 by 2028 and USD 76.9 billion by 2032 (CAGR ~9.15%). However, this is likely to include a broader definition of "feed phytogenics" than most conservative assessments (Data Bridge Market Research, 2025). Over the next 5–10 years, the PFA industry is expected to increase to mid-to-high single-digit CAGRs. Depending on adaptation, regulation, and raw material supply, the world market size is expected to double or more over the next 10 years.

11.6.2 Drivers of growth

According to Grand View Research, Inc. (2025), the growing number of in-depth R&D projects by different scientists and research organizations is the main driver boosting this market. These factors are focused on testing and investigating new phytogenic compounds for a variety of uses, including antibacterial properties, growth stimulation, antibiotic substitutes, and enhancing the quality of animal products. The demand for botanical feed additives is rising due to a number of convergent trends:

11.6.2.1 Antibiotic bans

Natural alternatives are in high demand because of global regulatory pressure to reduce or eliminate AGPs from livestock feed. Many countries have banned the use of antibiotics in animal feed because of the concerns surrounding antimicrobial resistance. Botanical feed additives offer a promising alternative, providing better health and performance without relying on antibiotics (MarkWide Research, 2025). Fruit pulp, guava, black pepper, turmeric, spices, and other things are being studied in a variety of animal species, including fish, poultry, and rabbits. A study published in the Brazilian Journal of Poultry Sciences in 2024 investigated the impact of *Citrullus colocynthis* (bitter apple or colocynth) fruit pulp (CCFP) in place of AGPs in the daily diet of broilers. Parameters such as growth performance, blood biochemistry, and intestinal microbial count were measured. The result demonstrated that while a diet containing CCFP decreased feed intake, overall growth was enhanced, and a healthy microbial balance was maintained. A Nigerian study investigated the use of turmeric as a potent alternative to antibiotics in poultry (Grand View Research, Inc., 2025). In November 2023, research published in the Translational Animal Science Journal found that turmeric (*Curcuma longa*) can exhibit antimicrobial, antiinflammatory, and immunomodulatory properties and enhance growth performance and gut health while lowering the overall incidence of diseases (Aderemi and Alabi, 2023).

11.6.2.2 Sustainability and environmental concerns

Sustainability has emerged as an essential aspect of modern livestock production, driven by worldwide demand for lessen environmental effects such as greenhouse gas emissions, nutrient pollution, and antibiotic overuse. PFAs are becoming more recognized as environmentally friendly alternatives because they can naturally improve gut health, lower methane emissions, and increase feed efficiency (Patra and Yu, 2012). Research shows that certain essential oils and plant extracts of thyme, oregano, cinnamon, garlic, horseradish, rhubarb, and

Frangula can regulate rumen fermentation, resulting in substantial decreases in methane production (Benchaar and Greathead, 2011). Furthermore, the drive toward antibiotic-free agriculture decreases the ecological concerns associated with antimicrobial residues entering soil and water bodies, promoting broader "One Health" objectives (Van Boeckel et al., 2019). As consumers and retailers push for greener livestock systems, producers increasingly utilize phytogenics to fit with eco-friendly certifications such as carbon reduction assurances and sustainable supply chain demands. Producers are attracted to botanical feed additives that may improve animal health, gut function, and feed conversion while potentially lowering waste, antibiotic dependence and environmental impact as intense livestock production is scrutinized for its effect on the environment (MarkWide Research, 2025).

11.6.2.3 Consumer demand for natural foods

Consumers' preference for natural, clean-label, and minimally processed foods has become one of the biggest market drivers influencing current animal production systems. According to surveys conducted in the EU, the United States, and Asia, consumers are increasingly associating "natural" and antibiotic-free animal products with greater safety, better nutrition, and enhanced animal welfare, which influence purchasing decisions (Asioli et al., 2017). This shift is particularly noticeable in the expansion of premium markets such as organic meat, hormone-free poultry, and antibiotic-free dairy, creating upstream pressure on feed manufacturers to replace synthetic or antibiotic-based feed additives with plant-derived alternatives. Ingredients familiarity and the avoidance of chemical-sounding components considerably influence purchase preferences, leading producers toward botanical, plant-based, and minimally processed formulations (Aschemann-Witzel et al., 2019). Consumers who are concerned about their health typically consider natural and organic foods as complementary to their dietary objectives, which reinforces their choice for goods that are advertised as natural, free from, or derived organically. This indicates a strong and rising WTP for food items that align with natural and clean-label standards, contributing directly to the market expansion of botanical-based feed additives.

11.6.3 Challenges to scalability

Scaling botanical feed additives remains challenging because of high manufacturing cost, fragmented supply chains, and instability of phytochemical compounds during processing. Cost challenges exist because essential oils, plant extracts, and standardized phytochemicals require complex cultivation, extraction, and quality control methods, making them more expensive than the traditional additives (Windisch et al., 2008). Costs also increase because of the necessity for standardized active ingredients and batch-to-batch uniformity, which are necessary for regulatory approval and worldwide market penetration (Zeng et al., 2015). Supply chain limitations further limit scalability. Botanical raw materials generally depend on seasonal harvests, geographic concentration, and changes in phytochemical content influenced by soil type, climate, and agronomic conditions (Hashemi and Davoodi, 2011). Large-scale feed companies that need steady volumes are in danger because of these variables, which result in inconsistent supply streams and variations in ingredient quality. A major technological problem is formulation stability. Many phytogenic compounds such as terpenes, phenolics, and flavonoids are volatile and degrade under heat, light, moisture, or during feed pelleting, resulting in reduced bioactivity and inconsistent efficacy (Zeng et al., 2015). Poor scalability needs microencapsulation and other advanced delivery techniques, which improve performance but drastically raise production costs (Windisch et al., 2008). The combination of high raw material prices, uncertain supply chains, and technological challenges makes it difficult for small and medium-sized companies to scale botanical additives competitively. Therefore, while demand for natural feed solutions continues to increase, overcoming these costs and stability constraints is vital for greater commercial adaptation.

11.7 Comparative analysis of market opportunities by region

11.7.1 Mature markets (EU, USA) versus emerging markets (Asia-Pacific, MENA)

11.7.1.1 European Union (High-science, high barrier market)

The EU has a mature, single-market regulatory framework with a centralized approval process for feed additives under Regulation (EC) No 1831/2003 and scientific risk assessment by EFSA's FEEDAP Panel. This ensures predictable, harmonized market access for authorized products, but the data requirements (such as identification, toxicity, residue, efficacy, and environmental evaluation) and expenses are substantial, benefiting the well-capitalized enterprises and generating a premium market for authorized botanical products. Firms that clear EFSA review obtain EU-wide market access and a significant credibility signal to purchasers (European Commission, 2003).

11.7.1.2 United States (Standard + flexible pathway)

The U.S. mixes federal authority (FDA/CVM) with industry standards (AAFCO). Two popular procedures for botanicals are GRAS determinations or Food-Additive Petitions. AAFCO ingredients recognition and FDA guidance assist in influencing labelling and state implementation. This offers various commercial paths (such as GRAS, AAFCO recognition, FAP) -permitting quicker entrance for items with strong history of use evidence but still imposing substantial scientific and regulatory criteria for novel botanicals (U.S. Food and Drug Administration, 2025).

11.7.1.3 *Asia-Pacific (Fast-growing but fragmented)*

Asia is diverse. China's MARA maintains national feed additive catalogues and has increased approval for plant extracts (catalog updates, periodic arrangements), especially after policies to reduce feed antibiotics, creating a large, rapidly growing market with somewhat faster but complex approval pathways compared to the EU. According to the MARA catalog updates and ASEAN guidelines, other APAC markets (such as Australia and ASEAN countries) vary from mature (Japan with well-established MAFF procedures) to developing (Southeast Asian States, with evolving guidance) markets, creating both enormous volume opportunities and regulatory fragmentation that exporters must manage.

11.7.1.4 *MENA and Africa (Important reliance and capacity gaps)*

Many MENA nations (GCC) and African countries rely substantially on imports of feed and additives and operate import-control frameworks and regional coordination efforts. Exporters can supply goods where domestic production is limited, but they must deal with inconsistent enforcement, different documentation requirements, and halal/traceability expectations, as regulatory capacity, lab infrastructure, and harmonized approval pathways are often weaker, as reported by GCC Guide and AU-IBAR RAFFS.

11.7.2 Role of export–import regulations in shaping competitiveness

Export–import regulations have a significant impact on competitiveness in the botanical feed additive sector by determining which companies can effectively access the global supply chain. Global sanitary and phytosanitary (SPS) regulations, such as the WTO SPS agreement and Codex Alimentarius standards, result in lower trade barriers, easier customs clearance, and quicker acceptance in international markets for exporters who show compliance with recognized international specifications. In contrast, nations with different national requirements such as distinctive testing, document formats, or legalization procedures increase costs and delay market entry (WTO, 1994;Codex Alimentarius Commission, 2023). Market access is further shaped by regulatory recognition and "positive list" systems such as Brazil's MAPA feed additive registration system, China's MARA Feed Additive Catalogue, and Regulation (EC) No 1831/2003. These systems prefer companies that invest in early dossier development, validated analytical methods, and GLP studies because only listed or formally registered additives may be imported or marketed (MAPA, 2019; European Commission, 2003; MARA, 2025). In regions such as the Middle East and Africa, import controls such as halal certification, certificates of analysis, and origin documentation can restrict importers lacking compliance capacity or reward importers with strong traceability and quality systems as described in

GCC regulatory guidance and AU-IBAR feed and fodder system frameworks. Overall, export–import regulations transform regulatory compliance into a competitive advantage: companies that can satisfy multijurisdictional standards (such as those in the United States, China, and the EU) are preferred, while smaller or poorly documented companies face delays, rejected shipments, and increased logistical risks.

11.7.3 Potential for harmonization of standards

The potential for harmonized standards in botanical feed additive regulation is increasing, supported by international anchors such as the Codex Alimentarius, JECFA scientific evaluations, and the WTO SPS agreement, which provide widely accepted risk-assessment frameworks and legal principles that countries can adopt to align national rules (WTO, 1994;FAO/WHO, 2022; Codex Alimentarius Commission, 2023). Although progress is still uneven and dependent on regional politics and technical infrastructure, regional initiatives such as the GCC's joint import-control guidelines and AU-IBAR's RAFFS program to build African feed-regulatory capacity demonstrate practical approaches to reduce duplication and improve mutual recognition (GCC Standardization Organization, 2014; AU-IBAR, 2022). Major challenges still exist, though; different legal approaches such as the EU's precautionary model against risk-based systems elsewhere limit regulatory convergence, and many nations lack laboratory capacity and risk-assessment competence to adopt harmonized standards (European Commission, 2003; AU-IBAR, 2022). Commercial interest may also resist harmonization if it raises compliance costs or exposes domestic sectors to more competitive imports, and WTO rules still allow scientifically justifiable national variances, preserving room for variety (WTO, 1994). Realistic harmonization pathways include mutual recognition agreements for analytical methods, wider adoption of Codex to avoid duplicative national assessments, and investment in regional testing laboratories to allow smaller economies to share scientific infrastructure and adopt more consistent standards (FAO/WHO, 2022).

11.8 Toward a harmonized global approach

International harmonization of botanical feed additive regulations demands alignment in definitions, safety testing techniques, and claims substantiation standards, as present global variations produce duplicated testing, inconsistent classifications, and challenges to trade. Definitions significantly range, such as between the EU's formal additive categories and the U.S. GRAS/AAFCO system, leading Codex to call for unified terminology to promote transparent, comparable risk assessment (Codex Alimentarius Commission, 2021). Safety testing frameworks differ as well. While U.S. and Chinese systems rely more on literature reviews or local verification, EFSA requires thorough tolerance, toxicology, and environmental studies, which lead to redundant or contradictory data requirements (EFSA, 2017; MARA, 2025). With internationally recognized risk-assessment principles, containment limits, and feed safety guidelines that many national authorities already use as a scientific benchmark, the Codex Alimentarius Commission offers the strongest foundation currently available for such harmonization (Codex Alimentarius Commission, 2021). In parallel, worldwide feed associations—including the International Feed Industry Federation (IFIF)-promote standardized quality assurance schemes and urge mutual recognition of safety data to simplify cross-border market access (International Feed Industry Federation, 2022). Further opportunities arise from integrating these initiatives into a larger One Health vision, since coordinated, cross-sector collaboration can support antimicrobial resistance mitigation, enhance animal welfare, and promote environmental and human health outcomes (FAO, WHO, & WOAH, 2022).

11.9 Conclusion

Global interest in botanical feed additives continues to rise, although misalignment between scientific data, regulatory constraints, and market expectations remains a major obstacle to expansion. For farmers and manufacturers, this misalignment raises developmental costs, hinders or delays market entry, and causes uncertainty. Uncertainty and development costs for industry are increased by region-specific approval, inconsistent testing procedures, and variability in botanical composition. However, increased demand for natural, sustainable livestock products, along with international momentum toward antibacterial reduction, creates tremendous prospects for harmonized progress. Advances in analytical chemistry, metabolomics, and standardized bioassays can increase evidence production, enabling regulators to adopt more harmonized, science-based approval processes. Policymakers can drive innovation by harmonizing definitions, safety criteria, and claim standards through Codex and One Health frameworks, while researchers advance standardized analytical techniques and field-relevant efficacy studies. For industries, scaling adaptation across developed and emerging markets will depend on increasing quality control, transparent sourcing, and active participation in global standard setting.

References

AAFCO, 2024. FDA's Final Guidance for AAFCO-Defined Ingredients. Association of American Feed Control Officials, Champaign, IL.

Aderemi, F.A., Alabi, O.M., 2023. Turmeric (Curcuma longa): an alternative to antibiotics in poultry nutrition. Transl. Anim. Sci. 7 (1). doi:10.1093/tas/txad133.

Ali, B.M., Ang, F., van der Fels-Klerx, H.J., Georgantzis, N., 2021. Consumer willingness to pay for plant-based foods produced using microbial applications to replace synthetic chemical inputs. PLoS One 16 (12), e0260488. doi:10.1371/journal.pone.0260488.

Aschemann-Witzel, J., Varela, P., Peschel, A.O., 2019. Consumers' categorization of food ingredients: do consumers perceive them as 'clean label' producers expect? An exploration with projective mapping. Food Qual. Preference 71, 117–128. doi:10.1016/j.foodqual.2018.06.003.

Asioli, D., Aschemann-Witzel, J., Caputo, V., Vecchio, R., Annunziata, A., Næs, T., Varela, P., 2017. Making sense of the "clean label" trends: a review of consumer food choice behavior and discussion of industry implications. Food Res. Int. 99 (Pt 1), 58–71. doi:10.1016/j.foodres.2017.07.022.

Babinszky, L., Oliveira, J., Santos, E.M., Payan-Carreira, R., 2021. Advanced Studies in the 21st Century Animal Nutrition. BoD–Books on Demand, Norderstedt, Germany.

Benchaar, C., Greathead, H., 2011. Essential oils and opportunities to mitigate enteric methane emissions from ruminants. Anim. Feed Sci. Technol. 166-167, 338–355. doi:10.1016/j.anifeedsci.2011.04.024.

Chauhan, B., Kumar, G., Kalam, N., Ansari, S.H., 2013. Current concepts and prospects of herbal nutraceutical: a review. J. Adv. Pharm. Technol. Res. 4 (1), 4–8. doi:10.4103/2231-4040.107494.

African Union–InterAfrican Bureau for Animal Resources (AU-IBAR), 2022. Resilient African Feed and Fodder Systems (RAFFS) Project. Available at: https://www.au-ibar.org/projects/resilient-african-feed-and-fodder-systems (accessed 26 April 2026).

CIRS Group, 2022. How to Search Positive Lists of Feed Additives and Feed Ingredients in China. CIRS Feed-Regulation Guidance. https://www.cirs-group.com/en/agrochemicals/how-to-search-positive-list-of-feed-additives-in-china (accessed 26 April 2026).

CIRS Group, 2024. Official consultation service for feed ingredient and feed additive approval in China. CIRS Feed Regulation Guidance. https://www.cirs-group.com/en/feed/official-consultation-service-for-feed-ingredient-and-feed-additive-approval-in-china (accessed 28 April 2026).

Codex Alimentarius Commission. (2021). Procedural Manual (28th ed.). Food and Agriculture Organization & World Health Organization.

Codex Alimentarius Commission. (2023). Codex Standards and Guidelines for Feed Safety. Food and Agriculture Organization & World Health Organization.

DAHD. (2023). Additional List of Animal Feed Additives/Premixes/Supplements for Import into India. Department of Animal Husbandry & Dairying.

Data Bridge Market Research. (2025). Global Feed Phytogenics Market Size, Share & Trends Analysis Report.

EFSA, FEEDAP Panel., 2017. Guidance on the assessment of the safety of feed additives for the target species. EFSA Journal 15 (10), e5021. https://doi.org/10.2903/j.efsa.2017.5021.

EFSA FEEDAP Panel, 2024. Guidance on the assessment of the efficacy of feed additives. EFSA J. 22 (7), e8856.

Bampidis, V., Azimonti, G., Bastos, M.D.L., Christensen, H., Durjava, M., Galobart, J.EFSA Panel on Additives and Products or Substances used in Animal Feed (FEEDAP), 2023. Guidance on the assessment of the safety of feed additives for the users. EFSA J. 21 (12), e8469.

Bampidis, V., Azimonti, G., Bastos, M.D.L., Christensen, H., Dusemund, B., Brozzi, R.EFSA Panel on Additives and Products or Substances used in Animal Feed (FEEDAP), 2021. Safety and efficacy of the feed additive consisting of Bacillus licheniformis DSM 28710 (B-Act®) for laying hens, minor poultry species for laying, poultry species for breeding purposes and ornamental birds (HuvePharma NV). EFSA J. 19 (3), e06449.

EMA (European Medicines Agency), 2020. Guideline on the Quality of Herbal Medicinal Products. European Medicines Agency, Amsterdam, Netherlands.

European Commission, 2003. Regulation (EC) No 1831/2003 of the European Parliament and of the Council on additives for use in animal nutrition. Off. J. Eur. Union L268, 29–43. https://eur-lex.europa.eu/legal-content/EN/TXT/?uri=CELEX:32003R1831.

FAO, WHO, & WOAH, 2022. One Health Joint Action Plan (2022–2026). Food and Agriculture Organization of the United Nations, Rome, Italy.

FAO/WHO, 2022. Joint FAO/WHO Expert Committee on Food Additives (JECFA) Evaluations: Principles and Methods. Food and Agriculture Organization of the United Nations, Rome, Italy.

European Medicines Agency, 2024. Final programming document 2024–2026 (EMA/426707/2023). Amsterdam: European Medicines Agency. https://www.ema.europa.eu/en/documents/corporate-report/final-programming-document-2024-2026_en.pdf (accessed 2 May 2026).

U.S. Food and Drug Administration, Center for Veterinary Medicine, 2024. CVM GFI #293: FDA enforcement policy for AAFCO-defined animal feed ingredients (Guidance for Industry No. 293). Silver Spring, MD: U.S. Food and Drug Administration. https://www.fda.gov/regulatory-information/search-fda-guidance-documents/cvm-gfi-293-fda-enforcement-policy-aafco-defined-animal-feed-ingredients (accessed 20 May 2026).

Fernández Miyakawa, M.E., Casanova, N.A., Kogut, M.H., 2024. How did antibiotic growth promoters increase growth and feed efficiency in poultry? Poult. Sci. 103 (2). doi:10.1016/j.psj.2023.103278.

Food and Agriculture Organization of the United Nations & African Union, 2007. African Model. Legislation for the Control of Transboundary Animal Diseases. Food and Agriculture Organization of the United Nations, Rome.

Food Safety and Standards Authority of India, 2019. Direction: Commercial Feeds/Feed Materials for Food-Producing Animals to Comply with Relevant BIS Standards. Food Safety and Standards Authority of India, New Delhi https://fssai.gov.in/upload/advisories/2019/12/5def78627ba0eDirection_Feed_Animal_BIS_10_12_2019.pdf.

Food Safety and Standards Authority of India. (2021). Extension of timeline for compliance with direction dated 27 January 2020 regarding compliance of commercial feeds/feed materials intended for meat and milk producing animals with relevant BIS standards. New Delhi: Food Safety and Standards Authority of India. https://fssai.gov.in/upload/advisories/2021/07/60eeaf0a1cb43Extension_TimeLine_Meat_milk_14_07_2021.pdf

Fortune Business Insights. (2024/2025). Feed Additives Market Size, Share & Trends Report. Pune: Fortune Business Insights. https://www.fortunebusinessinsights.com/feed-additives-market-104345

Future Market Insights. (2025). Phytogenic Feed Additives Market Size & Share Forecast 2025–2035. https://www.futuremarketinsights.com/reports/feed-phytogenic-market

Gadde, U., Kim, W.H., Oh, S.T., Lillehoj, H.S., 2017. Alternatives to antibiotics for maximizing growth performance and feed efficiency in poultry: A review. Anim. Health Res. Rev. 18 (1), 26–45. doi:10.1017/S1466252316000207.

GCC Standardization Organization, 2014. GCC Guide for Control on Imported Foods. Gulf Cooperation Council, Riyadh https://www.gso.org.sa/wp-content/uploads/2014/09/GCC-Guide-for-Control-on-Imported-Foods.pdf.

Grand View Research, Inc. (2025). Phytogenic Feed Additives Market to Reach USD 1.48 Billion by 2030. https://www.grandviewresearch.com/press-release/global-phytogenic-feed-additives-market

Grand View Research. Phytogenic Feed Additives Market Size | Industry Report (2024). San Francisco, CA. https://www.grandviewresearch.com/industry-analysis/phytogenic-feed-additives-market-report

Greathead, H., 2003. Plants and plant extracts for improving animal productivity. Proc. Nutr. Soc. 62 (2), 279–290. doi:10.1079/pns2002197.

Hallagan, J.B., Hall, R.L., Drake, J., 2020. The GRAS provision-the FEMA GRAS program and the safety and regulation of flavors in the United States. Food Chem. Toxicol. 138, 111236.

Hashemi, S.R., Davoodi, H., 2011. Herbal plants and their derivatives as growth and health promoters in animal nutrition. Vet. Res. Commun. 35 (3), 169–180. doi:10.1007/s11259-010-9458-2.

IFIF (International Feed Industry Federation), 2013. Comparison of Approval Process and Risk-Assessment Procedures for Feed Ingredients: Canada, European Union, and United States (Comparison Project). IFIF, Brussels https://ifif.org/wp-content/uploads/2018/06/IFIF-Comparison-Project.pdf.

International Feed Industry Federation, 2022. Annual report 2022: Global Feed Industry Priorities and Harmonization Efforts. International Feed Industry Federation, Brussels.

Kalra, E.K., 2003. Nutraceutical-definition and introduction. AAPS PharmSci 5 (3), 27–28. doi:10.1208/ps050325.

Making Better Policies for Food Systems. OECD Publishing, Paris (2021), doi: 10.1787/ddfba4de-en.

MAPA, 2019. Regulatory Requirements for Animal Feed Additives. Ministry of Agriculture, Livestock and Food Supply (Brazil), Brasília.

MARA, 2025. Catalogue of Feed Additives and Feed Additive Products. Ministry of Agriculture and Rural Affairs of China, Beijing.

MarkWide Research, 2025. Phytogenic Feed Additives Market Analysis: Industry Size, Share, Research Report, Insights, COVID-19 Impact, Statistics, Trends, Growth and Forecast 2025–2034. https://markwideresearch.com/phytogenic-feed-additives-market/ (accessed 2 April 2026).

U.S. Department of Agriculture, Foreign Agricultural Service, 2023. Guidelines for Applications for Plant-Extracted Feed Additives (GAIN Report No. CH2023-0029). Beijing: USDA Foreign Agricultural Service. https://apps.fas.usda.gov/newgainapi/api/Report/DownloadReportByFileName?fileName=Guidelines+for+Applications+for+Plant-Extracted+Feed+Additives_Beijing_China+-+People%27s+Republic+of_CH2023-0029.pdf (accessed 2 May 2026).

Patra, A.K., Yu, Z., 2012. Effects of essential oils on methane production and fermentation by, and abundance and diversity of, rumen microbial populations. Appl. Environ. Microbiol. 78 (12), 4271–4280. doi:10.1128/AEM.00309-12United States.

Rastogi, S., Pandey, M.K., Prakash, J., Sharma, A., Singh, G.N., 2015. Veterinary herbal medicines in India. Pharmacogn. Rev. 9 (18), 155–163. doi:10.4103/0973-7847.162140.

Pandey, A.K., Kumar, P. and Saxena, M.J. (2019). 'Feed additives in animal health', in Gupta, R.C., Srivastava, A. and Lall, R. (eds.) Nutraceuticals in Veterinary Medicine. Cham: Springer International Publishing, pp. 345–362. https://doi.org/10.1007/978-3-030-04624-8_23 (Accessed: 26 May 2026).

Rychen, G., Aquilina, G., Azimonti, G., Bampidis, V., Bastos, M.L, Bories, G., Chesson, A., Cocconcelli, P.S, Flachowsky, G., Gropp, J.ü, Kolar, B., Kouba, M., López-Alonso, M., Puente, S.Ló, Mantovani, A., Mayo, B., Ramos, F., Saarela, M., Villa, R.E, Wallace, R.J, Wester, P., Anguita, M., Galobart, J., Innocenti, M.L, Martino, L., 2017. Guidance on the assessment of the safety of feed additives for the target species. EFSA J. 15 (10). doi:10.2903/j.efsa.2017.5021.

Shahbakht, R.M., Nomi, Z.A., Kabeer, S.W., Anwar, M.Z., Shahid, S., Imtiaz, B., Younus, G., Muneer, M.H., Usman, M., Rajput, S.A., 2024. The use of phytogenic feed additives in animal nutrition in the Pakistan scenario', in Abbas. In: Akhtar, T., Asrar, R., Khan, A.M.A., Saeed, Z. (Eds.), Complementary and Alternative Medicine: Feed Additives. Unique Scientific Publishers, Faisalabad, pp. 326–334.

Steiner, T. and Syed, B. (2015). 'Phytogenic feed additives in animal nutrition', in Máthé, Á. (ed.) Medicinal and Aromatic Plants of the World: Scientific, Production, Commercial and Utilization Aspects. Dordrecht: Springer Netherlands, pp. 403–423. https://doi.org/10.1007/978-94-017-9810-5_20

Tedesco, D.E.A., Cagnardi, P., 2019. Regulatory guidelines for nutraceuticals in the European Union. In: Nutraceuticals in Veterinary Medicine. Springer International Publishing, Cham, pp. 793–805.

Trivedi, A., Shukla, P., Mishra, R., Maurya, A., Srivastava, M., 2025. Insight of herbal veterinary medicines: a review article. J. Drug Discov. Health Sci. 2 (01), 47–52. doi:10.21590/jddhs.02.01.07.

U.S. Department of Agriculture, Foreign Agricultural Service. (2018). Food and Agricultural Import Regulations and Standards — Narrative: Argentina. Buenos Aires.

U.S. Food & Drug Administration. Botanical Drug Development: Guidance for Industry. (CDER guidance; 2004/updated pages). Silver Spring, MD, USA

U.S. Food and Drug Administration, Center for Veterinary Medicine. (2024). CVM GFI #293 — FDA Enforcement Policy for AAFCO-Defined Animal Feed Ingredients. Silver Spring, MD, USA

Van Boeckel, T.P., Pires, J., Silvester, R., Zhao, C., Song, J., Criscuolo, N.G., Laxminarayan, R., 2019. Global trends in antimicrobial resistance in animals in low-and middle-income countries. Science 365 (6459), eaaw1944.

Windisch, W., Schedle, K., Plitzner, C., Kroismayr, A., 2008. Use of phytogenic products as feed additives for swine and poultry1. J. Anim. Sci. 86 (suppl_14), E140. doi:10.2527/jas.2007-0459.

U.S. Food and Drug Administration. (2025). Importing animal food. Silver Spring, MD: U.S. Food and Drug Administration. https://www.fda.gov/animal-veterinary/import-exports/importing-animal-food (Accessed: 26 April 2026).

WTO, 1994. Agreement on the Application of Sanitary and Phytosanitary Measures (SPS Agreement). World Trade Organization, Geneva, Switzerland.

Wu, W., Zhang, A., van Klinken, R.D., Schrobback, P., Muller, J.M., 2021. Consumer trust in food and the food system: a critical review. Foods 10 (10). doi:10.3390/foods10102490.

Zeng, Z., Zhang, S., Wang, H., 2015. Xiangshu Piao, Essential oil and aromatic plants as feed additives in non-ruminant nutrition: a review. J. Anim. Sci. Biotechnol. 6 (1). doi:10.1186/s40104-015-0004-5.

Zhou, J.Y. (2023). 'Domestic feed additives replacing antibiotics', China Daily. https://global.chinadaily.com.cn/a/202303/23/WS641bbce4a31057c47ebb61a1.html (Accessed: 26 April 2026).

Chapter 12

Challenges in functional phytofeed implementation

Mohammad Mehedi Hasan Khan[a], Abu Darda[a], Mohamed E. Abd El-Hack[b] and Shamima Sultana[a]

[a] *Department of Biochemistry and Chemistry, Sylhet Agricultural University, Sylhet, Bangladesh,* [b] *Department of Industrial Pharmacy, College of Pharmaceutical Sciences and Drug Manufacturing, Misr University for Science and Technology (MUST), Giza, Egypt*

12.1 Introduction

The livestock industry undergoes a burning urgency to establish sustainable, effective production systems while meeting the rising world demand for animal protein. Antibiotics persist in the body, which later may pose a threat to animals. Therefore since 2006, the European Union has banned the use of antibiotic growth promoters, along with a similar tendency toward this regulatory shift globally, phytogenic feed additives (PFAs) have offered an attractive option in improving animal welfare and performance by supplying beneficial bioactive ingredients from plants. These phyto-feed–based systems have significant benefits, such as antimicrobial activity against infectious bacteria, antioxidant properties for better meat quality, immunomodulatory effects for increasing disease resistance, and environmental advantages in terms of lower greenhouse gas emissions (Karásková et al., 2015; Nastoh et al., 2024; Biswas and Kim, 2025). Different types of spices, herbs, and essential oils are attracting the attention of researchers, as they help animal digestion and are beneficial for their health. Herbs worked as antioxidants and antibacterials. Essential oils suppress bacteria and fungus and enhance rumen fermentation. Plant extract increases digestive capacity and releases gases (Odhaib et al., 2021).

Scientific research progressively demonstrates that PFAs, including polyphenols (phenolic compounds), essential oils, saponins, isoprenoids, and alkaloids, can increase energy conversion by as much as 4%, decrease ammonia emissions by 40%, and support the integrity of the intestinal barrier while improving growth performance. In addition, these natural substitutes increase consumer concerns about food safety, environmental conservation, and animal welfare, which may earn a premium in the future international markets where quality is in great demand (Latek et al., 2022). Effective partnership between researchers

Plant-Derived Feed Additives and Animal-Based Foods: Linking Animal Health, Product Quality, and Consumer Perceptions. DOI: https://doi.org/10.1016/B978-0-443-45446-2.00005-3

and stakeholders is a key conditions that determine the long-term viability of phytotechnology, which ethnologically examines the effectiveness of solutions to large-scale pollution problems, including social and financial sustainability. However, there is a big challenge between research and the implementation of phytoremediation.

Despite this significant evidence base and market potential, the deployment of functional phytofeed is limited, with market penetration far lower than anticipated over a decade ago. The gap between scientific potential and commercial realization is a consequence of complex, interrelated challenges across the various sectors of the feed value chain. These hurdles are not just technical challenges but reflect systemic problems that need to be addressed in an integrated manner in scientific, regulatory, economic and social dimensions (Yildiz et al., 2024). As a result, phytofeed can be established as a cost-effective safe alternative not only in research but also commercially. The feed value system of the European Union is not protein self-sufficient. Legumes (soybean, pea, faba bean, and lentil) are the source of phytofeed (Balázs et al., 2021).

The objectives of this chapter are threefold: First, to systematically analyze the major challenges hindering widespread implementation of functional phytofeed systems; second, to analyze the potential key connections between specific challenges; and third, to propose evidence-based paths to overcome these barriers. This analysis can be used as a guideline for stakeholders seeking to develop sustainable phytogenic livestock nutrition systems.

12.2 Scientific and biological challenges

12.2.1 Variability in phytochemical composition

The inherent heterogeneity in phytochemical composition represents one of the most fundamental challenges. It limits the predictable efficacy of functional phytofeed additives. This variability manifests across multiple dimensions by creating significant difficulties for standardization and consistent biological activity (Al-Mijalli et al., 2022).

Plant origin and genetic factors: Different plants grow in different agroecological zones, and as a result, secondary metabolites, such as alkaloids, tannins, and saponins, differ in species. Research demonstrates that phytochemical profiles can vary significantly even within the same species because of genetic diversity. The comparative study on the essential oils of Mentha piperita from various Moroccan regions showed a considerable compositional variability; e.g., carvone contents as high as 70.25% in oil from the region of Azrou were significantly lower elsewhere. Likewise, extracts of *Agrimonia eupatoria* demonstrated that the phytochemical profile is more related to the species scale than to the genus level, where concentrations in secondary metabolites may differ up to 300% between closely related taxa (Loučková et al., 2024).

Harvest timing and seasonal variations: Temporal regulation of phytochemical production adds complexity because the concentration of bioactive compounds changes during plant development. Yields and composition of essential oils may vary 50-200% according to time of harvest, with ratios of monoterpenes to sesquiterpenes changing greatly between early vegetative and flowering stages. Environmental conditions also affect the genetic variability of the secondary metabolite biosynthesis pathways, which are also influenced by climate (drought stress, temperature fluctuations, photoperiod length) (El-Feky and AbdelRahman, 2025). With harvest time, fiber concentration (Acid detergent fiber (ADF), Acid detergent lignin (ADL), and Neutral detergent fiber (NDF)) increases, and digestibility and relative feed value (RFV) decrease. Therefore cultivar selection, harvest stage determination, and processing methods are essential to ensure the production and quality of sorghum fodder.

Processing method dependencies: The postharvest processing may play an important role in the retention and bioactivity of phytochemicals. Different processing techniques, such as hydrodistillation, steam distillation, and solvent extraction, provide different compound profiles from the same plant material. They also vary in volatile components losses (15%–60% depending on processing parameters). Similarly, drying methods strongly influence phytochemical stability, with freeze-drying resulting in approximately 85%–95% of original bioactivity compared to only 40%–60% retention when conventional heat drying is used (Sasidharan et al., 2011). Chopping, drying, and other processing methods also play a role in determining the digestibility and intake potential of forage .

Inconsistency in bioactivity: The compositional variation leads to variable biological actions. Antimicrobial efficacy from in vitro studies can vary 10-fold between the batches of the same phytogenic extract, while antioxidant activity exhibits identical variability patterns. This inconsistency lowers confidence in phytogenic products and complicates dose-response relationships, which are critical for regulatory approval as well as commercial use (Loučková et al., 2024). In addition, moisture content, processing methods, and the type of feed used also affect the efficacy and digestibility of bioactive compounds (Fung et al., 2018). This variability indicates that inconsistent bioactivity of bioactive compounds is a significant challenge in the production of animal feed from different sources, which directly affects feed formulation and livestock performance. Therefore it is essential to take this variability into account when feed for livestock nutrition.

12.2.2 Complexity of mechanisms

The biological complexity underlying the action of PFAs creates significant challenges for understanding, predicting, and optimizing their effects in livestock systems (Latek et al., 2022).

Multiple overlapping pathways: Phytogenic compounds exert their activities on different biological targets, such as antimicrobial effectiveness on pathogenic bacteria, immunomodulation through cytokine modulation, antioxidant activities by free radical scavenging, and metabolic effects in support of nutrient utilization efficiency. Essential oils can act simultaneously in antimicrobial, antioxidant, and immunomodulatory pathways. For example, an essential oil can destroy microbial cell walls (antimicrobial), reduce oxidative stress (antioxidant), and balance host immune responses (immunomodulatory). This multifaceted overlapping mechanism makes essential oils complex but effective (Lorenzo et al., 2014). Thymol and Carvacrol monoterpenes are found in plants such as oregano, thyme, and mint and have potent antioxidant effects. Thymol or carvacrol-containing essential oils, for instance, arrest bacterial membrane integrity but simultaneously activate the host antioxidant enzyme systems and alter the microbial gut flora composition (Sathishkumar et al., 2025).

Synergistic and antagonistic interactions: Because of their multi-component character, phytogenic extracts carry out a complex network of interactions among different compounds with either synergistic or antagonistic effects on certain biological activities. Studies of combined essential oil combinations have shown that some binary and tertiary formulations can increase the antimicrobial potency by up to 500% compared with their single component, whereas other combinations show reduced activity. These interactions are dose-related and species-dependent, which makes any attempts to optimize even more complex (El-Zaiat et al., 2023). Some bioactive compounds in essential oils affect the host immune system. They can regulate both proinflammatory and antiinflammatory responses, which helps in both infection control and tissue protection (Bhavaniramya et al., 2019).

Difficulty in establishing causal effects: The multitargeting mechanisms of action make it more challenging to establish causal relationships between specific phytochemical components and observed biological effects. This mechanistic uncertainty prevents the development of formulated standardized products. It also complicates the regulatory assessments that require clear cause-and-effect relationships (Zollers et al., 2024). The same essential oil is used as a dietary supplement in one country and as an herbal medicine in another, because regulatory policies differ between countries. According to clinical trial data, adverse effects of herbal products have been confirmed in many cases (Lorenzo et al., 2014). So, comparative identification and scientific decision-making have become important for regulatory agencies.

12.2.3 Animal-to-animal variability

Another critical challenge is the considerable variability between animals in responding to the PFAs, limiting uniform efficacy and commercial applicability.

Species, breed, and age differences: The response patterns differ remarkably between animal species, and ruminants exhibit more or less sensitivity than monogastric animals because of the microbial metabolism of bioactive substances in the rumen. Ruminants (cattle, sheep, and goats) may differ in grazing patterns, leader-follower behavior, and food neophobia (Sibbald et al., 2009).

Foraging strategies, diet selection, and toxin-handling capacity vary across species; for example, goats are more capable of browsing, while cows are primarily restricted to grazing (Bergvall et al., 2011).

Between breeds in a species, there are genetic variations that affect drug metabolism pathways, intestinal physiology, or immune response. Anatomical and physiological differences, such as age-dependent alterations in tissue architecture or function of digestive enzymes; altered hormone and immune responses; gut barrier integrity; and metabolic capacity, are further factors for variation, with young animals sometimes being more responsive to phytogenic interventions (Semwogerere et al., 2023). Breed-specific morphological and physiological traits (e.g., body size, rumen capacity, and growth potential) influence feeding behavior (Løvendahl and Munksgaard, 2016). Young ruminants (calves, lambs, and kids) are initially dependent on milk, then become accustomed to solid feed through social learning (Miller-Cushon and DeVries, 2015). Weaning age, time to start grain intake, and novel feed acceptance—all depend on age (Neave et al., 2018).

Health status dependencies: Animals' baseline health status has a significant impact on the PFA (PFAs) efficacy. Stressed (e.g., disease-challenged) animals typically respond to immunomodulatory or antioxidant compounds, but non-stressed (healthy) ones may fail to exhibit appreciable responses. This different responsiveness makes the product's economic validation in a commercial context more difficult (Biswas and Kim, 2025). Stress, fearfulness, food neophobia, and coping style these all influence feeding behavior (Van Reenen et al., 2005). Sick or weak animals may lose out in social competition, and feed intake decreases, resulting in reduced growth potential (Llonch et al., 2018). Cortisol response, immune status, handling stress—all directly linked with feeding pattern, which corresponds to health condition (Daniels et al., 2007).

Gut microbiota diversity: Most notably, variations in gut microbiota composition between individuals cause large differences in the metabolism and bioavailability of phytochemicals. The colon microflora is an important source of bioactive compound transformation, and also certain bacterial populations may potentially increase or promote reduction in phytochemical bioactivity via metabolic conversion routes. Studies showed that animals having different microbiota profiles can differ three- to fivefold in their response to the same phytogenic treatment. This clearly illustrates the importance of precision nutrition solutions based on individual microecosystem signatures (Abeltino et al., 2024).

12.3 Technological and formulation challenges

Livestock feed formulation is crucial for production efficiency and cost reduction, but balancing nutrient requirements accurately across species, breeds, ages, and health conditions is a major challenge. Climate change (drought, heat stress, water scarcity) is reducing feed availability and utilization, while feed costs account for 60%–80% of the total cost of livestock production. Alternative proteins (plant, insect meal, etc.) are needed but are difficult to integrate into the formulation. Traditional methods (Pearson's square, least-cost formulation, etc.) are cost-efficient but often create nutrient imbalances. Advanced techniques (linear/non-linear optimization, AI, DSS) are more accurate but have limitations in terms of data requirements, computational cost, and field-level adoption. Sensor-based precision feeding and AI may be future solutions, but digital infrastructure and farmer education remain major challenges.

12.3.1 Extraction and standardization

Extraction is an important step in livestock feed, as it allows the detection of various harmful substances such as mycotoxins, pesticide residues, and veterinary drug residues in the feed. Studies have shown that the complex matrix of feed materials (e.g., barley, maize, soy, wheat bran, and silage) makes it difficult to extract the target analyte accurately, for which special solvent systems are used. On the other hand, since feed formulation varies by species and age, it is not possible to standardize it globally. Therefore specific validation guidelines are followed in the feed industry of different countries. For example, the German accreditation body (DAkkS) has proposed multiresidue method validation for livestock feed, where satisfactory performance of at least 75% of the analytes must be ensured. As a result, extraction helps to separate toxins and drug residues in livestock feed, and standardization ensures the safety of livestock and the security of the food chain by managing the formulation variability of feed (Steiner et al., 2020). The production of such standardized, high-quality PFAs requires the most advanced extraction and standardization technologies, but these are still difficult and costly to apply in large quantities at commercial scale (Sasidharan et al., 2011).

Reproducible active compounds: Reproducible analysis is crucial when using new materials in animal feed management. Achieving batch-to-batch consistent concentrations of bioactive compounds is a basic technical challenge. Conventional extraction processes are also subject to variable yields because of sensitivity to process parameters such as temperature, pressure, composition of solvent, and length of extraction time. Modern extraction techniques such as supercritical fluid extraction, ultrasound, and microwave-assisted methods provide better reproducibility but demand large capital funding and very specialized, skilled personnel support (Ahamed et al., 2025).

Analytical standardization challenges: The complex multicomponent phytogenic extracts interfere with the analytical standardization. Although methods of analysis such as gas chromatography-mass spectrometry (GC-MS) and high-performance liquid chromatography (HPLC) provide detailed compositional determination. But establishing relevant bioactivity markers is still challenging. Several regulatory frameworks require standardization of certain marker compounds, which implies without a direct correlation to biological activity, which disconnects analytical specifications from functional performance (Sasidharan et al., 2011).

Plant cell cultures are currently being used to produce recombinant proteins because they can undergo complex post-translational modifications and are safe for humans, which is an advantage over microbes or whole plants. However, there are challenges in terms of purity, stability, and batch-to-batch consistency.

Purity: Secreted proteins can be easily purified by culture, but intracellular proteins require cell disruption, which increases the risk of contamination. Protein purity can be increased using affinity chromatography and tag-based purification methods.

Stability: Secreted proteins are susceptible to proteolysis, aggregation, and degradation. Stabilizing agents such as Polyethylene glycol (PEG), Bovine serum albumin (BSA), Dimethyl sulfoxide (DMSO), and methods of targeting proteins to the endoplasmic reticulum help to increase stability. Batch-to-Batch consistency: Differences in cell growth, genetic drift, and culture conditions can cause variations in protein production. Cryopreservation and proper cell banking are essential to ensure repeatability (Hellwig et al., 2004).

Quality control infrastructure: Implementing proper comprehensive quality control (QC) systems for PFAs often requires sophisticated, high-end analytical capabilities. And that is often beyond the scope of most of the traditional feed manufacturers. The requirements for gaseous phase detection in real-time monitoring of volatile compounds, stability testing in different environmental storage conditions, and bioactivity assay make it relatively complicated and a cost burden (Bagade et al., 2022).

12.3.2 Bioavailability and delivery

Poor bioavailability represents one of the most important technological barriers preventing photobiotic feed additives from attaining commercial success.

Solubility and stability limitations: Many bioactive phytochemicals are insoluble in water and chemically unstable during conventional feed processing and storage. In particular, essential oils are very volatile and are prone to oxidation, with activity loss of 30%–70% often seen when exposed to high temperatures (80–90°C) during pelleting. Similar problems are also faced by polyphenolic compounds with degradation rates that rapidly increase upon exposure to light, oxygen, and alkaline conditions (Sharma et al., 2019).

First-pass metabolism and efflux: Even if phytochemicals are effectively delivered to the gastrointestinal tract, a significant proportion may experience rapid metabolism and active efflux, restricting systemic or local tissue levels. The hepatic first-pass effect can remove 60%–90% of absorbed compounds before they reach the target tissues, and efflux transporters, including P-glycoprotein, expel compounds out of intestinal cells (Hu et al., 2023).

Advances in encapsulation technologies: Technological developments in microencapsulation and nano-encapsulation have brought valuable measures to overcome bioavailability. Lipid-based matrix entrapping systems that are employed to encapsulate volatile agents may protect such volatile agents against influences during feeding processing and can permit controlled release in the GI tract. Enhanced formulations using cyclodextrins, liposomes, and polymer-based delivery systems have been shown to increase the bioavailability of multiple phytogenic compounds up to two- to fivefold (Nutrition, 2021).

Nanotechnology applications: Emerging trends in nanotechnology, such as nanoparticles, nanoemulsions, and nanostructured lipid carriers, have given a potential proposition regarding the improved phytochemical delivery. They can increase solubility 10–100 times and also serve as delivery systems for targeted release to tissues and offer them protection from degradation. Nevertheless, the regulatory uncertainty of nanomaterials in feed applications is a bottleneck to their incorporation (Usman et al., 2025). Nanotechnology is used for the delivery of nutrients or therapeutics to livestock. Gibbs et al. 1999 showed that the spicy chilling method encapsulates fat-soluble minerals (zinc, iron, and iodine) and vitamin A. Encapsulation protects bioactive compounds and helps increase the stability and bioavailability of core minerals. Nanolevel encapsulation enhances the effectiveness of enzymes, nutrients, and therapeutic agents. Encapsulation prevents interaction with other compounds—proteins, amino acids. An advanced strategy for feed delivery is to use nanoscale technology, where the active ingredients are encapsulated and released slowly and in a controlled manner. Nanotechnology is being used in the animal health sector to improve the bioavailability of diagnostic tests and therapeutics. Nano selenium, nano chromium, and nanominerals are more effective than conventional sources even in low doses. It preserves nutrients in a way that prevents nutrient degradation due to feed processing and storage, the rumen environment, and external environmental factors. Nutrients reach specific organs and are released without degradation. Nanominerals (Zn, Se, and Cu) show greater efficacy at lower doses. Encapsulates nutrients and bioactive compounds to protect them from feed processing, storage, and the rumen environment. Encapsulation helps to release nutrients to a specific organ or size.

12.3.3 Integration into feed processing

The use of phytogenic products as additives in existing feed manufacturing schemes presents significant challenges and consequences in terms of both product efficacy and production efficiency.

Heat sensitivity during pelleting: The Standard feed pelleting process is conducted at temperatures of about 80–90°C and pressures that can degrade heat-sensitive phytochemicals. From essential oils, a loss rate above 50% for monoterpenoids on average is a common condition during pelleting. This, in turn, requires process changes that can have an adverse effect on products or costly protective systems (Biswas and Kim, 2025).

Compatibility with other feed ingredients: Phytogenic compounds might interact with other feed materials such that the bioactivity would be decreased or processing problems may occur. Interactions with minerals can result in complex formation and decreased solubility, whereas reactions with proteins can produce undesirable flavors or decrease the bioavailability of nutrients. Understanding and managing these interactions is difficult and requires extensive compatibility testing (Bagade et al., 2022). Mixing other ingredients can be difficult. For example, if the ingredients are incompatible, it is difficult to achieve proper mixing and uniformity. For example, certain minerals reacting with enzymes can reduce enzyme activity, making uniform nutrient distribution impossible. Random particle movement is required to avoid layering or segregation. It depends on factors such as particle size, density, moisture content, and hygroscopic nature, as well as compatibility. Incompatible particles may lead to clumping or segregation, resulting in nutrient heterogeneity. Proper mixing, uniformity, and compatibility with other feed ingredients are directly related to each other. Mixing efficiency, uniform nutrient distribution, and heat-sensitive additives are all affected by incompatible ingredients (Coop and Holmes, 1996).

Homogeneous distribution challenges: Uniform distribution of the PFAs over feed batches is technically challenging, particularly for lipophilic compounds, which may aggregate or separate during blending operations. Poor uniform distribution in variable dosing among animals may result in unpredictable biological responses (Hu et al., 2023).

12.4 Economic and market challenges

The cost of new technologies and changes to reduce greenhouse gases is increasing, which is affecting the cost of livestock production. Feed prices are becoming volatile and uncertain because of fluctuations in global trade in soybeans, maize, oilseeds, etc. Demand for livestock products is growing rapidly in developing countries, but demand in developed countries is stable. Market opportunities and pressures in international trade in both feed and livestock products are affecting livestock production and markets, especially in smaller countries. Consumer and societal concerns about animal welfare, food safety, environmental protection, etc., affect the market. Stringent laws and regulations make it difficult for livestock producers to enter the market and create barriers to exports and imports. Consumer and societal concerns about animal welfare, food safety, environmental protection, etc. affect the market. Stringent laws and regulations make it difficult for livestock producers to enter the market and create barriers to exports and imports. In phytofeed production, raw material supply scarcity is a major challenge, as medicinal plants or herbal ingredients are often seasonal and subject to limited availability, leading to market inefficiency and price fluctuation. The adoption of phytofeed largely depends on the supply chain, farmer awareness, and distribution. In phytofeed, issues such as quality control, standardization of digestibility, and farmer or industry perceptions act as significant market barriers.

12.4.1 Cost of raw materials and processing

Economic challenges limit functional phytofeed additives from being more extensively applied, with cost factors spreading over all stages in terms of raw material purchase and final product sale.

Competition with food and pharmaceutical industries: The base materials for high-quality PFAs generally compete with uses of higher value in the human foodstuff and pharmaceutical sector. The price of raw materials is increasing day by day. The demand in the market is increasing. While culinary products such as oregano, thyme, and rosemary that synthesize essential oils demand a high price in the food-flavoring industry, standardized herbal extracts are competing with nutraceutical uses. This competition results in raw material costs for phytogenic substances being multiple factors higher than those of conventional feed additives, while some of the phytogenic substances even cost 5–20 times more than synthetics (Yildiz et al., 2024).

Processing cost premiums: Compared to simple chemical synthesis, the sophisticated extraction and purification processes for stable PFAs are associated with high-cost premiums. Enriching processes such as the use of supercritical fluid extraction, standardized solvent extraction, and advanced purification processes may lead to production costs 200%–500% higher than for the manufacture of conventional feed additives. The requirement for special dedicated analytical method development and analytical testing, as well as stability studies and quality control tests also a cost burden (Singh et al., 2018). To use costs after other phytobiotic plant extracts in broiler feed, a high-quality extract must be produced. This means that active pharmaceuticals (alkaloids, saponins, flavonoids, etc.) must be highly purified or standardized. High-quality extract requires extraction solvent, purification, concentration, quality control, stability testing, and processing costs such as energy, labor, and equipment. The more purified or standardized the extract, the higher the production cost. Using high-quality extracts in feed formulation increases the overall feed cost. Therefore, this additional expense needs to be considered when evaluating growth performance and conducting cost–benefit analysis. The high production cost is primarily due to the expenses associated with solvents, purification processes, and quality control/testing (Løvendahl and Munksgaard, 2016).

Economic performance analysis: Comprehensive economic evaluations of phytogenic feed additives (PFAs) have reported inconsistent findings regarding their profitability. For example, a study on broiler production showed that although dietary supplementation with 2% PFAs achieved growth performance comparable to that of antibiotic growth promoters, the benefit–cost ratio was significantly lower because of the higher cost of the phytogenic ingredients. While PFAs have been reported to improve feed productivity by approximately 10–15% through enhanced production performance, these gains were often offset by a 20–30% increase in ingredient costs, ultimately resulting in lower overall economic efficiency than conventional feeding systems (Singh et al., 2018).

12.4.2 Scalability issues

The step from successful small-scale trials to commercial application can be difficult, which frequently affects the economic practicability of phytogenic feed systems.

Supply chain limitations: A considerable number of the more promising phytogenic ingredients are obtained from wild-harvested or small-scale agricultural cultivation practices that would be a critical challenge for supporting commercial-scale volumes. There is a worldwide shortage of premium-quality standardized plant extracts due to seasonal availability and variations in quality, which pose significant concerns about reliable supply. It is a high investment to establish reliable supply chains of specialty botanical material in agricultural infrastructure and processing capacity (Joseph and Hall, 2025).

Manufacturing scale-up challenges: It is difficult to scale up laboratory and pilot plant production to commercial level owing to scaling issues of extraction, processing, and formulation. When it comes to upscaling, the process conditions that may work at lab scale can be too expensive or inapplicable for industrial production. Commercial-scale phytogenic processing sites require capital for commercial-style facilities, which may be more than $10–50 million, which is a tremendous amount to make an investment (Joseph and Hall, 2025).

Infrastructure for distribution and storage: Phytogenic additives frequently need to be stored under specific conditions to remain stable and bioactive, and as a result, cold chain management or controlled atmosphere storage systems (which add costs) are required. Volatility and chemical instability of numerous bioactive agents impose shelf-life terminations, which can create erratic inventory management and wastage.

12.4.3 Market acceptance by farmers

Acceptance by the producer is the ultimate factor for commercial success, although farmers have not widely adopted PFAs despite demonstrating advantages.

Risk-benefit perception: Livestock producers work on thin profit margins and have low risk tolerance, especially for feed additives that can impact animal performance. The variable, and at times subtle, effects of phytogenic leave too much to "chance," a level of uncertainty that many operations will not accept given the reliability of conventional systems. If performance improvements are not assured, it is common for those who produce animals to select existing feeding programs instead of new ones (Singh et al., 2018). Increased mortality and reduced feed conversion efficiency and reproductive rates cause economic losses for farms (Dijkhuizen and Morris, 1997). Outbreaks of epidemic diseases have proven to be a real economic threat not only to individual farmers or the associated sector but also to regional and national economies (Tomassen et al., 2002).

Economic return requirements: Farmers generally require a very strong (clear economic profitable) incentive to adopt new nutritional feed technologies, but the value provided by phytogenic may be challenging to quantify in terms of economics. Better respect for animal welfare, decreased negative impact on the environment, and process on product quality may not directly reflect measurable economic benefit in the actual form of the market. The premium price of products from animals fed phytogenic diets is not high enough to encourage producers to adopt it (Janjić et al., 2022).

Technical support and knowledge requirements: Utilization of PFAs demands extensive technical support and education for the producer, which not many suppliers are able to cover. The difficulty in optimizing phytogenic programs for each production system, environmental factors and management implications requires an ongoing technical service, which also increases system cost (Joseph and Hall, 2025).

12.5 Regulatory and policy challenges

12.5.1 Lack of harmonized standards

The PFA regulatory framework is far from being harmonized between countries, which poses major impediments to international marketing and market access. Food regulations often vary from country to country and are not even the same between neighboring countries (WHO 2015). Although many countries and international organizations have declared the need for harmonization, it has not been implemented in practice. The process of harmonization is very complex and slow, which hinders progress. Nutritional feeding standards are typically formulated in accordance with a defined range of ambient temperatures. These standards are not always applicable in typical or low-input systems. Standards are not globally harmonized, and animals are forced to adapt to region-specific high-input systems, and in a low-input system, animal feed is inevitably adapted. Standards are applied differently according to different systems. The main reason for low animal production in the low-input system is not feed shortage but socioeconomic considerations. Scientists and policymakers often do not understand farmers' production goals, resulting in standards being applied inconsistently (Coop and Holmes, 1996). Information in tropical regions. Data in the temperature region cannot be applied directly (Torres Acosta et al., 2004).

Difference between EU, USA, and Asia in approval processes: The United States of America emphasizes postmarket surveillance and manufacturer responsibility for product safety. Product approval in India is relatively easy; the registration process is simple. In the EU, strict premarket approval and scientific validation are required before a new nutraceutical dietary supplement can be brought to the market. Regulatory differences hinder international trade and market integration of nutraceutical products. Compared to less premarket approval, emphasis, and postmarket monitoring before bringing new products to market, the USA process is different from the EU or Asia. Premarket approval, scientific validation, or rigorous approval and scientific testing are carried out before being brought to the market. Approval is more difficult in the EU than in the USA and Asia (EFSA, Europe, Gaisfort et al. 2001). The approval process for nutraceuticals and functional foods varies from country to country (Gaisford et al., 2001). There is no standard harmonization at the international level (Kaisford and Kerr, 2001).

International trade barriers: Regulatory divergences are posing considerable obstacles to trade in phytogenics. It could be that products approved in one territory need to undergo full reassessment in others, doubling the cost and complexity. The absence of mutual recognition agreements among major markets creates the need for going through difficult approval processes more than once, reducing efficient access to global markets and increasing developmental costs (Varol, 2023).

12.5.2 Safety and efficacy documentation

Regulatory authorities require detailed, extensive safety and efficacy documentation, which is particularly challenging for phytogenic products because of their complex and varying composition with a variable nature.

Toxicological assessment requirements: PFAs should be subjected to detailed toxicological studies, including acute exposure assessment as well as chronic exposure, reproductive toxicology, and bioaccumulation. The multicomponent nature of plant extracts adds greater complexity to such assessments, with potential interactions between components producing toxicological profiles distinct from the individual constituents. Also, the inherent variability in phytochemical composition requires a safety assessment that is based on such ranges rather than single chemical entities (Mantovani et al., 2022).

Long-term safety studies: The need to conduct long-term safety studies constitutes a challenge in the case of phytogenic additives, as chronic effects require prolonged feeding trials that are costly and time-consuming. These studies would likely be over several growing cycles and should include testing for potential residues on animal products intended for consumption by humans (Mantovani et al., 2022).

12.6 Environmental and sustainability challenges

12.6.1 Sourcing of plant materials

The environmental sustainability of the phytogenic used has to be measured using responsible sourcing approaches that face various difficulties. The environmental impact of manufacturing PFAs goes beyond raw material production, which includes energy-intensive processing operations. Increasing amounts of plant-based materials are being used to ensure feed supply in animal husbandry. Traditional feeds such as cereals, oil cakes, and meals are widely used. But these are not always suitable in tropical regions. Because climate and technology are different (1985). Nonconventional feed resources, especially plant-based materials, are being used to reduce pressure on conventional feed resources and ensure livestock feed sustainability (Ben Salem et al., 2004).

Energy-intensive extraction: Techniques: Advanced extraction methods such as supercritical fluid extraction, steam distillation, and microwave-assisted extraction require substantial energy inputs that contribute to the carbon footprint of phytogenic products. Life cycle assessments indicate that extraction and processing can account for 30%–50% of the total carbon footprint of finished PFAs(Feed & Additive 2024).

Overharvesting and biodiversity concerns: Overexploitation in livestock production refers to the overexploitation of natural resources such as pasture, fodder, plants, or the genetic resources of established breeds. Due to overuse, a limited number of commercial breeds are used to the exclusion of local breeds, which, while increasing production, causes loss of diversity and poses a serious threat to long-term sustainable livestock systems. Human population growth is the main cause of biodiversity loss (Crist, 2019). Because of population growth, the demand for food and products is increasing. This results in excessive grazing, deforestation, and overharvesting of feed plants. This leads to a decrease in biodiversity. Humans have encroached on wildlife habitats and are destroying ecosystems. Overexploitation of natural vegetation for animal feed, deforestation to create pastures, and overuse of agroindustry by-products reduce biodiversity (Cafaro et al., 2022). Harvesting Medicinal and Aromatic Plants from the wild to be used as PFAs may put pressure on the natural population. Several useful phytogenic species, such as wild oregano and thyme, and other medicinal herbs, are under population pressure because of escalating commercial needs. Sustainable gathering practices mean wild supplies are managed, and the availability of this product may be limited (Bature et al., 2024).

Cultivation sustainability: The extensive cultivation of plants for use as PFAs clashes with food production regarding arable land, water, and further resources. Water needs of plants producing essential oils can be high (e.g., two to three times as much irrigation as conventional feed crops). Moreover, the specific cultivation needs of high-quality phytogenic materials may include intense management requirements that might have an impact on the environment (All About Feed, 2022).

Competition with human food use: Several plants used in PFAs are also food crops or ingredients made from food crops, which could be argued will create some competition between use for efficacy and the need to feed humans. For example, Herbs (oregano, rosemary, and turmeric), spices (cinnamon, ginger, and garlic), essential oils (from eucalyptus, peppermint, etc.), plant extracts (saponins from yucca, tannins from chestnut) and other well-known phytogenic compounds are valued more in the food market than feed ingredients so production might be directed to the food rather than feed side (Joseph and Hall, 2025).

12.6.2 Carbon footprint of extraction and processing

The farm stage is the most important step in determining the carbon footprint of livestock production, as 70%–90% of total GHG emissions occur before the farm gate. Methane (CH_4) is the main contributor, which is produced due to enteric fermentation and manure management, while nitrous oxide (N_2O) comes from manure and fertilizer, and carbon dioxide (CO_2) comes from fuel combustion, agrochemical production, and land use change (de Vries and de Boer, 2010). The type of livestock system affects the carbon footprint; for example, pasture-based extensive beef systems have higher methane emissions, which can account for 60%–70% of the total carbon footprint, whereas intensive grain-fed beef systems have lower methane emissions due to feed digestibility (Cederberg et al., 2009). Life cycle assessment is used to calculate carbon footprints, with the cradle-to-farm gate approach being the most widely accepted. In addition, changes in land use and soil carbon also affect the carbon footprint, as productive grassland or cereal land can provide opportunities for carbon sequestration, which alters the emissions equation (Nguyen et al., 2010). In the case of dairy products, the FAO reports that the global average milk carbon footprint is approximately 2.4 kg CO_2 eq./kg, 93% of which occurs at the cradle-to-farm gate.

12.6.2.1 *Energy-intensive techniques*

In livestock and agricultural production, many steps are energy-intensive. These energy-intensive steps result in higher CO_2 and other greenhouse gas (GHG) emissions, increasing the total carbon footprint (Cormos et al., 2018). For example: Feed production and crop processing—fertilizer, machinery, irrigation, and fuel-intensive harvesting processes require a lot of energy, which increases CO_2 emissions. For example: Feed production and crop processing—fertilizer, machinery, irrigation, and fuel-intensive harvesting processes require a lot of energy, which increases CO_2 emissions. Integrated industrial processes: Modelling, simulation, and energy integration tools are used for CO_2 capture and utilization in power generation, cement, iron & steel, and chemical industries, which help determine the carbon footprint of the entire production chain (Cormos et al., 2018).

12.6.3 Waste management

Livestock waste refers to animal excreta, bedding material, washing water, hair, feathers, and other debris. According to the 19th Livestock Census of India, a total of 512 million animals in the country produce 1095 million metric tons of excreta annually. In modern animal husbandry, waste recycling and proper management are important to produce manure, feed, energy, and reduce environmental pollution (Westendorf et al., 2000). If livestock waste is not properly managed, it can increase greenhouse gas emissions, ammonia emissions, water pollution, and disease spread. However, proper management helps increase soil fertility, food production, fuel production (biogas), and protect animals and human health. Livestock waste can be mainly divided into solid (manure, food waste, and bedding material) and liquid (urine and washing water). The collection methods are also of two types: Separate collection (solid and liquid are separated) and flushing method (flushing together with water).

12.7 Consumer and sociocultural challenges

12.7.1 Perception of "herbal-fed" animal products

The consumers' perception and acceptance of animal products originating from livestock-fed phytogenic additives have multifaceted impacts that affect market acceptability and economic sustainability.

Skepticism about effectiveness: Despite increasing consumer interest in natural and sustainable food production, skeptical opinions are present regarding the potency of plant-based feed additives compared with conventional systems. According to consumer polls, only 30%–40% of consumers believe that herbal feed additives are as effective as the conventional alternatives, but 60%–70% of them express a preference for more natural methods of production. This skepticism results in market resistance, which reduces premiums for products originating from animals fed phytogenic diets (Stampa et al., 2020).

Knowledge gaps in consumers: There are substantial knowledge gaps about the mechanism and advantages of PFAs among consumers, as most have little or no idea how to distinguish between types of natural feed additives. The knowledge of consumers about the terms "phytogenic," "botanical," or "herbal" feed additives is limited, showing that they often associate these products with human medicinal herbal extracts or supplements. This absence of comprehension leads to market underdevelopment and hinders consumer education efforts (O'brien et al., 2023).

Associtation with product quality: Consumer perception of animal products derived from herb-fed animals differs markedly in terms of characteristics such as taste, nutritive value, and safety. While some consumers link herbal feeding with better animal welfare and product quality, others fear the development of off-flavors or adulterated product attributes. Research into preferences for grass-fed beef, for instance, shows that perceptions of "natural" production methods among consumers do not always coincide with real aesthetic preferences when tested through a blind tasting (Bukachi et al., 2021).

12.7.2 Cultural and regional preferences

Food habits are related to the geographical environment, weather, socioculture, and lifestyle of the region (Brouwer et al., 2021). What types of animal foods a person will or will not eat are deeply influenced not only by health or economic reasons, but also by their religious beliefs and cultural practices (Kittler et al., 2011). These foods may be a luxury or a sign of social status, while for others, they are an essential part of daily life. People's dietary patterns are deeply related to their biophysical context, such as climate, geography, and resource availability, and to sociocultural context, such as religion, culture, and tradition. Culture and religious beliefs (such as Muslims not eating pork, Hindus/Buddhists abstaining from meat) play an important role in shaping eating habits. Regional environment (weather, crop production) determines which foods will be predominant. As a result, nature and culture combine to form regional animal husbandry practices (Popkin, 1999). Using local feed increases production costs, so farmers will only do it if there is a subsidy or if consumers are willing to pay a higher price.

12.7.2.1 Acceptance varies by tradition and dietary culture

The concept "acceptance varies by tradition and dietary culture" clearly highlights the cultural and traditional influence on food acceptance and eating habits. Various studies show that people's food choices are deeply influenced not only by biological needs but also by culture, religion, social values, and customs. For example, religious prohibitions such as halal in Islam, kosher in Judaism, and vegetarianism in Hinduism and Buddhism make it clear that foods are acceptable in one tradition and culture, while forbidden or less valuable in another. Similarly, traditional foods such as sushi in Japan, pasta in Italy, and corn in Mexico—these are not just foods but symbols of cultural heritage. While these foods are highly accepted locally, they may not have the same value or acceptance in other regions. Geography and environment also influence food acceptability. For example, the warm climate and fertile soil of Mediterranean countries have made the Mediterranean diet popular there. On the other hand, seafood is more acceptable in coastal areas, while grain-based foods are predominant in inland areas (Willett et al., 1995). In addition, globalization and modernization have transformed food culture. The spread of Western fast-food culture has been called "McDonaldization" or "Coca-colonization," which has often altered or eliminated local traditional food habits. This demonstrates that cultural acceptance of food is not static but rather constantly changing because of time, modernity, and global influences.

12.7.2.2 Acceptance varies by tradition and dietary culture

Meiselman et al. (1996) stated that food is almost always part of a social process, and cultural/social influence plays a major role in dietary choices. Rozin et al. (2004) and Verbeke et al. (2009) have shown that consumers prefer "natural" foods, which is an extension of social/cultural belief systems. Traditional food innovation acceptance largely depends on this social-cultural perception. In the framework, perceived benefits and costs, novelty, and maintenance of traditional character are crucial determinants. Social/behavioral influences are reflected here (Guerrero et al., 2009).

12.8 Overcoming challenges: Pathways forward

12.8.1 Scientific solutions

Progress in the field of functional phytofeed increasingly demands engineered, technically advanced scientific solutions to the inherent problems of variability, complexity, and idiosyncratic response.

Multiomics approaches: The combination of genomics, transcriptomics, metabolomics, and microbiomics nomenclature enables unique possibilities in the elucidation and prediction of the effects of PFAs. Multiomics technologies can be used to generate biomarkers for predicting individual animal responsiveness to particular phytogenics, opening the possibility of precision nutrition where efficacy is maximized and costs are minimized. Such technologies can also help to reveal molecular mechanisms of how phytochemicals are bioactive, so that more targeted and effective formulations can be developed.

Microbiome science integration: The importance of the gut microbiota in phytochemical metabolism and bioactivity presents opportunities for a microbiome-informed phytogenic nutrition approach. Advanced microbiome profiling may reveal microbial signatures related to maximal phytogenic response, and probiotic coapplications could improve the bioavailability of phytochemicals and their effectiveness. In addition to their role in plant-microbiome relationships, knowledge of microbiome-phytochemical interactions can be employed for designing cooperative combinations that capitalize on microbial metabolism by potentiating bioactive compound activity (Mills et al., 2019).

Precise nutrition platforms: Development of precision nutrition technologies that integrate individual animal features in addition to environmental factors and microbiome data could enhance the use of PFA in feed. Such platforms can apply artificial intelligence and machine learning techniques to predict appropriate phytogenic formulations and dosing solutions for individual production environments. Real-time monitoring devices can support a proactive approach to developing more optimal phytogenic nutrition regimens.

Standardized bioassays: Standardized in vitro and in vivo bioassays to determine the effectiveness of PFAs could be a value addition to product development and quality control. These assays should contain relevant bioassays such as antimicrobial activity, antioxidant behavior, immunomodulation, and gut barrier integrity. Scalable formulation protocols will allow comparison between products and assist in regulatory review processes.

12.8.2 Technological innovations

New delivery technologies and processing techniques are solving some of the technical obstacles to overcoming the effectiveness and economic feasibility of PFAs.

Advanced encapsulation systems: The bioavailability and stability issues can be addressed by next-generation encapsulation technologies using complex delivery systems. Matrix confinement of sensitive compounds in lipid carriers, polymer microspheres, and cyclodextrin complexes can prevent them from being damaged or destroyed during feed manufacturing and be delivered to specific tissues in a controlled manner. These systems can increase bioavailability from 2–5 times that of pure forms, allowing much lower doses to be used for maximum benefit and value.

Nanotechnology applications: A range of nanodelivery systems has emerged, allowing new possibilities to improve the bioavailability and targeting of phytochemical extracts. Various advanced drug delivery systems, such as nanoparticles, nanoemulsions, and nanostructured lipid carriers, can enhance solubility 10–100 times, accompanied by protection against degradation and targeted tissue oscillation. Controlled release systems at the nano level can keep drugs at a therapeutic level for long durations without exceeding toxic doses and may reduce dosing intervals and resource utilization.

Smart feed technologies: Incorporation of smart delivery systems that exploit environmental stimuli, such as pH, temperature, or enzymatic activity, to modulate the accessibility of phytogenic compounds in tissues can enhance their release at the site of action. Such responsive systems can increase bioavailability while reducing degradation during feeding, processing, and storage. The time-release formulations allow for continuous delivery of the bioactive compound, which persists throughout grow-outs.

Process optimization: Advanced feed manufacturing processes, including low-temperature pelleting, steam conditioning optimization, and postpelleting application technologies, can reduce the losses of bioactive compounds during processing. Process analytical technologies may offer in-process monitoring of phytochemical stability during processing, which can lead to prompt corrections to preserve product quality.

12.8.3 Policy and regulation improvement

Changes in livestock production have increased the need to improve policies and regulations related to livestock feeding to reduce negative impacts on the environment and society (Rauw et al., 2023). The overuse of antibiotics is closely linked to the feeding system. Large-scale hog production farms (CAFOs) use about three times more antibiotics in feed and water than smaller farms. Therefore, it is essential to develop policies and regulations to reduce the incidence of unnecessary antibiotic use in feed and water . Government efforts to address the externalities of livestock production have largely focused on controlling water pollution, as excess animal waste and other diffuse pollutants from agricultural activities are entering water bodies at a large scale. Animal waste is polluting water, and there is widespread use of antibiotics on a large scale. Addressing this problem requires policy reforms, such as waste management regulations, food-related regulations, and antibiotic use controls.

12.8.3.1 Global harmonization, risk-based frameworks

Global harmonization and risk-based frameworks are currently of great importance in protecting food safety and public health. The World Health Organization (WHO) and the Food and Agriculture Organization (FAO) are using a risk-based approach to food-related health risk management. This risk analysis consists of three parts—risk assessment, risk management, and risk communication the main purpose of which is to ensure public health protection. Risk assessment is essentially a scientific process that analyzes the effects of potential pathogens or harmful substances in food. It involves a series of steps including hazard identification, hazard characterization, exposure assessment, and risk characterization (Alimentarius, 1999).

12.8.4 Economic strategies

Breeding and animal health improvement projects have not always been effective in increasing farmers' income or production. The multifaceted uses of animals and the complex interrelationships of biological, technological, and social factors require a systems-based approach to maximize resource use. Systems-based management reduces costs, increases profits, and ensures sustainable production. To sustainably increase crop and livestock production while preventing environmental degradation. The government needs to show a long-term commitment to supporting rural people. Economic empowerment of small farmers and women livestock keepers strengthens the livestock economy. Different strategies for different types of systems: Pastures, small farms, poor farms, and commercial farms. Adopting the right strategy ensures cost reduction, increased production, and profits. Animal health programs are an important part of a systems-based approach(Kaasschieter et al., 1992). European family farms and livestock farms have repeatedly faced economic pressures, social and historical challenges, although their production capacity is low and should not be sustained solely on economic criteria (Jervell et al., 2003). Rising production costs and inflation of certain inputs (e.g., feed increased by 31%, energy increased by 64%) have increased the financial pressure on livestock farms(Paseges, 2013). Lack of cash flow on farms and failure to receive timely payments from the dairy industry have also caused financial crises (Karelakis et al., 2013). Structural constraints on farms, such as small size, lack of modern technology, dependence on subsidies, and low levels of cooperation, further complicate this problem (Karanikolas and Martinos, 2012). Labor management is an important part of economic strategy. Sheep and goat farms deal with risk through the reorganization of labor (Karelakis et al., 2013).

12.8.4.1 Valorization of agricultural by-product

Reducing food losses and food waste is the first step in the valorization of agricultural residues. According to FAO, losses are identified in the production and supply chain (Mohan and Long, 2021). Extracting phytonutrients and bioactive compounds from inedible parts or agricultural residues and using them for food additives, nutraceuticals, therapeutics, and cosmetics is a prime example of valorization (Majerska et al., 2019). The antioxidant activity of some tropical fruit residues can reduce lipid peroxidation and can be used in food packaging or preservation (Rakariyatham et al., 2020). Pectin obtained from carrot residues increases antioxidant capacity, which is an example of valorization (Ben-Othman et al., 2020). Cocoa bean shell residues can be used in innovative and functional foods to prevent lipid oxidation, which is an example of valorization (Rojo-Poveda et al., 2020). Extracting bioactive compounds from olive mill wastewater to produce food preservation and functional ingredients (Alfano et al., 2018). Extracts obtained from broccoli seed residues have been used as food preservatives to reduce fungal contamination of cereals (Thery et al., 2020).

12.8.4.2 Incentives for sustainable feed adoption

Government incentives encourage farmers and stakeholders to use sustainable practices. Incentive and control systems influence farmer behavior and decisions. Government incentives and policies are important in both areas to reduce environmental impacts. An integrated approach that combines incentives and punishments is more effective. Awareness and technical issues are the main challenges. Incentives are used to encourage farmers to adopt environmentally friendly or sustainable agricultural/farm practices. Farmers' preferences, financial status, awareness, and cultural characteristics mediate the effect of incentives. Conserving natural resources and protecting the environment by adopting environmentally friendly methods. Incentives encourage farmers to adopt practices that bring environmental and economic improvements. Short-term profit or production growth, long-term environmental or sustainable impact are the main motivations. More research is needed on the relationship between different incentives, farmer acceptance, and outcomes. Governments or institutions provide economic incentives to encourage farmers to engage in sustainable agriculture. Short-term economic incentives influence the perception of intrinsic motivation and positive outcomes in the long term. Sustainable adoption has been explained by combining incentives, risk perception, and behavioral control. Adopting sustainable agriculture conserves the environment, increases productivity, and increases farmer resilience (Gagaoua et al., 2024).

12.9 Conclusion

The use of phytofeed additives with functional properties in animal production is limited by several intricate and multi-faceted constraints, which are interconnected across scientific, technological, economic, regulatory, environmental, and social aspects. Although such obstacles are formidable, they are not impossible to overcome—and important advances are underway in overcoming major limitations through creative research, technological advancement, and policy action.

The scientific hurdles to fit the bill for phytochemical variation and biological complexity demand further investments in methods of standardization, mechanistic studies, and precision nutrition. The combination of multiomics platforms, microbiome science, and artificial intelligence provides unprecedented opportunities to explore/understand/optimize applications of PFAs to individual animals and production systems.

Over the last few decades, advances in technology of delivery systems, processes, and quality control have helped overcome bioavailability and stability issues, which for long periods have posed restrictions to PFA efficacy. Novel encapsulation technologies, nanotechnology-based applications, and smart delivery systems offer opportunities to achieve pharmaceutical-level bioavailability and targeting, which, if exploited, could potentially revolutionize the commercialization of phytogenic products.

References

Abeltino, A., Hatem, D., Serantoni, C., Riente, A., Maria De Giulio, M., De Spirito, M., De Maio, F., Maulucci, G., 2024. Unraveling the gut microbiota: implications for precision nutrition and personalized medicine. Nutrients 16 (22), 3806. doi:10.3390/nu16223806.

Ahamed, M.S., Asghar, A., Hayat, N., Muqaddas, M., Bashir, L., Saif, S., Shahriar, S.Md.S., Riaz, R., Ashraf, M.A, Khalid, B., Sadia, A., Moeid Khan, M., Riaz, T., 2025. Optimizing bioactive compounds extraction: advanced food safety protocols for enhanced bioavailability and processing stability. S. Asian J. Life Sci. 13. doi:10.17582/journal.sajls/2025/13.58.65.

Alfano, A., Corsuto, L., Finamore, R., Savarese, M., Ferrara, F., Falco, S., Santabarbara, G., De Rosa, M., Schiraldi, C., 2018. Valorization of olive mill wastewater by membrane processes to recover natural antioxidant compounds for cosmeceutical and nutraceutical applications or functional foods. Antioxidants 7 (6), 72. doi:10.3390/antiox7060072.

All About Feed. (2022). Phytogenics: sustainability under challenging conditions. All about feed. https://www.allaboutfeed.net/specials/phytogenics-sustainability-under-challenging-⟨?PMU ?⟩conditions/.

Al-Mijalli, S.H., Mrabti, N.N., Ouassou, H., Sheikh, R.A., Abdallah, E.M., Assaggaf, H., Bakrim, S., Alshahrani, M.M., Awadh, A.A.A., Qasem, A., Attar, A., Lee, L.H., Bouyahya, A., Goh, K.W., Ming, L.C., Mrabti, H.N., 2022. Phytochemical variability, in vitro and in vivo biological investigations, and in silico antibacterial mechanisms of Mentha piperita essential oils collected from two different regions in Morocco. Foods 11 (21). doi:10.3390/foods11213466.

Bagade, S., Patil, D.D., Shirkhedkar, A., 2022. Standardization of herbal bioactives. In: Herbal Bioactive-Based Drug Delivery Systems. Academic Press, Cambridge, MA, pp. 393–407.

Balazs, B., Kelemen, E., Centofanti, T., Vasconcelos, M.W., Iannetta, P.P.M., 2021. Policy interventions promoting sustainable food-and feed-systems: A Delphi study of legume production and consumption. Sustainability 13 (14), 7597. https://doi.org/10.3390/su13147597.

Balázs, B., Mooney, P., Nováková, E., Bastin, L., Arsanjani, J.J., 2021. Data quality in citizen science. In: The Science of Citizen Science. Springer International Publishing, Hungary, pp. 139–157. doi:10.1007/978-3-030-58278-4_8.

Bature, I., Xiaohu, W., Ding, X., 2024. The roles of phytogenic feed additives, trees, shrubs, and forages on mitigating ruminant methane emission. Front. Vet. Sci. 11. doi:10.3389/fvets.2024.1475322.

Ben-Othman, S., Jõudu, I., Bhat, R., 2020. Bioactives from agri-food wastes: present insights and future challenges. Molecules 25 (3), 510. doi:10.3390/molecules25030510.

Ben Salem, H., Nefzaoui, A., Ben Salem, L., 2004. Spineless cactus (Opuntia ficus indica f. inermis) and oldman saltbush (Atriplex nummularia L.) as alternative supplements for growing Barbarine lambs given straw-based diets. Small Rumin. Res. 51 (1), 65–73. doi:10.1016/s0921-4488(03)00186-x.

Bergvall, U.A., Schäpers, A., Kjellander, P., Weiss, A., 2011. Personality and foraging decisions in fallow deer, Dama dama. Anim. Behav. 81 (1), 101–112. doi:10.1016/j.anbehav.2010.09.018.

Bhavaniramya, S., Vishnupriya, S., Saleh Al-Aboody, M., Vijayakumar, R., Baskaran, D., 2019. Role of essential oils in food safety: antimicrobial and antioxidant applications. Grain Oil Sci. Technol. 2 (2), 49–55. doi:10.1016/j.gaost.2019.03.001.

Biswas, S., Kim, I.H., 2025. A thorough review of phytogenic feed additives in non-ruminant nutrition: production, gut health, and environmental concerns. J. Anim. Sci. Technol. 67 (3), 497–519. doi:10.5187/jast.2025.e26.

Brouwer, I.D., van Liere, M.J., de Brauw, A., Dominguez-Salas, P., Herforth, A., Kennedy, G., Lachat, C., Omosa, E.B., Talsma, E.F., Vandevijvere, S., Fanzo, J., Ruel, M., 2021. Reverse thinking: taking a healthy diet perspective towards food systems transformations. Food Secur. 13 (6), 1497–1523. doi:10.1007/s12571-021-01204-5.

Bukachi, S.A., Ngutu, M., Muthiru, A.W., Lépine, A., Kadiyala, S., Domínguez-Salas, P., 2021. Consumer perceptions of food safety in animal source foods choice and consumption in Nairobi's informal settlements. BMC Nutrit. 7 (1). doi:10.1186/s40795-021-00441-3.

Cafaro, P., Hansson, P., Götmark, F., 2022. Overpopulation is a major cause of biodiversity loss and smaller human populations are necessary to preserve what is left. Biol. Conserv. 272, 109646. doi:10.1016/j.biocon.2022.109646.

Cederberg, C., Flysjö, A., Sonesson, U., Sund, V., & Davis, J. (2009). Greenhouse gas emissions from Swedish consumption of meat, milk and eggs 1990 and 2005. SIK Report No. 794. Gothenburg, Sweden: SIK – The Swedish Institute for Food and Biotechnology.

Codex Alimentarius Commission. (1999). Guidelines for the production, processing, labelling and marketing of organically produced foods (CAC/GL 32-1999). Rome, Italy: Food and Agriculture Organization of the United Nations and World Health Organization.

Coop, R.L., Holmes, P.H., 1996. Nutrition and parasite interaction. Int. J. Parasitol. 26 (8–9), 951–962. doi:10.1016/s0020-7519(96)80070-1.

Cormos, A.M., Dinca, C., Petrescu, L., Andreea Chisalita, D., Szima, S., Cormos, C.C., 2018. Carbon capture and utilisation technologies applied to energy conversion systems and other energy-intensive industrial applications. Fuel 211, 883–890. doi:10.1016/j.fuel.2017.09.104.

Crist, E., 2019. Abundant Earth: Toward an Ecological Civilization. University of Chicago Press, Chicago, IL doi:10.7208/chicago/9780226596945.001.0001.

Daniels, K.J., Donkin, S.S., Eicher, S.D., Pajor, E.A., Schutz, M.M., 2007. Prepartum milking of heifers influences future production and health. J. Dairy Sci. 90 (5), 2293–2301. doi:10.3168/jds.2005-881.

de Vries, M., de Boer, I.J.M., 2010. Comparing environmental impacts for livestock products: a review of life cycle assessments. Livest. Sci. 128 (1–3), 1–11. doi:10.1016/j.livsci.2009.11.007.

Dijkhuizen, A.A., Morris, R.S., 1997. Animal Health Economics: Principles and Applications. University of Sydney, Sydney, Australia.

Di Lorenzo, C., Santos, A.D, Colombo, F., Moro, E., Dell'Agli, M., Restani, P., 2014. Development and validation of HPLC method to measure active amines in plant food supplements containing Citrus aurantium L. Food Control 46, 136–142. doi:10.1016/j.foodcont.2014.05.017.

El-Feky, A.M., AbdelRahman, M.A., 2025. Harvest time and soil–plant relationship effects on phytochemical constituency and biological activities of *Psidium guajava* L. leaves. Sci. Rep. 15 (1), 25907. https://doi.org/10.1038/s41598-025-11323-0.

El-Zaiat, H.M., Ku-Vera, J.C., Soltan, Y.A., 2023. Editorial: Natural phytochemicals to enhance animal productivity and health status with low greenhouse gas emissions. Front. Vet. Sci. 10. doi:10.3389/fvets.2023.1280611.

Feed & Additive, 2024. Potential of phytogenic additives to reduce environmental footprint. Feed Addit. https://www.feedandadditive.com/potential-of-phytogenic-additives-to-reduce-environmental-footprint/.

Fung, F., Wang, H.S., Menon, S., 2018. Food safety in the 21st century. Biomed. J. 41 (2), 88–95. doi:10.1016/j.bj.2018.03.003.

Gagaoua, M., Das, A.K., Fu, Y., Dib, A.L., Nanda, P.K., 2024. Meat by-products as a source of bioactive peptides and functional ingredients: regulatory and safety barriers to valorization. Curr. Opin. Green Sustain. Chem. 47. doi:10.1016/j.cogsc.2024.100910.

Gaisford, J.D., Hobbs, J.E., Kerr, W.A., Perdikis, N., Plunkett, M.D., 2001. The Economics of Biotechnology. Edward Elgar Publishing, Cheltenham, UK.

Gaisford, J.D., Kerr, W.A., 2001. Economic Analysis for International Trade Negotiations. Edward Elgar Publishing, Cheltenham, UK.

Guerrero, L., Guàrdia, M.D., Xicola, J., Verbeke, W., Vanhonacker, F., Zakowska-Biemans, S., Sajdakowska, M., Sulmont-Rossé, C., Issanchou, S., Contel, M., Scalvedi, M.L., Granli, B.S., Hersleth, M., 2009. Consumer-driven definition of traditional food products and innovation in traditional foods. a qualitative cross-cultural study. Appetite 52 (2), 345–354. doi:10.1016/j.appet.2008.11.008.

Hellwig, S., Drossard, J., Twyman, R.M., Fischer, R., 2004. Plant cell cultures for the production of recombinant proteins. Nat. Biotechnol. 22 (11), 1415–1422. doi:10.1038/nbt1027.

Hu, Y., Lin, Q., Zhao, H., Li, X., Sang, S., McClements, D.J., Long, J., Jin, Z., Wang, J., Qiu, C., 2023. Bioaccessibility and bioavailability of phytochemicals: Influencing factors, improvements, and evaluations. Food Hydrocoll. 135. doi:10.1016/j.foodhyd.2022.108165.

Janjić, J., Savić, K.Š., Marković, R., Šefer, D., Nedić, D., Đurić, S., Vejnović, B., Mirilović, M., 2022. Influence of phytobiotics in feed on the cost-effectiveness of broiler production during fattening. Meat Technol. 63 (1), 51–58. doi:10.18485/meattech.2022.63.1.5.

Jervell, A.M., Jolly, D.A., 2003. Beyond Food: Towards a Multifunctional Agriculture. Norwegian Agricultural Economics Research Institute, Oslo, Norway.

Joseph, M., Hall, S., 2025. Challenges and trends in animal nutrition and feed industry. Feed Addit. https://www.feedandadditive.com/challenges-and-trends-in-animal-nutrition-and-feed-industry/.

Kaasschieter, G.A., de Jong, R., Schiere, J.B., Zwart, D., 1992. Towards a sustainable livestock production in developing countries and the importance of animal health strategy therein. The Veterinary Quarterly 14 (2), 66–75. https://doi.org/10.1080/01652176.1992.9694333.

Karanikolas, P., Hernández, P.A., Martinez-Gomez, V., Arnalte-Mur, L., Galli, F., Rivera, M., Prosperi, P., Goussios, G., Oikonomopoulou, E., Fonseca, A., 2021. Food system integration of olive-oil-producing small farms in Southern Europe. Glob. Food Sec. 28, 100499.

Karásková, K., Suchý, P., Straková, E., 2015. Current use of phytogenic feed additives in animal nutrition: a review. Czech J. Anim. Sci. 60 (12), 521–530. doi:10.17221/8594-cjas.

Karelakis, C., Abas, Z., Galanopoulos, K., Polymeros, K., 2013. Positive effects of the Greek economic crisis on livestock farmer behaviour. Agron. Sustain. Dev. 33 (3), 445–456. doi:10.1007/s13593-013-0136-y.

Kittler, P.G., Sucher, K.P., Nelms, M., 2011. Food and Culture (6th ed.). Cengage Learning, Belmont, CA.

Latek, U., Chłopecka, M., Karlik, W., Mendel, M., 2022. Phytogenic compounds for enhancing intestinal barrier function in poultry–a review. Planta Med. 88 (03/04), 218–236. doi:10.1055/a-1524-0358.

Llonch, P., Troy, S.M., Duthie, C.A., Somarriba, M., Rooke, J., Haskell, M.J., Roehe, R., Turner, S.P., 2018. Changes in feed intake during isolation stress in respiration chambers may impact methane emissions assessment. Anim. Prod. Sci. 58 (6), 1011–1016. doi:10.1071/AN15563.

Loučková, A., Neugebauerová, J., Křížkovská, B., Zlechovcová, M., Šebelová, K., Lipov, J., Viktorová, J., Hajšlová, J., 2024. Variability in phytochemical profiles and antimicrobial activity of Agrimonia Taxa. SSRN doi:10.2139/ssrn.5006737.

Løvendahl, P., Munksgaard, L., 2016. An investigation into genetic and phenotypic variation in time budgets and yield of dairy cows. J. Dairy Sci. 99 (1), 408–417. doi:10.3168/jds.2015-9838.

Majerska, J., Michalska, A., Figiel, A., 2019. A review of new directions in managing fruit and vegetable processing by-products. Trends Food Sci. Technol. 88, 207–219. doi:10.1016/j.tifs.2019.03.021.

Mantovani, A., Aquilina, G., Cubadda, F., Marcon, F., 2022. Risk-benefit assessment of feed additives in the One Health perspective. Front. Nutr. 9. doi:10.3389/fnut.2022.843124.

Meiselman, H.L., MacFie, H.J.H., 1996. Food choice, acceptance and consumption. Blackie Academic & Professional, London, UK.

Miller-Cushon, E.K., DeVries, T.J., 2015. Invited review: development and expression of dairy calf feeding behaviour. Can. J. Anim. Sci. 95 (3), 341–350. doi:10.4141/cjas-2014-163.

Mills, S., Stanton, C., Lane, J.A., Smith, G.J., Ross, R.P., 2019. Precision nutrition and the microbiome, part I: current state of the science. Nutrients 11 (4). doi:10.3390/nu11040923.

Mohan, A., Long, J.M., 2021. Valorization of wastes and by-products from the meat industry. In: Valorization of Agri-Food Wastes and By-Products. Academic Press, Cambridge, MA, pp. 457–474.

Nastoh, N.A., Waqas, M., Çınar, A.A., Salman, M., 2024. The impact of phytogenic feed additives on ruminant production: A review. J. Anim. Feed Sci. 33 (4), 431–453. doi:10.22358/jafs/191479/2024.

Nguyen, T.L.T., Hermansen, J.E., Mogensen, L., 2010. Environmental consequences of different beef production systems in the EU. J. Cleaner Prod. 18 (8), 756–766. doi:10.1016/j.jclepro.2009.12.023.

O'brien, K.D., Baker, C.N., Bush, S.A., Wolf, K.J., 2023. The meat of the matter: the effect of science-based information on consumer perception of grass-fed beef. J. Appl. Commun. 107 (4). doi:10.4148/1051-0834.2496.

Odhaib, S.A, Amiri, F.N, Altemimi, M.T, Imran, H.J, Alidrisi, H.A, Mohammed, M.J, Mansour, A.A, 2021. Development of the first health-related quality of life questionnaires in Arabic for women with polycystic ovary syndrome (Part II): dual-center validation of PCOSQoL-47 and PCOSQoL-42 questionnaires. Cureus 13 (9). doi:10.7759/cureus.18060.

Paseges, 2013. The Greek agricultural economy: Annual report 2013. Athens, Greece.

Popkin, B.M., 1999. Urbanization, lifestyle changes and the nutrition transition. World Dev. 27 (11), 1905–1916. doi:10.1016/S0305-750X(99)00094-7.

Rakariyatham, K., Zhou, D., Rakariyatham, N., 2020. Fereidoon Shahidi, Sapindaceae (Dimocarpus longan and Nephelium lappaceum) seed and peel by-products: potential sources for phenolic compounds and use as functional ingredients in food and health applications. J. Funct. Foods 67, 103846. doi:10.1016/j.jff.2020.103846.

Rauw, W.M., Gomez-Raya, L., Star, L., Øverland, M., Delezie, E., Grivins, M., Hamann, K.T., Pietropaoli, M., Klaassen, M.T., Klemetsdal, G., Gil, M.G., Torres, O., Dvergedal, H., Formato, G., 2023. Sustainable development in circular agriculture: an illustrative bee legume poultry example. Sustain. Dev. 31 (2), 639–648. doi:10.1002/sd.2435.

Rojo-Poveda, O., Barbosa-Pereira, L., Zeppa, G., Stévigny, C., 2020. Cocoa bean shell—a by-product with nutritional properties and biofunctional potential. Nutrients 12 (4), 1123. doi:10.3390/nu12041123.

Rozin, P., Spranca, M., Krieger, Z., Neuhaus, R., Surillo, D., Swerdlin, A., Wood, K., 2004. Preference for natural: Instrumental and ideational/moral motivations, and the contrast between foods and medicines. Appetite 43 (2), 147–154. https://doi.org/10.1016/j.appet.2004.03.005.

Sasidharan, S., Chen, Y., Saravanan, D., Sundram, K.M., Yoga Latha, L., 2011. Extraction, isolation and characterization of bioactive compounds from plants' extracts. Afr. J. Tradit. Complement. Altern. Med. 8 (1), 1–10.

Sathishkumar, G., Felix, N., Ranjan, A., Uma, A., Rajalakshmi, K., 2025. Exploring the impact of selected functional feed additives on growth performance, nutrient utilization, enzyme activities and immune gene expression of striped murrel (Channa striata) juveniles. Probiotics Antimicrob. Proteins doi:10.1007/s12602-025-10598-z.

Semwogerere, F., Chikwanha, O.C., Katiyatiya, C.L.F., Marufu, M.C., Mapiye, C., 2023. Bioavailability of bioactive phytochemicals in selected tissues and excreta from goats fed hempseed cake (Cannabis sativa L.) finisher diets. Trop. Anim. Health Prod. 55 (4). doi:10.1007/s11250-023-03676-3.

Sharma, B., Manepalli, N., 2019. Phytogenics: an innovative new trend in animal feed. MarketResearch.com Blog. https://blog.marketresearch.com/phytogenics-a-surprising-new-trend-in-animal-feed.

Sibbald, A.M., Erhard, H.W., McLeod, J.E., Hooper, R.J., 2009. Individual personality and the spatial distribution of groups of grazing animals: an example with sheep. Behav. Processes. 82 (3), 319–326. doi:10.1016/j.beproc.2009.07.011.

Singh, V.B., Singh, V.K., Tiwari, D., Gautam, S., Ruma, D., Singh, S.P., Chaturvedi, S., Singh, P., 2018. Effect of a phytogenic feed additive-supplemented diet on economic efficiency and cost of production of broiler chickens. Int. J. Curr. Microbiol. Appl. Sci. 7 (Special Issue-7), 2831–2837.

Stampa, E., Schipmann-Schwarze, C., Hamm, U., 2020. Consumer perceptions, preferences, and behavior regarding pasture-raised livestock products: a review. Food Qual. Preference 82, 103872. doi:10.1016/j.foodqual.2020.103872.

Steiner, D., Krska, R., Malachová, A., Taschl, I., Sulyok, M., 2020. Evaluation of matrix effects and extraction efficiencies of lc–ms/ms methods as the essential part for proper validation of multiclass contaminants in complex feed. J. Agric. Food Chem. 68 (12), 3868–3880. doi:10.1021/acs.jafc.9b07706.

Technical Team, EW Nutrition, 2021. Encapsulation: how a modern phytogenic feed additive makes all the difference. EW Nutrition. https://ew-nutrition.com/en-uk/encapsulation-modern-phytogenic-feed-additive/.

Thery, T., Lynch, K.M., Zannini, E., Arendt, E.K., 2020. Isolation, characterisation and application of a new antifungal protein from broccoli seeds – New food preservative with great potential. Food Control 117. doi:10.1016/j.foodcont.2020.107356.

Tomassen, F.H.M., de Koeijer, A., Mourits, M.C.M., Dekker, A., Bouma, A., Huirne, R.B.M., 2002. A decision-tree to optimise control measures during the early stage of a foot-and-mouth disease epidemic. Prev. Vet. Med. 54 (4), 301–324. doi:10.1016/s0167-5877(02)00053-3.

Torres Acosta, J.A., De Almeida Engler, J., Raes, J., Magyar, Z., De Groodt, R., Inzé, D., De Veylder, L., 2004. Molecular characterization of Arabidopsis PHO80-like proteins, a novel class of CDKA;1-interacting cyclins. Cell. Mol. Life Sci. 61 (12), 1485–1497. doi:10.1007/s00018-004-4057-4.

Usman, I., Sana, S., Afzaal, M., Imran, A., Jaffar, H.M., Sukhera, S., Munir, M., Asghar, A., Urugo, M.M., 2025. Phytochemicals in plant food waste: Innovative insight and implications. J. Agric. Food Res. 20. doi:10.1016/j.jafr.2025.101779.

Van Reenen, C.G., O'Connell, N.E., Van Der Werf, J.T.N., Korte, S.M., Hopster, H., Jones, R.B., Blokhuis, H.J., 2005. Responses of calves to acute stress: individual consistency and relations between behavioral and physiological measures. Physiol. Behav. 85 (5), 557–570. doi:10.1016/j.physbeh.2005.06.015.

Varol, B., 2023. Overcoming regulatory hurdles for plant-based feed additives. Feed Addit. https://www.feedandadditive.com/overcoming-regulatory-hurdles-for-plant-based-feed-additives/.

Verbeke, W., Scholderer, J., Lähteenmäki, L., 2009. Consumer appeal of nutrition and health claims in three existing product concepts. Appetite 52 (3), 684–692. https://doi.org/10.1016/j.appet.2009.03.007.

Westendorf, M.L., 2000. Food Waste to Animal Feed. Iowa State University Press, Ames, IA https://doi.org/10.1002/9780470290217.

Willett, W.C, Sacks, F., Trichopoulou, A., Drescher, G., Ferro-Luzzi, A., Helsing, E., Trichopoulos, D., 1995. Mediterranean diet pyramid: a cultural model for healthy eating. Am. J. Clin. Nutr. 61 (6), 1402S. doi:10.1093/ajcn/61.6.1402S.

Yildiz, D., 2024. Phytogenic feed additives and global market status. Feed Addit. https://www.feedandadditive.com/phytogenic-feed-additives-and-global-market-status/.

Zollers, B., Hüttinger, C., Payo Lewis, L., 2024. The road to regulatory approval of animal nutrition and health products. Feed Addit. https://www.feedandadditive.com/the-road-to-regulatory-approval-of-animal-nutrition-and-health-products/.

Chapter 13

Future directions at the feed-food-health nexus

Ahmed I. Elsherbeni[a], **Ayman A. Swelum**[b], **Elwy A. Ashour**[c] **and Mohamed E. Abd El-Hack**[d]

[a]*Animal Production Systems Research Department, Animal Production Research Institute, Agricultural Research Center, Giza, Egypt,* [b]*Department of Theriogenology, Faculty of Veterinary Medicine, Zagazig University, Zagazig, Egypt,* [c]*Poultry Department, Faculty of Agriculture, Zagazig University, Zagazig, Egypt,* [d]*Department of Industrial Pharmacy, College of Pharmaceutical Sciences and Drug Manufacturing, Misr University for Science and Technology (MUST), Giza, Egypt*

13.1 Introduction

In recent decades, diet has been observed to strongly influence both mental health and cognitive performance through numerous biological and psychological pathways. This has led to the emergence of a new class of nutritional innovations known as neuro-herbal foods (NHFs), which utilize a combination of active plant and nutrient components and microbiome modifiers to improve mood and cognition (Borrego-Ruiz and Borrego, 2025; Tuigunov et al., 2025). These foods represent an advancement in nutrition science that integrates nutritional psychology, pharmacology, and microbiome research (Gómez-Pinilla, 2008; Berding et al., 2021).

With the increasing prevalence of mental health disorders and the difficulty in accessing or accepting pharmaceutical treatments, a nutritional approach can provide a complementary and preventative strategy within the framework of "food as medicine" (Gowrishankar and Bruchas, 2019; Fekete et al., 2025). The scientific basis for these foods rests on converging evidence demonstrating that foods rich in rebiotics and plant compounds can influence nerve signaling, synapses, and inflammatory pathways along the gut–brain axis (Luna and Foster, 2015; Cryan et al., 2019). Polyphenols, omega-3 fatty acids (OM_3FAs), probiotics, and adaptogenic herbs such as *Bacopa monnieri* and *Panax ginseng* have the potential to modulate and enhance neurotransmitter systems and reduce oxidative stress (González, 2023; Lorca et al., 2023; Chowdhury, 2024). These recent findings have led to the development of a range of neuroprotective foods,

Plant-Derived Feed Additives and Animal-Based Foods: Linking Animal Health, Product Quality, and Consumer Perceptions. DOI: https://doi.org/10.1016/B978-0-443-45446-2.00014-4

such as fortified meats and fermented dairy products, as cognitive enhancers (Clark et al., 2020; Liu et al., 2022).

Neurofood innovation combines the sensory and cultural dimensions of food with scientific rigor, manifesting in two main branches: "Mood-enhancing meats" and "cognitive cooking." The former refers to animal-based foods fortified with neurotrophic factors or plant extracts to modulate mood, such as animal or poultry products rich in probiotics and OM_3FAs (Modlinska and Pisula, 2018; Liu et al., 2022). "Cognitive cooking" focuses on plant-based systems that include fermented foods and polyphenol-rich herbs to enhance immunity, memory, and executive functions (Lorca et al., 2023). These categories reveal how new nutritional concepts can be adapted to different dietary systems and cultures, emphasizing that neurofoods are holistic systems closely linked to everyday cooking and scientific evidence, rather than simply commercial products.

The international regulatory environment presents both opportunities and challenges for integrating such products into general dietary recommendations, particularly regarding evidence documentation, ensuring public safety, and establishing clear labeling to facilitate the inclusion of probiotics, OM_3FAs, and micronutrients. However, the fact that NHFs are hybrid products, falling between medicines, supplements, and foods, makes their regulation more complex. Therefore there is an urgent need to create coherent frameworks that balance innovation, consumer trust, scientific credibility, and general cultural acceptance, given the regulatory uncertainty, the wide variation in evidence, and the risk of exaggerated claims (Arai, 2021).

This chapter addresses the scientific and ethical foundations of NHFs and aims to clarify terms such as "mood-enhancing meats" and "cognitive cuisine." It also attempts to summarize the latest evidence, review current regulations, and explore the possibilities of incorporating these foods into national guidelines. Overall, this chapter positions NHFs within the framework of evidence-based and sustainable "food-as-medicine" as a potential contribution to improved mental health.

13.2 Emerging concepts: Mood-meat, cognitive cuisine

The concept of "neurophytonutrients" emerges from the convergence of microbiome science, phytopharmacology, and nutritional psychology, utilizing specific components and formulations to enhance mood, brain health, and various cognitive functions (Tuygunov et al., 2025). Both "mood meats" (meat products designed to influence mood) and "cognitive kitchens" (meal plans designed for cognitive function) exemplify the potential for innovation and development within this field.

Currently, growing research demonstrates the crucial role of modifiable diet in mental health and cognition. Studies on plant-based cognitive enhancers and nutritional psychology reveal that a range of dietary components can influence the gut–brain axis, reduce neuroinflammation, and modulate neurotransmitter systems (Gómez-Pinilla, 2008; Borrego-Ruiz, 2025). This has given rise to the concept of neurotrophic foods, or foods designed to support neurological performance.

"Cognitive kitchen" and "mood meat" represent two complementary approaches within this field. Mood meat refers to animal products modified to affect mood by enhancing micronutrient density (such as iron, B vitamins, and zinc), along with OM_3FAs, probiotics, and herbal extracts. While conclusive clinical evidence is lacking, preliminary evidence suggests that consumers are willing to pay for foods that claim to offer various mood-enhancing benefits, aligning with the growing demand for such foods that combine enjoyment with health benefits (Modlinska and Pisula, 2018). Cognitive kitchen comprises a set of meal plans aimed at improving short-term cognitive performance and protecting brain health by focusing on polyphenol-rich foods, fermented foods containing probiotics, and memory-enhancing plant extracts. These strategies can be implemented in scientific institutions, various work environments, nursing homes, and for personal nutrition. While preclinical and early clinical findings support these interventions, larger, longer-term, and controlled studies are still needed to translate them into environmentally appropriate cooking practices (Lorca and Lorca, 2023). The mechanisms behind these approaches are becoming clearer. Polyphenols and OM_3FAs reduce neuroinflammation and oxidative stress through their antiinflammatory and antioxidant properties (Borrego-Ruiz, 2025); probiotics and prebiotics improve gut microbiota populations with subsequent effects on brain signaling (Chowdhury, 2024); and cognitive phytostimulants such as ginseng and *Bacopa monnieri* (BM) exhibit neuroprotective and cholinergic activities (Lorca et al., 2023). Together, Mood Meat and Cognitive Cuisine establish the ability of NHFs to innovate interdisciplinarily by bringing together food technology, nutrition, sensory science, psychobiology, and food science to develop next-generation dietary systems that promote psychological and cognitive well-being; thus previous mechanistic insights support the viability of this approach.

13.3 Integration into national dietary guidelines

National dietary guidelines translate the latest nutrition science into public health recommendations. Historically, guidelines have included food categories and nutrients proven safe and beneficial at the population level (e.g., fruit and vegetable recommendations and saturated fat limitations) (Herforth et al., 2019; Dean et al., 2024a). The advent of products marketed specifically to influence mood or cognitive function (which are now referred to as NHFs) compels the authors of the guidelines to reconsider the evidence-based and regulatory criteria used for previous innovations such as probiotics, OM_3FAs, and the fortification of staple foods (Fig. 13.1).

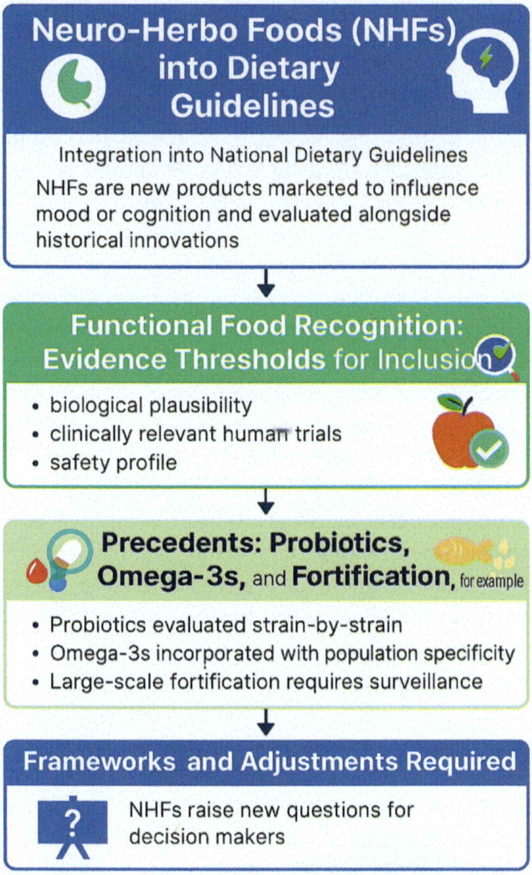

FIGURE 13.1 The incorporation of NHFs into national dietary guidelines: Supporting data, established practices, and legislative frameworks. *NHFs*, Neuroherbal foods.

13.4 Functional food recognition: Evidence thresholds for inclusion

Three pillars should support the inclusion of guidelines: Biological plausibility (mechanistic rationale), consistent data from human trials demonstrating clinically relevant benefit in the target demographic, and a safety profile acceptable to likely consumers (Ezeanochie et al., 2024). When evaluating causation claims, decision makers (such as ministries of health, national nutrition councils, or guidelines committees) typically seek systematic reviews and meta-analyses of randomized controlled trials (RCTs), supported where possible by dose-response data and repeatable biomarkers (Brambilla et al., 2020; Howitt and Taylor, 2020). When RCTs are limited, strong mechanistic data and high-quality observational studies can help direct research objectives and offer tentative recommendations with caution. Probiotics and other functional substances have established procedures that align with these concepts.

13.5 Precedents: Probiotics, omega-3s, and fortification, for example

Many national dietary guidelines have avoided classifying probiotics as a single category, despite international expert groups emphasizing the need for strain characterization, safety assessment, and effective human trials before supporting any health claims. This caution stems from the considerable variability among strains and the inconsistency of evidence, making the handling of probiotic strains a matter of prudence and reliance on evidence alone (European Food Safety Authority EFSA, 2024).

OM_3FAs represent another example. Based on rigorous systematic reviews, reputable bodies have recommended specific levels of EPA, DHA, and DPA for certain life stages and conditions, while others permit conditional health claims where reliable standards are met (Tuscher et al., 2019; Lewis et al., 2025). OM_3FAs, for example, illustrate how a widely available nutrient with reliable biomarkers can be included in guidelines with age-specific differentiation (Walmar, 2025a).

Population-level nutritional fortification represents a third model: The World Health Organization and other organizations recommend fortification when there is a proven deficiency, technical feasibility, economic viability, and robust safety monitoring systems. This should be accompanied by legal measures or incentives to ensure equitable access, as well as public health monitoring to control dosage and safety (Bashir et al., 2024; Ashraf, 2025; Deman et al., 2025; Soni et al., 2025) (Fig. 13.1).

13.6 Neuroherbal foods and population mental health

13.6.1 Role in mental well-being promotion strategies

We increasingly recognize dietary interventions as components of mental health promotion and prevention efforts. Connections exist between specific nutrients and dietary patterns (such as omega-3, B vitamins, and diets rich in polyphenols) and mood/cognitive outcomes, and clinical trials in nutritional psychiatry are expanding (Ortega et al., 2022; Raza et al., 2025). NHFs may be valuable additions to population strategies, particularly when intended for higher-risk populations with nutritional deficiencies that affect neurotransmitter synthesis or inflammation, as long as claims are evidence-supported and incorporated into larger lifestyle and clinical care pathways (Grajek et al., 2022). The following are some possible roles: (1) improving food programs at work and in schools to promote cognitive function at periods of high demand (Daniels et al., 2022), (2) integrating into perinatal nutrition guidelines where micronutrient sufficiency affects the mood of the mother (Naaz and Muneshwar, 2023; Dudley, 2025), and (3) providing dementia-risk reduction components within senior nutrition programs (Anderson et al., 2025; Matton et al., 2025). Moreover, NHFs should be positioned as complementary, evidence-based elements within comprehensive frameworks for mental health promotion, rather than as stand-alone therapies for psychiatric diseases (Grajek et al., 2022).

13.6.2 Challenges in standardizing intake recommendations

Standardizing the diverse forms of dietary supplements presents numerous scientific and practical challenges. The wide variety of formulations, including plant extracts, probiotic strains, and omega-3 levels, precludes the development of standardized dosage recommendations (Rao et al., 2024). Furthermore, underlying individual characteristics, such as nutrient deficiencies, gut microbiome composition, and medication use, influence response rates and safety, highlighting the need for tailored guidelines rather than a one-size-fits-all approach (Shukla et al., 2024; Rodriguez et al., 2025). Regulatory differences and interactions between foods, supplements, and therapeutic foods also affect permissible claims and stringent regulatory requirements. Consequently, the safety of individuals in cases of overdose and potential interactions, such as those between herbs and drugs, must be considered (Beeb, 2023).

13.6.3 Practical framework for guideline integration

Incorporating natural health foods into national dietary guidelines requires a practical, evidence-based approach (FAO/WHO, 2006; Hill et al., 2014). First, expert committees should adopt multitiered classification systems—as is done with probiotic and micronutrient guidelines—to rank recommendations based on the quality and reliability of RCTs and rigorous systematic reviews, distinguishing between claims supported by strong evidence and those with less robust evidence (U.S. Department of Health and Human Services & U.S. Department of Agriculture, 2020; NIH Office of Dietary Supplements, 2023). Second, active ingredients (such as probiotic strains and omega-3 dosages) must be precisely identified and characterized, including their stability and bioavailability (Calder, 2018; Roe et al., 2022). Third, priority should be given to safety assessments, interventions, and monitoring of long-term effects, as demonstrated by the cases of iodine and folic acid fortification (Olson et al., 2021). Fourth and finally, equity and fair access must be considered, as expensive health foods may exacerbate health inequalities if supportive policies are not in place (Allen et al., 2006; Beal et al., 2017). Field trials and phased implementation in settings such as workplaces, schools, and universities can provide realistic data on feasibility, efficacy, and safety before widespread adoption. This promotes the evaluation and integration of natural health foods (Ashwell et al., 2020).

13.6.4 Multidisciplinary research models

Over the past 10 years, there has been a sharp increase in interest in foods and food ingredients that affect brain health, whether directly through neuroactive phytochemicals, microbiome-mediated processes, or nutrient-regulated neurotrophic pathways. Research goals centered on diet–brain interactions are described by terms such as "FFS, nutraceuticals," and "probiotics" (Karantonis, 2025). Intentionally created foods that contain herbal elements (such as adaptogens or botanicals high in polyphenols) or isolates obtained from herbs with the goal of enhancing cognitive, emotional, or neurodevelopmental outcomes are known as NHFs (Grosso et al., 2023; Ingegneri et al., 2025). Neuroherbo foods research necessitates integrated, multidisciplinary models that span laboratory, animal, clinical, and population sciences due to the multiple mechanisms involved, which include neurochemical, immunological, endocrine, and microbial pathways. Translational pipelines connecting mechanistic indicators to validated psychometric outcomes are needed, according to recent reviews that highlight the critical role of nutrition, microbiome, and brain interactions (Berding et al., 2021; Serrano et al., 2025; Slykerman et al., 2025).

13.6.5 Cross-disciplinary collaboration

Food science (formulation, stability, and bioavailability), veterinary and animal sciences (preclinical efficacy and safety in pertinent species), neuroscience (mechanistic and biomarker assays), and psychology/psychiatry (measuring cognitive and affective outcomes) are all areas of expertise needed for NHFs development and evaluation (Makkar et al., 2020).

Institutions should put in place integrated research platforms with interoperable data standards (metadata schemas, ontologies) and governance procedures that handle technological, ethical, and legal issues to facilitate long-term collaboration. Pooled datasets (pharmacokinetic profiles, microbiome sequences, and psychometric batteries) are made possible by shared platforms, which also allow for reuse for repeatability checks and secondary analysis. Shared platforms enable pooled datasets (e.g., pharmacokinetic profiles and microbiome sequences) and permit reuse for secondary analyses and reproducibility checks (Barker et al., 2022; Behan et al., 2023).

Developing NHF-specific common data elements is essential. Suggested shared databases and common data elements (CDE) domains include food/formulation descriptors (botanical source, batch metadata, phytochemical fingerprint, and stability) (Naravane, 2024; Cui et al., 2025), dosing and dietary context (matrix, coingested foods) (Russell and Kingsley, 2014; De Castro Cogle et al., 2024), pharmacokinetics and metabolomics (plasma metabolites, brain-penetrant compounds) (Alves et al., 2024; Rahban et al., 2024), microbiome taxonomic and functional profiles, inflammatory and neurotrophic biomarkers (Bakir-Gungor et al., 2022; Onisiforou and Spyrou, 2022; Yuan et al., 2025), and harmonized psychometric instruments. Standardized CDEs accelerate meta-analyses, power calculations for large-scale trials, and regulatory dialogues. Practical precedents exist in probiotic/prebiotic literature where microbiome and psychometric outcomes have been jointly reported (Misera et al., 2021).

13.6.6 Systematic trial designs

Validated psychometric tools and objective biochemical measurements should be incorporated into integrated evaluation frameworks to substantially illustrate the neurocognitive and affective effectiveness of NHFs. Using validated measures for mood (e.g., Patient Health Questionnaire-9 [PHQ-9]; Kroenke et al., 2001), depression (Beck Depression Inventory-II; Beck et al., 1996), and anxiety (Generalized Anxiety Disorder-7 [GAD-7]; Spitzer et al., 2006), psychometric endpoints offer standardized assessment of psychological and cognitive domains. Neuropsychological tests used to assess memory, control, and processing speed can be employed to evaluate cognitive performance. Examples of such tests include the National Institutes of Health (NIH) Toolbox and the Cambridge Automated Battery of Neuropsychological Tests (CANTAB) (Weintraub et al., 2013; Langley et al., 2023). Furthermore, recording changes in mood and cognitive function in a natural, real-time environment enhances the credibility of environmental economics (Shiffman et al., 2008; Schneider et al., 2020). Numerous studies have demonstrated that measuring peripheral biomarkers associated with neuroinflammation, neuroplasticity, and stress physiology provides an automated understanding of processes related to neurodegenerative stress. These include cortisol diurnal patterns, neurotrophic factors such as brain-derived neurotrophic factor (BDNF), which indicate neuroplastic adaptations, and serum and plasma levels of inflammatory cytokines, e.g., interleukin-6 (IL-6) and tumor necrosis factor-α (TNFα) (Duman and Monteggia, 2006; Sen et al., 2008).

Metabolomic analyses enable detection of NHFs-derived metabolites and host cometabolites influencing neural signaling (Nicholson et al., 2012), while gut microbiome profiling reveals diet–microbiota–brain interactions that may mediate cognitive and emotional outcomes (Luna and Foster, 2015; Cryan et al., 2019). Harmonizing psychometric and biochemical endpoints across studies will allow meta-analytic synthesis and facilitate the identification of reproducible biomarker signatures associated with NHFs' intake, thereby strengthening translational validity and public-health relevance.

13.7 Food–pharma convergence

A "food–pharma convergence"—a process where the pharmaceutical industry's formulation techniques, discovery pipelines, and regulatory policies progressively overlap with those of food science and nutrition—has emerged in the

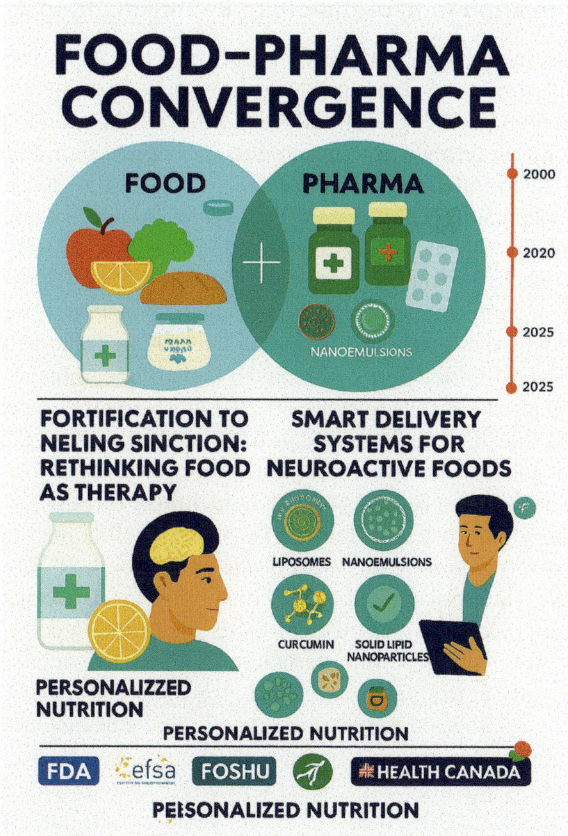

FIGURE 13.2 NHFs: The convergence of foods and pharmacology. *NHFs*, Neuroherbal foods.

global health sector during the past 20 years. This change reflects the increasing understanding that diet-derived bioactive that alter molecular targets that were previously thought to be only pharmacological can have an impact on neuropsychiatric disorders and chronic diseases (Weenen et al., 2013; Herforth et al., 2019) (Fig. 13.2). Product innovation (nutraceuticals, medical foods, and functional beverages), regulatory experimentation (such as the US medical foods category and Europe's novel foods framework), and translational science that connects the microbiome, nutrient metabolism, and neural signaling are all examples of this convergence (Dean et al., 2024a; Fekete et al., 2025). New therapeutic and preventive options are promised by this integration, but it also calls into question long-standing distinctions between "drugs" and "food," as well as between medical intervention and culinary culture.

13.7.1 Fortification to neurofunction: Rethinking food as therapy

Beyond conventional nutrient fortification (iron-enriched meat, vitamin D milk, etc.), functional meat and dairy products now have specific bifunctionality meant to affect mood and cognition. Modern innovations include probiotic- or postbiotic-containing fermented dairy products, designed to modulate the gut–brain axis and neuroinflammatory responses; omega-3-fortified milk and yoghurt, linked to decreased depressive symptoms and enhanced cognitive flexibility; and tryptophan-enriched dairy proteins, which function as serotonin precursors and may impact sleep and regulation (Cryan et al., 2019; Clarke et al., 2020; González, 2023). Similar patterns can be seen in "functional meat" projects, compounds, which investigate the neuroprotective and antiinflammatory benefits of enriching meat with bioactive compounds, including carnosine, taurine, and OM_3FAs (Liu et al., 2022; Zhang et al., 2023). Only a small number of products have undergone thorough RCTs that can fulfil pharmaceutical-grade evidence criteria, notwithstanding the molecular plausibility shown by in vitro and animal models. Hence, rather than focusing on risk reduction or therapy, the majority are limited to claims about structure/function or general well-being (e.g., "supports emotional balance") (Kušar et al., 2021; Neale and Tapsell, 2022).

On the other side, the United States distinguishes between conventional foods, dietary supplements, and medical foods under the Federal Food, Drug, and Cosmetic Act, which permits structure/function statements but forbids disease claims without drug approval. In contrast, the European Union's Regulation (EC) No. 1924/2006 requires human clinical substantiation for authorized health claims. Preapproved functional labeling based on compositional and efficacy data is possible with Japan's FOSHU system, which is a hybrid model (Katan et al., 2020; Arai, 2021). In addition, similar efforts to control this regulatory overlap are demonstrated by other jurisdictions, such as Australia's Food–Medicine Interface guidelines and Canada's Natural Health Products framework (Health Canada, 2022; Therapeutic Goods Administration TGA, 2023). To support reliable health claims and the safe market adoption of mood-supporting FFs, harmonized international standards and transparent evidence hierarchies are necessary. However, evidence heterogeneity, population variability, and dietary confounders continue to complicate data interpretation (Nieman and Trumbo, 2021).

13.7.2 Smart delivery systems for neuroactive foods

Poor aqueous solubility, instability during gastrointestinal digestion, and limited blood–brain barrier permeability continue to impede the translation of neuroactive phytochemicals, such as curcumin, resveratrol, gamma-aminobutyric acid analogs, and alkaloids, into effective FFs (Parvin et al., 2025a). Encapsulation and nanocarrier technologies have been incorporated into food-pharma design in response to these bioavailability issues, which aim to improve compound stability, absorption, and target specificity while maintaining regulatory safety and sensory quality. Numerous incorporation methods include solid lipid particles and composites that protect highly sensitive ingredients from oxidation and mitigate undesirable flavors; liposomes and nanoemulsifiers that enhance the solubility of hydrophobic polyphenols (Goktaş et al., 2020); and polymers such as chitosan, alginates, and whey protein matrices, which may impart expectorant properties (Madureira et al., 2022; Klojdová et al., 2023a; Klojdová et al., 2023b). From a food safety perspective, these techniques can exhibit pharmacological behavior similar to low-dose therapies when incorporated into processed foods (de Azeredo et al., 2021). Recently, these products have been evaluated through a multitiered pathway that includes both physicochemical characterization, in vitro digestion models, Caco-2 cell permeability tests, and blood–brain barrier cultures, followed by human clinical trials that measure these biochemical, microbial, and psychological responses (Su and Kang, 2020; Parvin et al., 2025b). Population-level fortification is also being improved in parallel with all these developments. In line with the concept of temporal nutrition, smart foods go beyond simply adding elements such as iodine, folic acid, or vitamin D; they rely on data-driven personalization by employing nutritional genetics data or microbiome analysis to adjust bioavailability to individuals' metabolic and circadian rhythms (Ordovás et al., 2023; Faienza et al., 2024a, 2024b). With all of the above, the classification of these enhanced foods as dietary supplements or medicines remains unclear, and disparities in access may lead to increased rates of inequality in personal nutrition, posing ethical and regulatory challenges (Monteiro et al., 2022; Rzymski et al., 2023).

13.7.3 Cultural compatibility and ethical governance in neuroherbal foods

The cultural, sensory, and ethical dimensions of the food industry influence its sustainability and public acceptance. While technological innovations improve everyday meals, consumer awareness remains focused on familiarity, taste, and the impression of naturalness, which often outweighs interest in advertised health benefits (Verbeek et al., 2021; Baker et al., 2022a; Baker et al., 2022b). Acceptance increases significantly when bioactive compounds are incorporated into products that align with local dietary identities—such as probiotic yogurt in many Mediterranean countries, fermented soybeans and herbal teas in East Asian countries, or fortified bread in many Middle Eastern countries (Rosin, 2020; Savvaidis et al., 2022; Ray et al., 2024). Conversely, potent foods may blur the line between therapy and nutrition, appearing alien or unfamiliar and raising numerous concerns about safety and commercial motives (Bogan, 2021;

Chen et al., 2023). A practical framework suggests a balance between scientific advancements, traditions, and social values through a design that includes chefs, ethnographers, sensory experts, and impartial consumer panels (Schnitler et al., 2021). Furthermore, distributive equity, autonomy, ethical consent, and cultural compatibility must be considered, as these foods, unlike medicines, are widely consumed without sufficient awareness of their bioactive content. Folic acid fortification is one historical example that demonstrates both the advantages for public health and the risk of overconsumption among some subgroups (Brouwer et al., 2019; Abdul Manan and Salmerón, 2025). Such worries are heightened when neuroactive substances are found in common meals, especially for vulnerable groups such as youngsters, expectant mothers, or people on medication. Therefore before enacting extensive fortification or bioactive enrichment policies, responsible innovation necessitates population-specific risk–benefit analyses, postmarket neurobehavioral surveillance, transparent labeling of all functional additions, and organized public participation (Baker et al., 2022b; Chen et al., 2023). In the end, as important as the societal legitimacy of food-pharma products is their biochemical efficacy, assuring cultural compatibility and ethical accountability (Fig. 13.2).

13.8 Recommendations for policymakers, industry, and researchers

Consumers, researchers, and businesses are becoming more interested in foods and herbs that affect brain function, whether through micronutrient support for neurotransmitter synthesis, gut–brain axis modulation, or neuroprotective phytochemicals. These days, probiotic/prebiotic formulations, fortified products, and traditional and experimental botanicals (such as turmeric/curcumin, sage, and Bacopa) are frequently marketed with claims of mood or cognitive benefits (Giromini et al., 2021; Altaf, 2024). However, scientific evidence varies from small randomized trials to mechanistic lab work, legal frameworks for nutrition and health claims vary greatly, and industry labeling practices can deceive consumers when claims are unsupported (Todt and Luján, 2017; Ashwell et al., 2023). To guarantee that neuroherbo foods provide genuine health benefits free from harm or false information, this combination calls into question the proper regulation, industry practices, and research priorities.

13.9 Policy frameworks

Regional and international regulatory frameworks already offer helpful guidelines for nutrition and health claims that can be modified for newly developed neurofunctional or NHF products. According to the Codex Alimentarius General Guidelines on Claims, all claims must be supported by evidence, and no food representation may be false, misleading, or deceptive (Grossklaus, 2009; Aggett et al., 2012; Boatwright et al., 2024). In addition, health claims must be supported by a solid and adequate body of scientific data, be in line with national nutrition and health policy, and refrain from deceiving consumers or negatively changing dietary patterns (FAO/WHO, 2004). For instance, a standardized framework was established in the European Union by Regulation (EC) No. 1924/2006: only approved claims may be used, nutrition and health claims must be supported by science, and submitted and approved claims are recorded in a public EU register; furthermore to providing regulatory guidelines on phrasing, satisfactory health claims, and circumstances (such as nutrient profiles) under which claims may be used, the European Union framework separates generic or ambiguous claims from those that have been properly examined (Laser; European Commission, 2007; Reuterswärd, 2007; Boer, 2021). On the other hand, the Food and Drug Administration (FDA) and the Federal Trade Commission (FTC) in the United States, on the other hand, make a distinction between health, nutrient-content, and structure/function claims (Evans et al., 2025; Tai et al., 2025).

Health claims necessitate substantial scientific consensus or petitioned approval, whereas structure/function claims can be made more broadly (as long as they are not deceptive) with the proper disclaimers, but there is a chance that consumers will misinterpret them (e.g., claims must avoid implying treatment or disease cure) (Lindsey, 2009; Martirosyan and Stratton, 2023; Wang, 2024). To better protect consumers, US regulatory initiatives have more recently suggested modernizing front-of-pack nutrition labeling, revising the definition of a "healthy" label, and increasing transparency in claims (Brown, 2022; Marks et al., 2024; Muzzioli et al., 2025). Furthermore, the new One Health paradigm, which acknowledges the interdependencies between human, animal, and environmental health, proposes that policies should be consistent across domains when neuroactive bioactive compounds come from ecological or animal sources (e.g., welfare, residue safety, ecological sustainability, and translational validity of animal studies) (Dikšaitytė et al., 2021).

Therefore three guiding concepts should be prioritized while modifying current frameworks to accommodate NHFs. First, evidence proportionality makes sure that the quality, amount, and reproducibility of supporting data match the specificity and strength of health or cognitive claims. The Codex Alimentarius Guidelines for Use of Nutrition and Health Claims are in line with this (Ekkekakis and Tiller, 2023; FAO/WHO, 2004). Second, to reduce customer misunderstanding, transparency necessitates the clear disclosure of ingredient provenance, sustainability credentials, evidentiary limits, and potential safety or interaction caveats. In the food and nutraceutical industries, transparent labeling and traceable supply chains are becoming more and more important for customer trust and sustainability (Nazzaro, 2020a; Cao et al., 2023; Komala et al., 2023; Gupta et al., 2025). Third, caution and equity highlight the moral and societal obligation of regulators to guarantee that vulnerable groups (such as those with mental health disorders or low health literacy) are not disproportionately exposed to risk and to prevent harm, whether through nutrient displacement, herb–drug interactions, or deceptive "neuroenhancement" marketing (Verhagen and Van Lovren, 2016; Wies et al., 2021; Bhugra et al., 2022). Targeted modifications, such as a tiered claim taxonomy, public claim registries, and cross-sectoral cooperation, could build on existing precedents and allow neurofunctional food innovation to proceed responsibly while preserving public confidence, consumer protection, and transparency.

13.10 Industry roadmap

In the field of functional neurofoods, both the food and supplement sectors play a crucial role in shaping public health outcomes and consumer confidence. Therefore manufacturers must maintain high standards of transparency, sustainability, and reliable scientific verification, as such products may claim to influence mood, brain function, and cognitive performance. Functional claims, ingredient sources, and realistic sustainability reporting, including the publication of both positive and negative scientific research, must be verified (Niessen, 2020; HRS Communications 2025). Many functional neurofood ingredients, such as OM_3FAs, *Panax ginseng*, and *Bacopa monnieri*, are sourced from environmentally sensitive areas. Concerns about animal welfare or poaching practices undermine consumer confidence and environmental goals (FAO, 2019; WHO, 2022). Therefore companies should showcase sustainability certifications, disclose the geographical origins of these ingredients, and implement various mechanisms to protect the environment and biodiversity (Damico et al., 2020). Such measures not only ensure compliance with environmental, social, and governance (ESG) requirements but also support the growing "food as medicine" strategy, where consumers prefer ethical, research-backed, and socially responsible medical products (Nestlé, 2023).

On the other hand, consumer awareness plays a crucial role in assessing both the safety and efficacy of nonsteroidal antiinflammatory drugs (NSAIDs) once they are marketed. Misinterpretations of claims—such as exaggerating product benefits or presenting it as an alternative to proven treatments—can lead to decreased public trust and potentially serious health consequences (Verhagen and Van Lovren, 2016; Nazzaro et al., 2020a). Therefore relevant authorities must clearly articulate realistic and tangible expected outcomes, safe dosage limits, and appropriate user groups (e.g., adults versus children, or individuals with comorbidities). To promote health awareness, the differences between the conclusions of laboratory studies and clinical trials should be highlighted

(Grunert et al., 2018). These efforts can be supported through collaboration between public health institutions, academia, and industry. Digital platforms, community seminars, and cofunded campaigns can also help to highlight the potential benefits and risks of neuroprotective foods (European Food Safety Authority EFSA, 2021; HRS Communications 2025). Providing safety information, contraindications, and authoritative references from sources such as the US FDA and the World Health Organization through informational brochures and QR code-enabled packaging is also recommended. This will help reduce misuse, encourage informed decision-making, and promote the responsible practice of the "food as medicine" concept.

13.11 Research funding and priorities

Cross-disciplinary cooperation is necessary due to the intricate relationship between mental health, neurology, and nutrition. Public and commercial research funders should establish grant programs that specifically recognize collaborations between nutrition science, psychiatry, phytochemistry, veterinary medicine, and data science while enforcing strict conflict-of-interest governance to expedite development (Hampel et al., 2019; Darko et al., 2025). The potential benefits of using plant extracts and bioactive compounds have been translated from laboratory and theoretical research into real-world human trials, supported by funding programs focused on applied neuronutrition research (Ames, 2018). Early initiatives such as the European Union's Horizon program and the National Institutes of Health (NIH) have demonstrated that intersectoral collaboration enhances the reliability of research methodologies and promotes the application of findings (Karin et al., 2022; Sipido et al., 2022; Zafar, 2025). Furthermore, measures such as different data-sharing requirements, the use of previously established methodologies, and the promotion of open-source articles can increase both reproducibility and transparency (Azevedo et al., 2019; Ioannidis, 2019; Dluzen and Schmidt, 2023).

Providing empirical evidence that is both biologically grounded and behaviorally relevant is crucial for the future of functional foods (FFS). This requires extensive and rigorous applied research to measure the effects on mental and psychological health. Therefore funding should focus on three complementary research pathways: (1) Real-world case studies evaluating the efficacy, adherence, acceptability, and long-term safety of treatments (Gregg et al., 2023); (2) RCTs to assess clinically relevant neuropsychological outcomes, such as improved mood, stress tolerance, and cognitive function (McGorry et al., 2020); and (3) behavioral science research investigating how consumers understand and respond to claims related to neurological functions (Jacka et al., 2017; Benton, 2020).

The integration of preliminary findings, which often focus on neurochemical or microbiome mechanisms, into large-scale, well-regulated human studies remains crucial for clinical validation and regulatory approval (European Food Safety Authority (EFSA), 2021; Wubairis et al., 2023). Furthermore, incorporating biomarkers, digital cognitive assessment, and multigenomic data analysis can improve the correlation between consumption patterns and neurobehavioral outcomes (Schmidt et al., 2021; Linseisen et al., 2025). Beyond biological validation, research into behavioral nutrition is essential for understanding the impacts of language, cultural context, and media coverage on the perceived safety and efficacy of neuro-friendly foods. Overall consumer confidence can also be influenced by the marketing strategies employed for various products, whether as mood enhancers or complementary therapies (Nazzaro et al., 2020b; Achemann-Wetzel and Grünert, 2022). Social and cultural factors, such as dietary habits, the symbolic meanings of different herbs, and varying attitudes toward "medicinal" foods, must also be considered when designing experiments (Rosin et al., 2022). Different funding organizations and public–private partnerships can enhance the efficiency of scientific research. For example, priority can be given to projects that achieve significant improvements in the mental health of the general population, reduce absenteeism due to stress, or boost cognitive productivity, as suggested by Godfrey et al. (2022) .

13.12 Conclusion

Recently, the link between nutrition, brain health, and societal well-being has been significantly strengthened by the advent of neurotrophic herbal foods. These foods demonstrate how dietary patterns can be designed to produce tangible and noticeable effects on mood, cognitive abilities, and psychological well-being by combining nutritional science, neuroscience, herbal medicine, and innovative culinary techniques. However, ensuring the accuracy and reliability of claims regarding their neurotrophic effects, conducting thorough safety assessments, and guaranteeing equitable access to them remain crucial for their integration into public health programs. While the hierarchy of scientific evidence helps distinguish between promising and proven treatments, establishing clear regulatory frameworks, based on precedents such as probiotics, OM_3FAs, and fortified foods, can ensure the high degree of accuracy required for their inclusion in official dietary recommendations. Collaboration among nutritionists, neurologists, clinicians, behavioral scientists, and policymakers is essential for any future progress in this field. To avoid the pitfalls of excessive marketing and misinformation, sustainable promotion must be based on transparency and consumer education. Ultimately, the future of natural remedies hinges on demonstrating not only their biological efficacy but also their behavioral and social significance, transforming the concept of "mind food" into a tangible reality in the mental health of communities, grounded in scientific evidence and governed by sound ethical frameworks Table 13.1.

TABLE 13.1 Comparing new functional food types and their current regulations.

Category	Main features and examples	Regulatory framework	Policy frameworks
Herb-based functional foods (adaptogens, botanical extracts, and neuro-modulatory herbs)	Botanicals and bioactive substances derived from plants that have metabolic or neuropsychological advantages (e.g., curcumin, ginseng, and *Bacopa monnieri*)	The terms "foods" and "supplements" frequently overlap, categorized as either functional foods or nutraceuticals based on the claims made	EU: The Novel Foods and Health Claims Regulation governs botanicals (Turck et al., 2022). US: Only structure/function claims are permitted by the FDA (FDA, 2023). In India, botanicals are covered by the Nutraceutical Regulations of the FSSAI (FSSAI, 2022). There are still gaps in the evidence supporting neurofunctional claims (Bagchi et al., 2015; Ashwell et al., 2023)
Probiotic, prebiotic, and symbiotic-enriched foods	Living bacteria or oligosaccharides (such as *Lactobacillus*, *Bifidobacterium*, and inulin) that support gut–brain health	Established regulatory precedents; assessed for dosage, claim specificity, and strain safety	EU: Claims must be supported by strain-specific evidence (European Food Safety Authority EFSA, 2024,). Japan: If safety and effectiveness are established, permitted under FOSHU (Arai, 2021). US: No illness claims; permitted structure/function claims (Howitt et al., 2020b)

(*continued on next page*)

Fortified functional food, enriched with omega-3	Foods (such as omega-3 milk and bread supplemented with folate) that have been fortified with vitamins, minerals, and EPA/DHA	Well acknowledged in relation to nutritional reference values and fortification schemes	WHO/FAO: Based on the prevalence of deficiencies, recommend fortification (). Quantitative intake guidelines have been set in the United States and the EU (Tocher et al., 2019; Wollmar, 2025a; Wollmar, 2025). Overexposure and required fortification continue to raise ethical problems (Olson et al., 2021)
Cell-based protein or farmed beef	Meat produced from animal cells; sometimes known as "cultivated" or "clean" meat	Categorized as a novel food that needs a safety study before going on sale	Singapore: SFA gives initial approval (GFI, 2024). US: FDA-USDA joint supervision (FSIS, 2023). EU: governed by the European Commission, 2023. Novel Foods Regulation (EC) No. 2015/2283. Debates on ethics and labeling are still going on (Fekete et al., 2025)

(*continued on next page*)

TABLE 13.1 Comparing new functional food types and their current regulations—cont'd

Category	Main features and examples	Regulatory framework	Policy frameworks
Neuro-functional foods and NHFs	Functional foods or formulations (such as mood meat or cognitive cuisine) that target mood, cognition, or neuroprotection	There is not a specific category yet; it is assessed under the current regulations governing food, supplements, and medicinal foods	EU: EFSA evaluates individual health claims related to cognition (Dean et al., 2024b). US: According to the FDA (2023), only structure/function claims are permitted. Asia: New regulations are being proposed (China NHC draft, 2024). Inability to establish precise criteria for proof (Jacka et al., 2017; Lorca et al., 2023; Tuigunov et al., 2025).
Food supplements and nutraceuticals	Concentrated extracts of nutrients that are not food (tablets, capsules)	Distinct category under the law governing health products or dietary supplements	European Commission (2023); FDA (2023). Enforcement and claim monitoring deficiencies in developing nations (Azevedo et al., 2019; Nestle, 2023)

References

Abdul Manan, M., Salmerón, I., 2025. Progress in probiotic science: prospects of functional probiotic-based foods and beverages. Int. J. Food Sci. 2025 (1). doi:10.1155/ijfo/5567567.

Aggett, P.J., Hathcock, J., Jukes, D., Richardson, D.P., Calder, P.C., Bischoff-Ferrari, H., Nicklas, T., Mühlebach, S., Kwon, O., Lewis, J., Lugard, M.J.F., Prock, P., 2012. Nutrition issues in Codex: health claims, nutrient reference values and WTO agreements: a conference report. Eur. J. Nutr. 51 (1). doi:10.1007/s00394-012-0306-8, S1.

Allen, L., de Benoist, B., Dary, O., Hurrell, R., 2006. Guidelines on Food Fortification with Micronutrients. World Health Organization and Food and Agriculture Organization of the United Nations (FAO), Geneva, Switzerland.

Altaf, A., 2024. Functional foods and nutraceuticals as immunity boosters and in combating lifestyle diseases. Functional Foods and Nutraceuticals: Chemistry, Health Benefits and the Way Forward 363–383. doi:10.1007/978-3-031-59365-9_17.

Alves, L.F., Moore, J.B., Kell, D.B., 2024. The biology and biochemistry of kynurenic acid, a potential nutraceutical with multiple biological effects. Int. J. Mol. Sci. 25 (16). doi:10.3390/ijms25169082.

Ames, B.N., 2018. Prolonging healthy aging: longevity vitamins and proteins. Proc. Natl. Acad. Sci. 115 (43), 10836–10844. doi:10.1073/pnas.1809045115.

Anderson, N.D., D'Amico, D., Rotenberg, S., Addis, D.R., Gillen, J., Moore, D., Chertkow, H., 2024. Validation of a community-based approach toward personalized dementia risk reduction: the Kimel Family Centre for Brain Health and Wellness. J. Prev. Alzheimer's Dis. 11 (5), 1455–1466.

Arai, S., 2021. Functional food science in Japan: history and current perspectives. Biosci. Biotechnol. Biochem. 85 (1), 1–10.

Aschemann-Witzel, J., Grunert, K.G., 2022. Influence of food marketing claims on consumer perception and behavior: implications for functional foods. Trends Food Sci. Technol. 123, 258–270. doi:10.1016/j.tifs.2022.03.015.

Ashraf, S.A., 2025. Food fortification as a sustainable global strategy to mitigate micronutrient deficiencies and improve public health. Discover Food 5 (1). doi:10.1007/s44187-025-00512-5.

Ashwell, M., Hickson, M., Stanner, S., Prentice, A., Williams, C.M., 2023. Nature of the evidence base and strengths, challenges and recommendations in the area of nutrition and health claims: a position paper from the Academy of Nutrition Sciences. Br. J. Nutr. 130 (2), 221–238. doi:10.1017/S0007114522003300.

Ashwell, M., Lambert, J.P., Alles, M.S., Brzozowska, A., et al., 2020. How we will produce the evidence-based European guidance on personalized nutrition: a consensus roadmap. Genes Nutr. 15 (1), 1–12.

Bagchi, D., Preuss, H.G., Swaroop, A., 2015. Nutraceuticals and Functional Foods in Human Health and Disease Prevention. CRC Press, United States, pp. 1–692. doi:10.1201/b19308.

Baker, M.T., Han, J., Zhang, Y., 2022a. Consumer acceptance toward functional foods: determinants and implications for adoption. Front. Nutr. 9, 8835010. doi:10.3389/fnut.2022.883501.

Baker, P., Machado, P., Santos, T., Sievert, K., Backholer, K., Lawrence, M., Baker, P., Machado, P., Santos, T., Sievert, K., Backholer, K., Lawrence, M., 2022b. Ultra-processed foods and the corporate capture of nutrition science. Glob. Health 18 (1), 1–12.

Bakir-Gungor, B., Hacılar, H., Jabeer, A., Nalbantoglu, O.U, Aran, O., Yousef, M., 2022. Inflammatory bowel disease biomarkers of human gut microbiota selected via different feature selection methods. PeerJ. 10, e13205. doi:10.7717/peerj.13205.

Barker, M., Chue Hong, N.P., Katz, D.S., Lamprecht, A.L., Martinez-Ortiz, C., Psomopoulos, F., Harrow, J., Castro, L.J., Gruenpeter, M., Martinez, P.A., Honeyman, T., 2022. Introducing the FAIR Principles for research software. Sci. Data 9 (1). doi:10.1038/s41597-022-01710-x.

Bashir, K., Jan, K., Maurya, V.K., Shakya, A., 2024. Food Fortification: Trends and Technologies Food Fortification. CRC Press, India, pp. 1–405.

Beal, T., Massiot, E., Arsenault, J.E., Smith, M.R., Hijmans, R.J., 2017. Global trends in dietary micronutrient supplies and estimated prevalence of inadequate intakes. PLoS ONE 12 (4). doi:10.1371/journal.pone.0175554.

Beck, A.T., Steer, R.A., Brown, G.K., 1996. Manual for the Beck Depression Inventory-II. Psychological Corporation, San Antonio, TX.

Beebe, S., 2023. Herbal medicine regulation, adverse events, and herb-drug interactions. In: Memon, M.A., Xie, H. (Eds.), Integrative Veterinary Medicine. John Wiley & Sons, Hoboken, NJ, USA, pp. 79–84.

Behan, B., Jeanson, F., Cheema, H., Eng, D., Khimji, F., Vaccarino, A.L., Gee, T., Evans, S.G., MacPhee, F.C., Dong, F., Shahnazari, S., Sparks, A., Martens, E., Lasalandra, B., Arnott, S.R., Strother, S.C., Javadi, M., Dharsee, M., Evans, K.R., Nylen, K., Mikkelsen, T., 2023. FAIR in action: Brain-CODE - a neuroscience data sharing platform to accelerate brain research. Front. Neuroinform. 17. doi:10.3389/fninf.2023.1158378.

Benton, D., 2020. The influence of dietary status on the cognitive performance of children. Mol. Nutr. Food Res. 64 (18), 1901052. doi:10.1002/mnfr.201901052.

Berding, K., Vlckova, K., Marx, W., Schellekens, H., Stanton, C., Clarke, G., Jacka, F., Dinan, T.G., Cryan, J.F., 2021. Diet and the microbiota-gut-brain axis: sowing the seeds of good mental health. Adv. Nutr. 12 (4), 1239–1285. doi:10.1093/advances/nmaa181.

Bhugra, D., Tribe, R., Poulter, D., 2022. Social justice, health equity, and mental health. S. Afr. J. Psychol. 52 (1), 3–10. doi:10.1177/00812463211070921.

Boatwright, M., Lawrence, M., Carriedo, A., Slater, S., McCoy, D., Northcott, T., Baker, P., 2024. Understanding the politics of food regulation and public health: an analysis of Codex standard-setting processes on food labelling. Int. J. Health Policy Manag. 13, 8310. doi:10.34172/ijhpm.8310.

Boer, A., 2021. Fifteen years of regulating nutrition and health claims in Europe: the past, the present and the future. Nutrients 13 (5), 1725. doi:10.3390/nu13051725.

Azevedo, F., Parsons, S., Micheli, L., Strand, J.F., Rinke, E.M., Guay, S., Elsherif, M.M., Quinn, K.A., Wagge, J.R., Steltenpohl, C.N., Kalandadze, T., Vasilev, M.R., Oliveira, C.M., Aczel, B., Miranda, J.F., Baker, B.J., Galang, C.M.O., Pennington, C.R., Marques, T., Laverty, C., Liu, M., Weisberg, Y.J., Evans, T.R., Pownall, M., Clark, K., Albayrak-Aydemir, N., Westwood, S.J., Hartmann, H., FORRT, 2019. Introducing a Framework for Open and Reproducible Research Training (FORRT). OSF Preprints. https://doi.org/10.31219/osf.io/bnh7p.

Bogan, R.R., 2021. Food-as-medicine: an everyday strategy of health, Doctoral dissertation. City University of New York.

Borrego-Ruiz, A., Borrego, J.J., 2025. Nutritional psychiatry: a novel approach to the treatment of mental health disorders. Actas Esp. Psiquiatr 53 (2), 443–445. doi:10.62641/aep.v53i2.1920.

Brambilla, G., Greiner, M., Gundert-Remy, U., Heinemeyer, G., Neisel, F., ter Burg, W., 2020. Major principles and concepts of risk assessment. In: The Practice of Consumer Exposure Assessment. Springer International Publishing, Italy, pp. 5–54. doi:10.1007/978-3-319-96148-4_2.

Brouwer, I.A., Dusseldorp, M., West, C.E., 2019. Risks and benefits of mandatory folic acid fortification. Public Health Nutr. 22 (5), 924–932.

Brown, L.T., 2022. Food labeling disclosures and the commercial speech doctrine: Combating America's growing problem. U. Louisville L. Rev. 61, 589.

Calder, P.C., 2018. Very long-chain n -3 fatty acids and human health: fact, fiction and the future. Proc. Nutr. Soc. 77 (1), 52–72. doi:10.1017/s0029665117003950.

Cao, S., Xu, H., Bryceson, K.P., 2023. Blockchain traceability for sustainability communication in food supply chains: an architectural framework, design pathway and considerations. Sustainability 15 (18), 13486. doi:10.3390/su151813486.

Chen, J., Li, Y., Zhang, Z., 2023. Consumer trust and perception of medicalized foods: balancing health innovation and naturalness. Appetite 186, 106556.

Chowdhury, S.R., 2024. Psychobiotics and gut–brain axis—a new paradigm for improving mental health. In: Probiotics: A Comprehensive Guide to Enhance Health and Mitigate Disease. CRC Press, India, pp. 146–158. doi:10.1201/9781003452249-7.

Clarke, G., O'Neill, S.M., Cryan, J.F., Dinan, T.G., 2020. Probiotics and the brain–gut axis: mechanisms and clinical implications. Neurogastroenterol. Motil. 32 (1), e13730.

Cryan, J.F., O'riordan, K.J., Cowan, C.S.M., Sandhu, K.V., Bastiaanssen, T.F.S., Boehme, M., Codagnone, M.G., Cussotto, S., Fulling, C., Golubeva, A.V., Guzzetta, K.E., Jaggar, M., Long-Smith, C.M., Lyte, J.M., Martin, J.A., Molinero-Perez, A., Moloney, G., Morelli, E., Morillas, E., O'connor, R., Cruz-Pereira, J.S., Peterson, V.L., Rea, K., Ritz, N.L., Sherwin, E., Spichak, S., Teichman, E.M., van de Wouw, M., Ventura-Silva, A.P., Wallace-Fitzsimons, S.E., Hyland, N., Clarke, G., Dinan, T.G., 2019. The microbiota-gut-brain axis. Physiol. Rev. 99 (4), 1877–2013. doi:10.1152/physrev.00018.2018.

Cui, Z., Qi, C., Zhou, T., Yu, Y., Wang, Y., Zhang, Z., Zhang, Y., Wang, W., Liu, Y., 2025. Artificial intelligence and food flavor: how AI models are shaping the future and revolutionary technologies for flavor food development. Compr. Rev. Food Sci. Food Saf. 24 (1), 1541–4337. doi:10.1111/1541-4337.70068.

D'Amico, P., Di Vita, G., Monaco, L., 2020. Sustainability and consumer perception of nutraceuticals and functional foods. Sustainability 12 (24), 10415.

Daniels, S., Clemente, D.B.P., Desart, S., Saenen, N., Sleurs, H., Nawrot, T.S., Malina, R., Plusquin, M., 2022. Introducing nature at the work floor: A nature-based intervention to reduce stress and improve cognitive performance. Int. J. Hyg. Environ. Health 240. doi:10.1016/j.ijheh.2021.113884.

Dean, E., Xu, J., Jones, A.Y.M., Vongsirinavarat, M., Lomi, C., Kumar, P., Ngeh, E., Storz, M.A., 2024a. An unbiased, sustainable, evidence-informed Universal Food Guide: a timely template for national food guides. Nutr. J. 23 (1). doi:10.1186/s12937-024-01018-z.

Dean, M., Herforth, A., Fanzo, J., Dean, M., Herforth, A., Fanzo, J., 2024b. Translating nutrition science into dietary policy: lessons from global guidelines. Public Health Nutr. 27 (3), 401–414.

Darko, D., Kwekutsu, E., Idoko, I.P., 2025. Synergistic effects of phytochemicals in combating chronic diseases with insights into molecular mechanisms and nutraceutical development. Int. J. Innov. Sci. Res. Technol. 10 (3), 25–42. doi:10.5281/zenodo.14987704.

de Azeredo, H.M.C., Otoni, C.G., McHugh, T.H., 2021. Nanostructures in food systems: design, functionality, and applications. Trends Food Sci. Technol. 110, 603–616.

De Castro Cogle K., Anastasiadi M., Mohareb F.R., Rossi C., Kubo M., De Castro Cogle, K., Anastasiadi, M., Mohareb, F.R., Rossi, C., Kubo, M., 2024. An Artificial Intelligence Approach Towards Screening of the Food Matrix and Digestion Effects in Predicting Phytochemical Bioaccessibility, Doctoral dissertation. Cranfield University.

Dhiman, S., Kaur, S., Thakur, B., Singh, P., Tripathi, M., 2025. Nutritional enhancement of plant-based fermented foods: Microbial innovations for a sustainable future. Fermentation 11 (6), 346.

Dikšaitytė, A., Zaltauskaite, J., Kniuipyte, I., 2021. Cd-induced alteration in photosynthetic performance of rapes affecting Cd-phytoextraction performance under warmer climate. In: Global Challenges, An Emergency for Environmental Sciences–Abstract Book: SETAC Europe 31st Annual Meeting. Society of Environmental Toxicology and Chemistry Europe (SETAC), Europe, Brussels, Belgium, p. 362 3–6 May 2021 (Virtual).

Dluzen, D.F., Schmidt, M.H.M., 2023. Rigor and Reproducibility in Genetics and Genomics: Peer-reviewed, Published, Cited. Elsevier, United States, pp. 1–482. doi:10.1016/C2018-0-01901-5.

Dudley K.T., Dudley, K.T., 2025. The Association Between Maternal Nutrition, Perinatal Mental Health, and Birth Outcomes in Socioeconomically Disadvantaged Women, Doctoral dissertation. Walden University.

Duman, R.S., Monteggia, L.M., 2006. A neurotrophic model for stress-related mood disorders. Biol. Psychiatry 59 (12), 1116–1127. doi:10.1016/j.biopsych.2006.02.013.

Ekkekakis, P., Tiller, N.B., 2023. Extraordinary claims in the literature on high-intensity interval training: II. Are the extraordinary claims supported by extraordinary evidence? Kinesiol. Rev. 12 (2), 144–157. doi:10.1123/kr.2022-0003.

European Commission, 2023. Regulation (EU) 2015/2283 on Novel Foods. Consolidated Version). Publications Office of the European Union, Luxembourg.

European Food Safety Authority EFSA, 2021. Guidance on the scientific requirements for health claims related to the nervous system, including psychological functions. EFSA J. 19 (7), e06548. doi:10.2903/j.efsa.2021.6548.

Evans, M., Lewis, E.D., Antony, J.M., Crowley, D.C., Charrette, A., Guthrie, N., Blumberg, J.B., Reid, G., 2025. Revisiting the definition of 'Healthy' participants in substantiation of structure/function claims for dietary supplements. J. Diet. Suppl. 22 (1), 41–57. doi:10.1080/19390211.2023.2301383.

European Food Safety Authority (EFSA), 2024. Guidance for health claim applications. European Food Safety Authority, Parma, Italy.

Ezeanochie, C., Akomolafe, O.O., Adeyemi, C., 2024. Integrating phytomedicine research into clinical development pathways for novel therapies. Shodhshauryam Int. Sci. Refereed Res. J. 7 (4), 166–196. doi:10.32628/SHISRRJ247158.

Faienza, M.F., Giardinelli, S., Annicchiarico, A., Chiarito, M., Barile, B., Corbo, F., Brunetti, G., 2024a. Nutraceuticals and functional foods: a comprehensive review of their role in bone health. Int. J. Mol. Sci. 25 (11), 5873.

Faienza, M.F., Goffredo, R., Portaccio, M., 2024b. Precision nutrition and the future of food fortification. Front. Nutr. 11, 1442178.

FAO/WHO, 2004. Guidelines for Use of Nutrition and Health Claims. Food and Agriculture Organization of the United Nations /World Health Organization.

FAO/WHO, 2006. Probiotics in Food: Health and Nutritional Properties and Guidelines for Evaluation. FAO Food and Nutrition Paper No. 85. Food and Agriculture Organization of the United Nations (FAO) and World Health Organization (WHO, Rome, Italy.

FDA, 2023. Structure/Function Claims and Dietary Supplement Labeling Guidance. U.S. Food and Drug Administration.

Fekete, M., Lehoczki, A., Kryczyk-Poprawa, A., Zábó, V., Varga, J., Bálint, M., Fazekas-Pongor, V., Csípő, T., Rząsa-Duran, E., Varga, P., 2025. Functional foods in modern nutrition science: mechanisms, evidence, and public health implications. Nutrients 17 (13), 2153. doi:10.3390/nu17132153.

FSSAI. (2022). Nutraceutical and Functional Food Regulations, India.

Giromini, C., Omodei Zorini, F., Sandrini, S., Savoini, G., Buccioni, A., Invernizzi, G., 2021. Camelina sativa cake and Cynara cardunculus meal into the diet of dairy goats affects the antioxidant activity and levels of total phenolic compounds in their milk. J. Dairy Sci. 104 (suppl. 1), 196.

Godfrey, K.M., Clark, H., Cooper, C., 2022. Nutrition, mental health, and productivity: the need for integrated research policy. Public Health Nutr. 25 (5), 1078–1086. doi:10.1017/S1368980021003282.

Goktaş, A., Şahin, S., Özdemir, N., 2020a. Encapsulation technologies for bioactive food compounds: recent advances and applications. Food Chem. 322, 126761.

Gómez-Pinilla, F., 2008. Brain foods: the effects of nutrients on brain function. Nat. Rev. Neurosci. 9 (7), 568–578. doi:10.1038/nrn2421.

González, A., et al., 2023. Dairy-based probiotics and mental health: a systematic review. Nutrients 15 (4), 856 6.

Gowrishankar, R., Bruchas, M.R, Jacka, F.N., Berk, M., 2019. Food for thought: the current state of nutritional psychiatry. In: Curr. Opin. Behav. Sci., 26, pp. 146–156. doi:10.1016/j.cobeha.2019.01.002.

Grajek, M., Krupa-Kotara, K., Białek-Dratwa, A., Sobczyk, K., Grot, M., Kowalski, O., Staśkiewicz, W., 2022. Nutrition and mental health: a review of current knowledge about the impact of diet on mental health. Front. Nutr. 9. doi:10.3389/fnut.2022.943998.

Gregg, E.W., Patorno, E., Karter, A.J., Mehta, R., Huang, E.S., White, M., Patel, C.J., McElvaine, A.T., Cefalu, W.T., Selby, J., Riddle, M.C., Khunti, K., 2023. Use of real-world data in population science to improve the prevention and care of diabetes-related outcomes. Diabetes Care 46 (7), 1316–1326. doi:10.2337/dc22-1438.

Grossklaus, R., 2009. Codex recommendations on the scientific basis of health claims. Eur. J. Nutr. 48 (S1), 15–22. doi:10.1007/s00394-009-0077-z.

Grosso, C., Santos, M., Barroso, M.F., 2023. From plants to psycho-neurology: unravelling the therapeutic benefits of bioactive compounds in brain disorders. Antioxidants 12 (8), 1603. doi:10.3390/antiox12081603.

Grunert, K.G., Wills, J.M., Fernández-Celemín, L., 2018. Nutrition knowledge, and use and understanding of nutrition information on food labels among consumers in the UK. Appetite 76, 137–147.

Gupta, M.K., Balodi, V., Tripathi, A.D., Rizwana, A.A., Nirmal, N.P., 2025. The importance of labeling and certification: ensuring quality and transparency. In: Unleashing the Power of Functional Foods and Novel Bioactives. Elsevier, India, pp. 449–465. doi:10.1016/B978-0-443-28862-3.00022-4.

Hampel, H., Vergallo, A., Perry, G., Lista, S., 2019. The Alzheimer precision medicine initiative. J. Alzheimer's Dis. 68 (1), 1–24. doi:10.3233/JAD-181121.

Health Canada, 2022. Natural Health Products Regulations. Health Canada, Government of Canada, Ottawa, ON, Canada.

Herforth, A., Arimond, M., Álvarez-Sánchez, C., Coates, J., Christianson, K., Muehlhoff, E., 2019. A global review of food-based dietary guidelines. Adv. Nutr. 10 (4), 590–605. doi:10.1093/advances/nmy130.

Hill, C., Guarner, F., Reid, G., Gibson, G.R., Merenstein, D.J., Pot, B., Morelli, L., Berni Canani, R., Flint, H.J., Salminen, S., Calder, P.C., Sanders, M.E, 2014. The International Scientific Association for Probiotics and Prebiotics consensus statement on the scope and appropriate use of the term probiotic. Nat. Rev. Gastroenterol. Hepatol. 11 (8), 506–514. doi:10.1038/nrgastro.2014.66.

Howitt, C., Taylor, K., Howitt, C., Taylor, K., 2020a. The central role of food in promoting health. Cereal Foods World 65 (3). doi:10.1094/CFW-65-3-0025.

Howitt, C., Taylor, M., Howitt, C., Taylor, M., 2020b. Evaluating causality in functional food claims. Crit. Rev. Food Sci. Nutr. 60 (9), 1473–1488.

HRS Communications, 2025. Transparency and transformation in food as medicine. HRS Communications. https://www.hrscommunications.com/news/transparency-in-food-as-medicine/.

Ingegneri, M., Smeriglio, E., Zebbiche, Y., Cornara, L., Visalli, L., Smeriglio, A., Trombetta, D., 2025. The dark side of "Smart Drugs": cognitive enhancement vs. clinical concerns. Toxics 13 (4), 247. doi:10.3390/toxics13040247.

Ioannidis, J.P.A., 2019. Meta-research: why research on research matters. PLoS Biol., 17 (3), e2005759.

Jacka, F.N., O'Neil, A., Opie, R., Itsiopoulos, C., Cotton, S., Mohebbi, M., Castle, D., Dash, S., Mihalopoulos, C., Chatterton, M.L., Brazionis, L., Dean, O.M., Hodge, A.M., Berk, M., 2017. A randomised controlled trial of dietary improvement for adults with major depression (the 'SMILES' trial). BMC Med. 15 (1). doi:10.1186/s12916-017-0791-y.

Karantonis, H.C, 2025. Editorial board members' collection series: functional foods. Appl. Sci. 15 (11), 6058. doi:10.3390/app15116058.

Karin, S., David, F.O., Tom, V., Wolfgang, G., & Reinhilde, V., 2022. Fostering coherence in EU health research: strengthening EU research for better health, Doctoral dissertation Panel for the Future of Science and Technology (STOA), European Parliament.

Katan, M.B., De Roos, N.M., 2020. Promises and pitfalls of health claims on functional foods. Ann. Rev. Nutr. 40, 35–55.

Klojdová, I., Králová, M., Červeňová, M., Klojdová, I., Králová, M., Červeňová, M., 2023a. Biopolymeric nanocarriers for delivery of neuroprotective phytochemicals. J. Funct. Foods 107, 105766.

Klojdová, I., Mikulová, K., Sláma, P., Klojdová, I., Mikulová, K., Sláma, P., 2023b. Encapsulation: a strategy to deliver therapeutics and nutraceuticals. Pharmaceutics 15 (1), 10053789.

Komala, M.G., Ong, S.G., Qadri, M.U., Elshafie, L.M., Pollock, C.A., Saad, S., 2023. Investigating the regulatory process, safety, efficacy and product transparency for nutraceuticals in the USA, Europe and Australia. Foods 12 (2). doi:10.3390/foods12020427.

Kroenke, K., Spitzer, R.L., Williams, J.B.W., 2001. The PHQ-9: validity of a brief depression severity measure. J. Gen. Intern. Med. 16 (9), 606–613. doi:10.1046/j.1525-1497.2001.016009606.x.

Kušar, A., Pravst, I., Miklavec, K., 2021. Comparison of requirements for using health claims on foods across jurisdictions. Compr. Rev. Food Sci. Food Saf. 20 (5), 4335–4350.

Langley, C., Sahakian, B.J., Robbins, T.W., 2023. Cambridge neuropsychological test automated battery (CANTAB). In: The SAGE Handbook of Clinical Neuropsychology: Clinical Neuropsychological Assessment and Diagnosis. SAGE Publications Ltd, p. 435. doi:10.4135/9781529789539.n28.

Lewis, E., Steenson, S., Haslam, R.P., Mcdonald, E., Sharman, M., Traka, M., Stanton, A., Napier, J.A., Sweeting, A., Saleh, R.N.M., Hornberger, M., Givens, D.I., Calder, P.C., Minihane, A.M., 2025. Sustainable and available sources of omega-3 fatty acids for health: are the current dietary recommendations, food sources and legislation fit for purpose? Nutr. Res. Rev. doi:10.1017/S0954422425100127.

Lindsey, M., 2009. Dietary supplements and structure-function claims: the dysfunctional structure of current regulation. J. Food Law & Policy 5 (2), 1942–9762. doi:10.54119/jflp.mjnt6512.

Linseisen, J., Renner, B., Gedrich, K., Wirsam, J., Holzapfel, C., Lorkowski, S., Watzl, B., Daniel, H., Leitzmann, M., 2025. Data in personalized nutrition: Bridging biomedical, psycho-behavioral, and food environment approaches for population-wide impact. Advances in Nutrition 16 (7), 100377. https://doi.org/10.1016/j.advnut.2025.100377.

Liu, H., Zhang, Y., Xu, J., 2022. Nutritional and functional properties of bioactive meat compounds: implications for mental health. Food Chem. 375, 131879.

Lorca, C., Mulet, M., Arévalo-Caro, C., Sanchez, M.Á., Perez, A., Perrino, M., Bach-Faig, A., Aguilar-Martínez, A., Vilella, E., Gallart-Palau, X., Serra, A., 2023. Plant-derived nootropics and human cognition: a systematic review. Crit. Rev. Food Sci. Nutr. 63 (22), 5521–5545. doi:10.1080/10408398.2021.2021137.

Lorca, R., Lorca, R., 2023. Cognitive cuisine and plant-based nootropics: Translational evidence. Phytother. Res. 37 (2), 321–339.

Luna, R.A., Foster, J.A., 2015. Gut brain axis: diet microbiota interactions and implications for modulation of anxiety and depression. Curr. Opin. Biotechnol. 32, 35–41. doi:10.1016/j.copbio.2014.10.007.

Madureira, A.R., Pereira, C.I., Gomes, A.M.P., Pintado, M.E., Malcata, F.X., 2022. Proteins as encapsulation materials for bioactive compounds in food and nutraceutical applications. Food Hydrocoll. 131, 107782.

Makkar, R., Behl, T., Bungau, S., Zengin, G., Mehta, V., Kumar, A., Uddin, M.d.S., Ashraf, G.Md., Abdel-Daim, M.M., Arora, S., Oancea, R., 2020. Nutraceuticals in neurological disorders. Int. J. Mol. Sci. 21 (12), 4424. doi:10.3390/ijms21124424.

Marks, A.B., 2024. Defining "Healthy" on food labels: one step forward, two steps back. Ind. J. Glob. Leg. Stud. 31, 47.

Martirosyan, D., Stratton, S., 2023. Comparing the FDA's process for health claim submission and the FFC's guidelines for regulating functional foods. Funct. Foods Health Dis. 13 (9), 424–436. doi:10.31989/ffhd.v13i9.1228.

Matton, A., Stephen, R., Daniilidou, M., Barbera, M., Alanko, V., Ballin, M., Ford, J., Hemiö, K., Lehtisalo, J., Rocha, S.Ló, Mangialasche, F., Ngandu, T., Rosenberg, A., Saadmaan, G., Udeh-Momoh, C., Uusimäki, K., Solomon, A., Kivipelto, M., 2025. Mechanisms of interventions targeting modifiable factors for dementia risk reduction. Mol. Neurodegener. 20 (1), 75. doi:10.1186/s13024-025-00845-w.

McGorry, P.D., 2020. Research for reform. In: Uhlhaas, P.J., Wood, S.J. (Eds.), Youth Mental Health: A Paradigm for Prevention and Early Intervention. MIT Press, Cambridge, MA, USA, pp. 237–246.

Misera, A., Liśkiewicz, P.ł, Łoniewski, I., Skonieczna-Żydecka, K., Samochowiec, J., 2021. Effect of psychobiotics on psychometric tests and inflammatory markers in major depressive disorder: meta-analysis of randomized controlled trials with meta-regression. Pharmaceuticals 14 (10), 952. doi:10.3390/ph14100952.

Modlinska, K., Pisula, W., 2018. Selected psychological aspects of meat consumption—a short review. Nutrients 10 (9), 1301. doi:10.3390/nu10091301.

Monteiro, C.A., Cannon, G., Lawrence, M., 2022. Ethical implications of personalized nutrition: Between science, marketing, and inequality. Public Health Nutr. 25 (4), 947–957.

Muzzioli, L., Maddaloni, L., Pintavalle, M., Poggiogalle, E., Vincenzo, O.D, Migliaccio, S., Vinci, G., Donini, L.M, 2025. Toward multidimensional front-of-pack labels: integrating nutritional, environmental, and processing information. Nutrients 17 (14), 2258. doi:10.3390/nu17142258.

Naaz, A., Muneshwar, K.N, 2023. How maternal nutritional and mental health affects child health during pregnancy: a narrative review. Cureus 15 (11). doi:10.7759/cureus.48763.

Naravane, T. (2024). Harnessing Food Composition Data: Machine Learning Models to Predict Taste and Health Outcomes of Food Processing (Doctoral dissertation, University of California, Davis).

Nazzaro, C., Lerro, M., Marotta, G., 2020a. Assessing consumers' attitudes toward functional foods: a socio-economic perspective. Nutrients 12 (12), 3833.

Nazzaro, F., Fratianni, F., Coppola, R., Nazzaro, F., Fratianni, F., Coppola, R., 2020b. Functional foods: consumer attitudes, market trends, and regulatory challenges. Foods 9 (5), 534.

Neale, E.P., Tapsell, L.C., Neale, E.P., Tapsell, L.C., Probst, Y.C., 2022. Nutrition and health claims: consumer use and evolving regulatory challenges. Nutrients 14 (8), 1724. doi:10.1007/s13668-022-00422-3.

Nestle, M., 2023. Food politics and the ethics of nutraceutical marketing. Ann. Rev. Public Health 44, 19–35.

Nicholson, J.K., Holmes, E., Kinross, J., Burcelin, R., Gibson, G., Jia, W., Pettersson, S., 2012. Host-gut microbiota metabolic interactions. Science 336 (6086), 1262–1267. doi:10.1126/science.1223813.

Nieman, D.C., Trumbo, P., 2021. Evidence-based review of mood and cognitive benefits of functional foods. Annu. Rev. Nutr. 41, 1–22.

NIH Office of Dietary Supplements. (2023). Omega-3 fatty acids: health professional fact sheet. National Institutes of Health.

Nissen, S., 2020. Transparency in nutrition and health research: addressing conflicts of interest. Public Health Nutr. 23 (5), 768–771.

Olson, R., Gavin-Smith, B., Ferraboschi, C., Kraemer, K., 2021. Food fortification: the advantages, disadvantages and lessons learned. Nutrients 13 (4), 1118.

Onisiforou, A., Spyrou, G.M., 2022. Immunomodulatory effects of microbiota-derived metabolites at the crossroad of neurodegenerative diseases and viral infection: network-based bioinformatics insights. Front. Immunol. 13. doi:10.3389/fimmu.2022.843128.

Ordovás, J.M., Ferguson, L.R., Tai, E.S., Mathers, J.C., 2023. Personalised nutrition and health: time for implementation. Nat. Rev. Genet. 24 (1), 1–17.

Ortega, M.A., Fraile-Martínez, Ó., García-Montero, C., Alvarez-Mon, M.A., Lahera, G., Monserrat, J., Llavero-Valero, M., Gutiérrez-Rojas, L., Molina, R., Rodríguez-Jimenez, R., Quintero, J., De Mon, M.A., 2022. Biological role of nutrients, food and dietary patterns in the prevention and clinical management of major depressive disorder. Nutrients 14 (15). doi:10.3390/nu14153099.

Parvin, N., Aslam, M., Joo, S.W., Kumar Mandal, T., Parvin, N., Gupta, D., Reddy, M., 2025a. Nanophytomedicine: harnessing plant-derived compounds with nanocarriers for improved therapeutic utility. Molecules 30 (15), 3177. doi:10.3390/molecules30153177.

Parvin, S., Rahman, M.S., Alam, M.A., 2025b. Nanostructured delivery systems for neuroactive compounds: challenges and future directions. Food Res. Int. 180, 113569.

Rahban, M., Joushi, S., Bashiri, H., Saso, L., Sheibani, V., 2024. Characterization of prevalent tyrosine kinase inhibitors and their challenges in glioblastoma treatment. Front. Chem. 11. doi:10.3389/fchem.2023.1325214.

Rao, T.J.M., Kesharwani, R.K., Keservani, R.K., Sharma, A.K., 2024. Formulations, Regulations, and Challenges of Nutraceuticals. Apple Academic Press, India, pp. 1–356.

Ray, R.C., Paramithiotis, S., Thekkangil, A., Nethravathy, V., Rai, A.K., Martin, J.G.P., 2024. Food Fermentation and its relevance in the human history. In: Trending Topics on Fermented Foods. Taylor and Francis, India, pp. 1–57. doi:10.1007/978-3-031-72000-0_1.

Raza, M.L., Hassan, S.T., Jamil, S., Fatima, W., Fatima, M., 2025. Nutritional interventions in depression: the role of vitamin D and omega-3 fatty acids in neuropsychiatric health. Clin. Nutr. 45, 270–280. doi:10.1016/j.clnu.2025.01.009.

Reuterswärd, A.L., 2007. The new EC Regulation on nutrition and health claims on foods. Scand. J. Food Nutr. 51 (3), 100–106. doi:10.1080/17482970701685625.

Rodriguez, I., Huckins, L.M., Bulik, C.M., Xu, J., Igudesman, D., 2025. Harnessing precision nutrition to individualize weight restoration in anorexia nervosa. J. Eat. Disord. 13 (1). doi:10.1186/s40337-025-01209-x.

Roe, A.L., Boyte, M.E., Elkins, C.A., Goldman, V.S., Heimbach, J., Madden, E., Oketch-Rabah, H., Sanders, M.E., Sirois, J., Smith, A., 2022. Considerations for determining safety of probiotics: a USP perspective. Regul. Toxicol. Pharmacol. 136. doi:10.1016/j.yrtph.2022.105266.

Rozin, P., 2020. Cultural approaches to human food preference. Appetite 149, 104634.

Rozin, P., Fischler, C., Shields, C., Masson, E., 2022. The cultural shaping of food and health perceptions. Appetite 173, 105997. doi:10.1016/j.appet.2022.105997.

Russell, M., Kingsley, M., 2014. The efficacy of acute nutritional interventions on soccer skill performance. Sports Med. 44 (7), 957–970. doi:10.1007/s40279-014-0184-8.

Rzymski, P., Nowak, J., Poniedziałek, B., 2023. Personalized nutrition in the era of big data: promise and pitfalls. Trends Food Sci. Technol. 139, 55–65.

Savvaidis, I.N., Al Katheeri, A., Lim, S.H.E., Lai, K.S., Abushelaibi, A., 2022. Traditional foods, food safety practices, and food culture in the Middle East. In: Savvaidis, I.N., Osaili, T.M. (Eds.), Food Safety in the Middle East. Academic Press, Cambridge, MA, USA,pp, pp. 1–31.

Schmidt, F.M., Kluge, M., Stengel, A., 2021. Linking nutrition and neurobiology: emerging multi-omics approaches. Front. Neurosci. 15, 664137. doi:10.3389/fnins.2021.664137.

Schneider, S., Junghaenel, D.U., Gutsche, T., Mak, H.W., Stone, A.A., 2020. Comparability of emotion dynamics derived from ecological momentary assessments, daily diaries, and the day reconstruction method: observational study. J. Med. Internet Res. 22 (9). doi:10.2196/19201.

Schnettler, B., Miranda-Zapata, E., Orellana, L., Sepúlveda, J., 2021. Consumer involvement in co-creation of functional foods: a cross-cultural approach. Food Qual. Prefer. 90, 104153.

Sen, S., Duman, R., Sanacora, G., 2008. Serum brain-derived neurotrophic factor, depression, and antidepressant medications: meta-analyses and implications. Biol. Psychiatry 64 (6), 527–532. doi:10.1016/j.biopsych.2008.05.005.

Serrano, J.I., Cruz-Gil, S., Fernández, C.M., Fernández, L.P., Sobrino-Santos, A., Espinosa-Salinas, I., Maestre, C., Latasa, M.J., Cebrián, M., Ramos-Ruiz, R., de Molina, A.R., del Castillo, M.D., 2025. Gut microbiota profiles are associated with different spontaneous cortical activity in healthy older people. Sci. Rep. 15 (1). doi:10.1038/s41598-025-16090-6.

Shiffman, S., Stone, A.A., Hufford, M.R., 2008. Ecological momentary assessment. Ann. Rev. Clin. Psychol. 4, 1–32. doi:10.1146/annurev.clinpsy.3.022806.091415.

Shukla, V., Singh, S., Verma, S., Verma, S., Rizvi, A.A., Abbas, M., 2024. Targeting the microbiome to improve human health with the approach of personalized medicine: latest aspects and current updates. Clin. Nutr. ESPEN 63, 813–820. doi:10.1016/j.clnesp.2024.08.005.

Sipido, K., Fajardo, D., Vercruysse, T., Glänzel, W., Veugelers, R., 2022. Fostering coherence in EU health research: strengthening EU research for better health. Study for the Panel for the Future of Science and Technology (STOA), Directorate-General for Parliamentary Research Services (EPRS). European Parliament, Luxembourg https://doi.org/10.2861/711150.

Slykerman, R.F., Davies, N., Vlckova, K., O'Riordan, K.J., Bassett, S.A., Dekker, J., Schellekens, H., Hyland, N.P., Clarke, G., Patterson, E., 2025. Precision psychobiotics for gut–brain axis health: advancing the discovery pipelines to deliver mechanistic pathways and proven health efficacy. Microb. Biotechnol. 18 (1). doi:10.1111/1751-7915.70079.

Soni, R., Verma, D., Chopra, R., Singh, V., Goswami, D., 2025. Demystifying intricate factors of nutritional anemia beyond iron deficiency–a narrative review. Clin. Nutr. ESPEN 69, 745–764. doi:10.1016/j.clnesp.2025.08.034.

Spitzer, R.L., Kroenke, K., Williams, J.B.W., Löwe, B., 2006. A brief measure for assessing generalized anxiety disorder: the GAD-7. Arch. Intern. Med. 166 (10), 1092–1097. doi:10.1001/archinte.166.10.1092United States.

Su, S., Kang, P.M., Goktaş, Z., Demirci, B., Kaplan, M., 2020. Nanoencapsulation of phytochemicals: strategies to enhance stability and bioavailability. Pharmaceutics 12 (9), 837. doi:10.3390/pharmaceutics12090837.

Tai, F.M., Subedar, A.I., Harsoliya, M.H., Galiyara, M.R., 2025. Navigating nutraceuticals regulation: a comparative assessment of the USA and India. J. Generic Med. 21 (3), 94–111. doi:10.1177/17411343251362186.

Therapeutic Goods Administration (TGA), 2023. Food–Medicine Interface Guidance Tool (FMIGT). Australian Government Department of Health and Aged Care, Canberra, ACT, Australia.

Tocher, D.R., Betancor, M.B., Sprague, M., Olsen, R.E., Napier, J.A., 2019. Omega-3 long-chain polyunsaturated fatty acids, EPA and DHA: bridging the gap between supply and demand. Nutrients 11 (1). doi:10.3390/nu11010089.

Todt, O., Luján, J.L., 2017. Health claims and methodological controversy in nutrition science. Risk Anal. 37 (5), 958–968. doi:10.1111/risa.12665.

Tuigunov, D., Sinyavskiy, Y., Nurgozhin, T., Zholdassova, Z., Smagul, G., Omarov, Y., Dolmatova, O., Yeshmanova, A., Omarova, I., 2025. Precision nutrition and gut–brain axis modulation in the prevention of neurodegenerative diseases. Nutrients 17 (19), 3068. doi:10.3390/nu17193068.

Turck, D., Bohn, T., Castenmiller, J., de Henauw, S., Hirsch-Ernst, K.I., Knutsen, H.K., Maciuk, A., Mangelsdorf, I., McArdle, H.J., Naska, A., Peláez, C., Pentieva, K., Thies, F., Tsabouri, S., Vinceti, M., Bresson, J.L., Siani, A., 2022. Scientific advice related to nutrient profiling for the development of harmonised mandatory front-of-pack nutrition labelling and the setting of nutrient profiles for restricting nutrition and health claims on foods. EFSA J. 20 (4). doi:10.2903/j.efsa.2022.7259.

U.S. Department of Health and Human Services & U.S. Department of Agriculture, 2020. Dietary guidelines for Americans 2020–2025 (9th ed.). U.S. Government Printing Office.

Verbeke, W., Pérez-Cueto, F.J.A., Barcellos, M.D.D., Krystallis, A., Grunert, K.G., 2021. Consumer awareness and perception of functional foods: cross-national insights. Food Qual. Prefer. 92, 104212.

Verhagen, H., van Loveren, H., 2016. The assessment of the safety of functional foods: toward practical guidelines. Trends Food Sci. Technol. 56, 164–171.

Wang, C.K., 2024. Legal aspects of functional food and dietary supplements in medicinal foods around the world. In: Consumer Perceptions and Food. Springer Nature, Taiwan, pp. 143–180. doi:10.1007/978-981-97-7870-6_8.

Weenen, T.C., De Regt, M., Balder, H.F., 2013. Food–pharma convergence in medical nutrition: Evidence from technological and patent analyses. PLOS One 8 (12), e82609.

Weintraub, S., Dikmen, S.S., Heaton, R.K., Tulsky, D.S., Zelazo, P.D., Bauer, P.J., Carlozzi, N.E., Slotkin, J., Blitz, D., Wallner-Allen, K., Fox, N.A., Beaumont, J.L., Mungas, D., Nowinski, C.J., Richler, J., Deocampo, J.A., Anderson, J.E., Manly, J.J., Borosh, B., Havlik, R., Conway, K., Edwards, E., Freund, L., King, J.W., Moy, C., Witt, E., Gershon, R.C., 2013. Cognition assessment using the NIH Toolbox. Neurology 80 (Issue 11, Supplement 3), S54. doi:10.1212/WNL.0b013e3182872ded.

Wies, B., Landers, C., Ienca, M., 2021. Digital mental health for young people: a scoping review of ethical promises and challenges. Front. Digit. Health 3. doi:10.3389/fdgth.2021.697072.

Wollmar, M., 2025. Opportunities and Challenges with the Shift to Climate-Adapted Food Consumption: Balancing Nutrition, Climate Impact, and Acceptance in Public and Private Meals. Ph.D. Dissertation. University of Gothenburg, Gothenburg, Sweden.

Wopereis, S., Stroeve, J.H.M., Stafleu, A., Dragsted, L.O., 2023. Systems biology approaches for nutritional neuroscience. Adv. Nutr. 14 (4), 345–359. doi:10.1093/advances/nmad010.

Yuan, H., Zhou, J., Li, H., Kang, S., Park, S., 2025. A systems biology approach to memory health: integrating network pharmacology, gut microbiota, and multi-omics for health functional foods. Int. J. Mol. Sci. 26 (14), 6698. doi:10.3390/ijms26146698.

Zafar, A., 2025. Quantum computing in finance: regulatory readiness, legal gaps, and the future of secure tech innovation. Eur. J. Risk Reg. 16 (4), 1535–1566. doi:10.1017/err.2025.10050.

Zhang, R., Zhou, H., Li, C., 2023. Functional meat products: opportunities and challenges in food–pharma innovation. Trends Food Sci. Technol. 138, 70–82.

Glossary

Antibiotic Growth Promoters (AGPs) Substances previously added to animal feed to enhance growth and prevent infections by inhibiting pathogenic bacteria. Their use has been restricted because of concerns about antibiotic resistance and residues in food.

Antioxidants Compounds that inhibit oxidation and neutralize free radicals, protecting cells from oxidative damage.

Aquaculture The farming of aquatic organisms, including fish, shrimp, mollusks, and aquatic plants, for food, economic, or environmental purposes.

Bioavailability The proportion of a nutrient or bioactive compound that is absorbed and utilized by the body.

Cecum A pouch connected to the junction of the small and large intestines in poultry and other animals, serving as a primary site for microbial fermentation and SCFA production.

Dysbiosis An imbalance in the gut microbiota that can lead to disease or impaired digestive and immune functions.

Essential Oils (EOs) Concentrated hydrophobic liquids containing volatile compounds from plants, such as thymol, carvacrol, and eugenol, which have antimicrobial, antioxidant, and immunomodulatory properties.

Eubiotic Referring to a balanced gut microbiota that supports optimal health and digestive efficiency.

Feed Conversion Ratio (FCR) A measure of an animal's efficiency in converting feed mass into increased body mass.

Fermentation Modulation The process of altering microbial fermentation in the gut or rumen to enhance nutrient utilization, produce beneficial metabolites, or reduce harmful by-products such as methane.

Functional Feed Feed supplemented with bioactive compounds that provide benefits beyond basic nutrition, including improved immunity, gut microbiota balance, and productivity.

Gut Microbiota The community of microorganisms, including bacteria, viruses, fungi, and protozoa, residing in the gastrointestinal tract, playing key roles in digestion, immunity, and overall health.

Immunomodulation The process of regulating or modifying the immune system response through dietary or bioactive interventions.

Metabolomics The study of metabolites in biological systems to understand physiological states and responses to interventions.

Neuroactive Phytochemicals Plant-derived compounds capable of modulating neural and behavioral responses in animals, influencing stress, cognition, and overall neurological health.

Oxidative Stress An imbalance between reactive oxygen species (ROS) production and antioxidant defenses, leading to cellular damage.

Phytochemicals Bioactive compounds naturally occurring in plants, such as flavonoids, phenolics, and alkaloids, with potential health benefits.

Phytogenic Feed Additives (Phytofeed) Plant-derived substances such as herbs, spices, essential oils, polyphenols, saponins, and tannins, used in animal feed to improve gut health, immunity, and performance.

Prebiotics Nondigestible food components that selectively stimulate the growth or activity of beneficial microorganisms in the gut.

Precision Feeding Tailoring feed composition and delivery to the specific needs of individual animals or groups to optimize growth, health, and feed efficiency.

Quorum Sensing A bacterial communication mechanism that regulates gene expression, including biofilm formation and virulence, based on cell population density.

Ruminants Herbivorous animals, such as cattle, sheep, and goats, possessing a complex stomach with multiple compartments, including the rumen, facilitating fermentation of plant materials.

Saponins Plant-derived glycosides that exhibit antimicrobial and anti-protozoal activities; in ruminants, they can reduce methane emissions, while in poultry and aquaculture they improve gut health and disease resistance.

Short-Chain Fatty Acids (SCFAs) Fatty acids with fewer than six carbon atoms (e.g., acetate, butyrate, and propionate) produced by microbial fermentation of fibers, which serve as energy sources for gut epithelial cells and modulate inflammation.

Tannins Polyphenolic compounds in plants (condensed or hydrolysable) that can modify microbial activity, inhibit pathogens, and reduce methane production in ruminants.

Volatile Fatty Acids (VFAs) Organic acids produced during microbial fermentation, primarily in the rumen or cecum, including acetate, propionate, and butyrate, contributing to energy supply and gut health.

Index

Page numbers followed by "*f*" and "*t*" indicate, figures and tables respectively.

E

F

G

H

I

K

L

M

N

O

P

Printed and bound by CPI Group (UK) Ltd, Croydon, CR0 4YY
17/09/2026
02218533-0002